Advanced Signal Processing in Wearable Sensors for Health Monitoring

Advanced Signal Processing in Wearable Sensors for Health Monitoring

Editors

Maysam Abbod
Jiann-Shing Shieh

MDPI • Basel • Beijing • Wuhan • Barcelona • Belgrade • Manchester • Tokyo • Cluj • Tianjin

Editors
Maysam Abbod
Brunel University London
UK

Jiann-Shing Shieh
Yuan Ze University
Taiwan

Editorial Office
MDPI
St. Alban-Anlage 66
4052 Basel, Switzerland

This is a reprint of articles from the Special Issue published online in the open access journal *Sensors* (ISSN 1424-8220) (available at: https://www.mdpi.com/journal/sensors/special_issues/ signal_processing_wearable_sensors).

For citation purposes, cite each article independently as indicated on the article page online and as indicated below:

LastName, A.A.; LastName, B.B.; LastName, C.C. Article Title. *Journal Name* **Year**, *Volume Number*, Page Range.

ISBN 978-3-0365-3887-7 (Hbk)
ISBN 978-3-0365-3888-4 (PDF)

Contents

About the Editors

Maysam Abbod received his PhD in Control Engineering from The University of Sheffield, U.K., in 1992. He is currently a Reader of Electronic Systems with the Department of Electronic and Computer Engineering, Brunel University London, U.K. He has authored over 50 papers in journals, 9 chapters in edited books, and over 50 papers in refereed conferences. His current research interests include intelligent systems for modeling and optimization. He is a member of the IET (U.K.), and a Chartered Engineer (U.K.). He is currently working as an Associate Editor of the Engineering Application of Artificial Intelligence (Elsevier).

Jiann-Shing Shieh received his BSc and MSc degrees in Chemical Engineering from National Cheng Kung University, Taiwan, in 1983 and 1986, respectively, and his PhD in Automatic Control and Systems engineering from The University of Sheffield, U.K., in 1995. He is currently a Professor with the International Program in Engineering for Bachelor, a Joint Professor with the Department of Mechanical Engineering, Graduate School of Biotechnology and Bioengineering, and also serves as the Provost and Dean of the College of Engineering, Yuan Ze University, Taiwan. His research interests are focused on biomedical engineering, particularly in bio-signal processing, intelligent analysis and control, medical automation, pain model and control, critical care medicine monitoring and control, dynamic cerebral autoregulation research, and brain death index research.

sensors

Special Issue "Advanced Signal Processing in Wearable Sensors for Health Monitoring"

Maysam Abbod [1,*] **and Jiann-Shing Shieh** [2,*]

1   Department of Electronic and Electrical Engineering, Brunel University London, Uxbridge UB8 3PH, UK
2   Department of Mechanical Engineering, Yuan Ze University, Taoyuan 32003, Taiwan
*   Correspondence: maysam.abbod@brunel.ac.uk (M.A.); jsshieh@saturn.yzu.edu.tw (J.-S.S.)

Citation: Abbod, M.; Shieh, J.-S. Special Issue "Advanced Signal Processing in Wearable Sensors for Health Monitoring". *Sensors* **2022**, 22, 2189. https://doi.org/10.3390/s22062189

Received: 7 March 2022
Accepted: 8 March 2022
Published: 11 March 2022

Publisher's Note: MDPI stays neutral with regard to jurisdictional claims in published maps and institutional affiliations.

Wearable sensors are becoming very popular recently due to their ease of use and flexibility in recording data from home. They can range from simple adhesive sensors to more sophisticated, stretchable implants to monitor health or for diagnosis. The basic unit of a wearable sensor is the electrodes or wires, the power source, and the interface/communication unit, which can be a smartphone or other types of signal receivers. One of the most important features of a wearable sensor is flexibility: It has to flex, stretch and twist without straining the sensory part and maintain the quality of the measured signal. Most wearable sensors measure physiological signals and incorporate a real-time decision system to interpret the signal and detect symptoms or measure context awareness. Different system implementations and technological approaches are used for the design of the state-of-the-art in wearable biosensors: some have advantages and some have shortcoming. The literature of such techniques is surveyed in order to provide direction for future research improvements [1].

Technology is continually improving, making many tools and algorithms available to developers with diverse applications and connectivity. Wearable sensors are benefiting from the underlying versatile technologies enabling them to capture rich contextual information that deliver a legitimately personalized experience. The extensive and diverse classification of wearable devices with wireless communication, data processing and on-board classification is reported in a survey that highlights the challenges and future solutions in this field [2].

The market of wearable sensors is growing exponentially, with an annual growth rate of 20%. Moreover, the outbreak of COVID-19 had a tremendous impact on the evolution of wearable device, driven by the requirements of home sensing and diagnosis devices [2]. Many systems have been developed for different applications, such as protection for the elderly, health home monitoring, gait analysis, interactive media and animation that helps people become familiar with such technologies [3].

In this Special Issue, special attention is given to the AI technologies that are utilized in signal processing and diagnosis. Signals usually need filtering, conditioning and processing. Advanced algorithms are computationally intensive and require fast hardware. This can represent a major hurdle in developing such systems, and most developers revert to performing the heavy signal processing on advanced computing systems such as GPUs, computer clusters, cloud applications and edge computing. This serves wearable sensor very well, as information can be transferred via the Internet, providing sophisticated algorithms and high-speed processing power.

Papers published in this Special Issue are focused onto two subjects: health monitoring using biological signal (EEG, ECG) and physical health monitoring (movements). The subject of health monitoring focused on recording vital signs such as brain activities (EEG) and applications to the cardiovascular system (ECG, HR, PPG); this issue includes seven papers in this field. The application of EEG is rather difficult due to the need for good wearable sensors that can detect the signal reliably. However, filtration and signal conditions are some of the techniques that can be used to extract meaningful information from noisy signals. EEG can be used to monitor different activities; one particular example is drowsiness detection, which is very useful for drivers who might lose attention during driving [4]. On the other hand, the use of ECG signal processing using deep learning (DL) algorithms tend to be the latest in the field. One particular application is the monitoring of the hemodynamics using ECG and DL without the need for invasive sensing [5].

Other biological signal applications are mainly related to the cardiovascular system, such as the detection of heart rate and blood pressure using either facial expressions [6] or PPG sensors [7,8] respectively. Such measurements are not easy to realize with high accuracy, hence, there is a need for DL systems to process images or signals in order to obtain good accuracy. Finally, DL is also used for the detection heart rhythm anomalies, and a short survey is presented in [9] that looks at the different techniques utilizing wearable sensors. The final application is the use of ECG for the detection of myocardial infarction (MI), which is one of the most prevalent cardiovascular diseases. An LSTM network is used to detect MI based on ECG signal [10].

Physical health monitoring using wearable sensors is presented in two papers related to physical movements, such as the prediction of joint momentum for the purpose of predicting the force generated by the muscle using an ANN for the purpose of skeleton control [11]. Another related work presented in [12] is based on the detection of chewing event using EMG signals.

The last paper is a mini-review that addresses pain as a subjective feeling. The review presents the correlation between pain and stress, and the measurement approach uses wearable sensors. Various physiological signals (i.e., heart activity, brain activity, muscle activity, electrodermal activity, respiratory, blood volume pulse, skin temperature) as well as expression/behavior are listed as measurable signs using wearables sensors. Wearable sensors used for healthcare monitoring systems can detect pain and stress. As a consequence, pain leads to multiple symptoms such as muscle tension and depression; hence, integrating modern computing techniques with wearable sensor measurements can help in pain control [13].

Author Contributions: M.A. and J.-S.S. wrote the paper. These authors contributed equally to this work. All authors have read and agreed to the published version of the manuscript.

Institutional Review Board Statement: Not applicable.

Informed Consent Statement: Not applicable.

Data Availability Statement: Not applicable.

Conflicts of Interest: The authors declare no conflict of interest.

References

1. Pantelopoulos, A.; Bourbakis, N.G. A survey on wearable sensor-based systems for health monitoring and prognosis. *IEEE Trans. Syst. Man Cybern. Part C Appl. Rev.* **2010**, *40*, 1–12. [CrossRef]
2. Ometov, A.; Shubina, V.; Klus, L.; Skibińska, J.; Saafi, S.; Pascacio, P.; Flueratoru, L.; Gaibor, D.Q.; Chukhno, N.; Chukhno, O.; et al. A survey on wearable technology: History, state-of-the-art and current challenges. *Comput. Netw.* **2021**, *193*, 108074. [CrossRef]
3. Olson, J.S. A survey of wearable sensor networks in health and entertainment. *MOJ Appl. Bionics Biomech.* **2018**, *2*, 280–287. [CrossRef]
4. Stancin, I.; Frid, N.; Cifrek, M.; Jovic, A. EEG signal multichannel frequency-domain ratio indices for drowsiness detection based on multicriteria optimization. *Sensors* **2021**, *21*, 6932. [CrossRef]
5. Sadrawi, M.; Lin, Y.-T.; Lin, C.-H.; Mathunjwa, B.; Hsin, H.-T.; Fan, S.-Z.; Abbod, M.F.; Shieh, J.-S. Non-invasive hemodynamics monitoring system based on electrocardiography via deep convolutional autoencoder. *Sensors* **2021**, *21*, 6264. [CrossRef]
6. Lee, H.; Cho, A.; Whang, M. Fusion method to estimate heart rate from facial videos based on RPPG and RBCG. *Sensors* **2021**, *21*, 6764. [CrossRef] [PubMed]
7. Sadrawi, M.; Lin, Y.-T.; Lin, C.-H.; Mathunjwa, B.; Fan, S.-Z.; Abbod, M.F.; Shieh, J.-S. Genetic deep convolutional autoencoder applied for generative continuous arterial blood pressure via photoplethysmography. *Sensors* **2020**, *20*, 3829. [CrossRef] [PubMed]
8. Lee, J.; Kim, M.; Park, H.-K.; Kim, I.Y.; Lee, J. Motion artifact reduction in wearable photoplethysmography based on multi-channel sensors with multiple wavelengths. *Sensors* **2020**, *20*, 1493. [CrossRef] [PubMed]

9. Li, H.; Boulanger, P. A survey of heart anomaly detection using ambulatory electrocardiogram (ECG). *Sensors* **2020**, *20*, 1461. [CrossRef] [PubMed]
10. Chuang, Y.-H.; Huang, C.-L.; Chang, W.-W.; Chien, J.-T. Automatic classification of myocardial infarction using spline representation of single-lead derived vectorcardiography. *Sensors* **2020**, *20*, 7246. [CrossRef] [PubMed]
11. Xiong, B.; Zeng, N.; Li, Y.; Du, M.; Huang, M.; Shi, W.; Mao, G.; Yang, Y. Determining the online measurable input variables in human joint moment intelligent prediction based on the hill muscle model. *Sensors* **2020**, *20*, 1185. [CrossRef] [PubMed]
12. Zhang, R.; Amft, O. Retrieval and timing performance of chewing-based eating event detection in wearable sensors. *Sensors* **2020**, *20*, 557. [CrossRef] [PubMed]
13. Chen, J.; Abbod, M.; Shieh, J.-S. Pain and stress detection using wearable sensors and devices—A review. *Sensors* **2021**, *21*, 1030. [CrossRef] [PubMed]

sensors

Article

EEG Signal Multichannel Frequency-Domain Ratio Indices for Drowsiness Detection Based on Multicriteria Optimization

Igor Stancin, Nikolina Frid, Mario Cifrek and Alan Jovic *

Faculty of Electrical Engineering and Computing, University of Zagreb, Unska 3, 10000 Zagreb, Croatia; igor.stancin@fer.hr (I.S.); nikolina.frid@fer.hr (N.F.); mario.cifrek@fer.hr (M.C.)
* Correspondence: alan.jovic@fer.hr

Citation: Stancin, I.; Frid, N.; Cifrek, M.; Jovic, A. EEG Signal Multichannel Frequency-Domain Ratio Indices for Drowsiness Detection Based on Multicriteria Optimization. *Sensors* **2021**, *21*, 6932. https://doi.org/10.3390/s21206932

Academic Editors: Maysam Abbod and Jiann-Shing Shieh

Received: 30 August 2021
Accepted: 17 October 2021
Published: 19 October 2021

Publisher's Note: MDPI stays neutral with regard to jurisdictional claims in published maps and institutional affiliations.

Abstract: Drowsiness is a risk to human lives in many occupations and activities where full awareness is essential for the safe operation of systems and vehicles, such as driving a car or flying an airplane. Although it is one of the main causes of many road accidents, there is still no reliable definition of drowsiness or a system to reliably detect it. Many researchers have observed correlations between frequency-domain features of the EEG signal and drowsiness, such as an increase in the spectral power of the theta band or a decrease in the spectral power of the beta band. In addition, features calculated as ratio indices between these frequency-domain features show further improvements in detecting drowsiness compared to frequency-domain features alone. This work aims to develop novel multichannel ratio indices that take advantage of the diversity of frequency-domain features from different brain regions. In contrast to the state-of-the-art, we use an evolutionary metaheuristic algorithm to find the nearly optimal set of features and channels from which the indices are calculated. Our results show that drowsiness is best described by the powers in delta and alpha bands. Compared to seven existing single-channel ratio indices, our two novel six-channel indices show improvements in (1) statistically significant differences observed between wakefulness and drowsiness segments, (2) precision of drowsiness detection and classification accuracy of the XGBoost algorithm and (3) model performance by saving time and memory during classification. Our work suggests that a more precise definition of drowsiness is needed, and that accurate early detection of drowsiness should be based on multichannel frequency-domain features.

Keywords: drowsiness detection; EEG; frequency-domain features; multicriteria optimization; machine learning

1. Introduction

Drowsiness is the intermediate state between wakefulness and sleep [1]. Terms such as sleepiness or tiredness are used synonymously with drowsiness in related studies [2–4]. Although it is intuitively clear what drowsiness is, it is not so easy to determine exactly whether a person is in a drowsy state or not. The reason for this is the unclear definition of drowsiness. Some researchers define drowsiness as stage 1 sleep (S1) [5–9], which is also known as non-rapid eye movement 1 (NREM 1) sleep. Da Silveira et al. [10] used S1 sleep stage data in their research of drowsiness. Johns [11] claims that the S1 sleep stage is equivalent to microsleep (episodes of psychomotor insensitivity due to sleep-related wakefulness loss [12]), while drowsiness is stated to occur before S1 sleep, but it is not stated when it begins and what characterizes it. Researchers who do not use any of the aforementioned definitions of drowsiness typically use a subjective assessment of drowsiness, e.g., the Karolinska sleepiness scale [13]. In this paper, the term drowsiness is used as a synonym for the S1 sleep stage.

In a drowsy state, people are not able to function at the level required to safely perform an activity [14], due to the progressive loss of cortical processing efficiency [15]. Drowsiness is, therefore, a significant risk factor for human lives in many occupations, e.g., for air traffic controllers, pilots and regular car drivers [16]. According to the reports from NASA [17] and the National Transportation Safety Board [18], one of the main factors in road and air accidents is drowsiness. Gonçalves et al. [19] conducted a study across 19 European countries and concluded that in the last two years, 17% of drivers fell asleep while driving, while 7% of them had an accident due to drowsiness. The high frequency and prevalence of drowsiness-related accidents speak in favor of the development of early drowsiness detection systems, which is the subject of this paper.

Many researchers are trying to solve the problem of early detection of drowsiness in drivers. Balandong et al. [20], in their recent review, divided the techniques for detecting driver drowsiness into six categories: (1) subjective measures, (2) vehicle-based systems, (3) driver's behavior-based systems, (4) mathematical models of sleep–wake dynamics, (5) human physiological signal-based systems and (6) hybrids of one or more of these techniques. Currently, the most common techniques used in practice are vehicle-based systems [5], but these systems are mostly unreliable and depend largely on the driver's motivation to drive as well as possible [20].

Physiological signals are the promising alternative for reliable drowsiness detection [21]. The main problem with this approach is that these systems are often not easy to use and are intrusive to drivers [20]. Nevertheless, many researchers are working on small, automated and wearable devices [21–24], or on steering wheel devices [25,26] in order to overcome these obstacles. Techniques for detecting drowsiness based on physiological signals can be further subdivided according to the type of signal used, such as electroencephalogram (EEG) [27], electrooculogram (EOG) [28] or electrocardiogram (ECG) [29].

The most studied and applied physiological signal to detect drowsiness is the EEG. In this paper, frequency-domain features of the EEG signal are analyzed and two novel multichannel ratio indices for the detection of drowsiness are proposed. Besides the frequency-domain features, there are also other types of features: (1) nonlinear features [30], (2) spatiotemporal (functional connectivity) features [31] and (3) entropies [32]. These three groups of features have a lower frequency of use compared to the frequency-domain features, so in this paper, we focus only on frequency-domain features. Based on the recent review [33] of EEG-based drowsiness detection systems, 61% of the included papers used frequency-domain features, 38% used entropies, 10% used nonlinear features and 10% used spatiotemporal features (some papers used multiple groups of features, so the sum of the percentages is greater than 100%). This shows the difference in the use of drowsiness detection systems, and the difference is even greater in the general field of neurophysiological scientific papers. Although the three feature groups mentioned above are used less frequently, there are still a certain number of papers that include them, especially entropies.

Frequency-domain features estimate the power spectral density in a given frequency band. The bands typically used in the analysis of EEG signals are delta (δ, 0.5–4 Hz), theta (θ, 4–8 Hz), alpha (α, 8–12 Hz), beta (β, 12–30 Hz) and gamma (γ, >30 Hz). An increase in theta activity [34] and an increase in alpha activity [35] indicate drowsiness. An increase in the beta activity, however, is a sign of wakefulness and alertness [36]. There are several widely used frequency-domain ratio indices for detecting drowsiness. Eoh et al. [36] proposed the θ/α and β/α ratio indices, Jap et al. [37] proposed the $(\theta + \alpha)/\beta$, θ/β and $(\theta + \alpha)/(\alpha + \beta)$ ratio indices and da Silveira et al. [10] proposed the γ/δ and $(\gamma + \beta)/(\delta + \alpha)$ ratio indices. These ratio indices provide improvement in the detection of drowsiness compared to the frequency-domain features alone and are shown to correlate with drowsiness.

All these frequency-domain features and ratio indices are calculated from a single EEG channel, i.e., from a single brain region. In recent research, Wang et al. [38] showed that the significance of a decrease in delta and an increase in $(\theta + \alpha)/\beta$ indices depends on the brain region. This significant diversity of the correlation of features with drowsiness in different brain regions is the motivation for this research. Since all currently used frequency-domain features and ratio indices are based on a single channel (single brain region), this work aims to use the best distinguishing features of each brain region for the detection of drowsiness and to combine them into a single multichannel ratio index feature.

In our work, we use a computational method based on multicriteria optimization to extract the multichannel EEG-based frequency-domain ratio index features. This method allows us to discover new multichannel ratio indices that show improvements in the detection of drowsiness compared to single-channel ratio indices. Finally, with the use of machine learning models, we prove that multichannel indices detect drowsiness with higher accuracy, higher precision, reduced memory and faster computation compared to single-channel features.

In the Materials and Methods Section, we show the methodology of our work, including a description of the dataset, preprocessing and feature extraction methods used. Novel multichannel ratio indices and the multi-objective optimization method are also described there. In the Results Section, we present the results of our work, including statistical analysis, drowsiness prediction and computational properties of the proposed indices. In the Discussion Section, we discuss in more detail the topics covered in this paper. Finally, in the last section, we conclude the paper.

2. Materials and Methods

2.1. Dataset, Preprocessing and Feature Extraction

The data used in this paper were obtained from the PhysioNet portal [39], in particular from the 2018 PhysioNet computing in cardiology challenge [40]. The original dataset contains data records from 1985 subjects, and each recording includes a six-channel EEG, an electrooculogram, an electromyogram, a respiration signal from the abdomen and chest, airflow and oxygen saturation signals and a single-channel electrocardiogram during the all-night sleep. The records were divided into training and test sets of equal size. The sleep stages [41] of all subjects were annotated by clinical staff based on the American Academy of Sleep Medicine (AASM) manual for the scoring of sleep [42]. There are six types of annotations for different stages: wakefulness (W), stage 1 (S1), stage 2 (S2), stage 3 (S3), rapid eye movement (REM) and undefined.

In this research, we wanted to use a training set (992 subjects) to detect drowsiness. The officially provided way of acquiring the data is through torrent download, but we managed to download only 393 subjects completely, due to a lack of seeders. Of these 393 subjects, EEG signal recordings from 28 subjects were selected, based on the condition that each recording had at least 300 s of the W stage and, immediately after that, at least 300 s of the S1 stage. From each recording, a fragment of 600 s (300 s of W stage and 300 s of S1 stage) was used for analysis. In the original dataset, each EEG signal recording consists of six channels (*F3*, *F4*, *C3*, *C4*, *O1* and *O2*, based on the International 10/20 System), with a sampling frequency of 200 Hz. Table 1 shows the identification numbers of all the selected subjects. The subjects were divided into two groups, one group used for training of the model (16 subjects) and the other one for the test of the obtained models (12 subjects). The training set was used to obtain novel ratio indices (with the method described below) and the test set was used to check these novel indices on the unseen data.

Table 1. The identification numbers of all the selected subjects. The training set is in the upper part and the test set is in the lower part of the table.

tr03-0092	tr03-0256	tr03-0876	tr03-1389
tr04-0649	tr04-0726	tr05-1434	Tr05-1675
tr07-0168	tr07-0458	tr07-0861	tr08-0021
tr08-0111	tr09-0175	tr10-0872	tr13-0204
tr04-0653	tr07-0127	tr09-0453	tr13-0170
tr05-0028	tr08-0157	tr12-0255	tr13-0508
tr05-0332	tr09-0328	tr12-0441	tr13-0653

Before feature extraction, the EEG signal must be filtered. For this purpose, the DC component was removed from the signal and the signal was filtered with a Butterworth filter to remove high-frequency artifacts and low-frequency drifts. We used the sixth-order Butterworth filter, the low-cut frequency of 1 Hz and the high-cut frequency of 40 Hz. In the selected fragments of the recordings, there was an insignificant number of eye-related artifacts, so we decided not to use the independent component analysis for their removal in order to prevent potential information loss due to component removal.

The signals were divided into epochs to calculate features. The epochs were five seconds long with a 50% overlap between them. Frequency-domain features are often used in EEG signal analysis. These features were extracted from the power spectral density (PSD) of the signal. To obtain the PSD of the signal, Welch's method [43] was used. Welch's method is used more often than Fast Fourier transform in the field of EEG signal analysis since it produces PSD with lower variance. The standard frequency-domain features were calculated, i.e., delta (δ, 0.5–4 Hz), theta (θ, 4–8 Hz), alpha (α, 8–12 Hz) and beta (β, 12–30 Hz) bands. We also calculated the less frequently used frequency-domain features, i.e., gamma (γ, >30 Hz), sigma (σ, 12–14 Hz), low alpha (α1, 8–10 Hz) and high alpha (α2, 10–12 Hz) bands [44].

2.2. Novel Multichannel Ratio Indices

Ratios between frequency-domain features have often been used as new features in different areas of EEG signal analysis [10,36]. All these features have a simple mathematical formulation but often lead to an improvement in detection and reduction of dimensionality for drowsiness. Moreover, they are calculated based on a single channel only. The idea behind the novel indices we present in this work is to design the feature formulation in such a way that frequency-domain features from different channels can be combined. Figure 1 illustrates the difference between these two approaches. For simplicity of visualization, only four epochs, two channels (*F3* and *F4*) and three features per channel are shown in Figure 1.

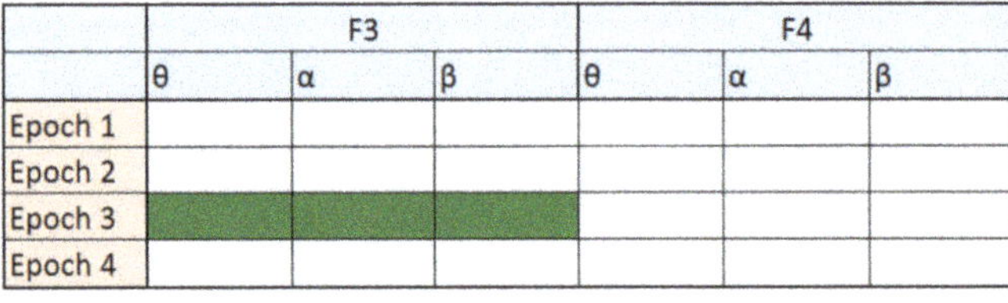

	F3			F4		
	θ	α	β	θ	α	β
Epoch 1						
Epoch 2						
Epoch 3	███	███	███			
Epoch 4						

	F3			F4		
	θ	α	β	θ	α	β
Epoch 1						
Epoch 2						
Epoch 3	███	███	███	███	███	███
Epoch 4						

Figure 1. A visualization of tables with features. The green color represents the possibilities for creating a ratio index, the first table (**top**) are the possibilities reported in the related work to create a single-channel ratio index, while the second table (**bottom**) are the possibilities explored in our novel multichannel approach.

We define a new index, I, for each epoch, e, which is calculated as a ratio of the feature values, $F(e)$, for all six channels in the epoch, e. In both the nominator and denominator, the feature value of each channel, j, is multiplied with a dedicated coefficient, C_{ij} or K_{ij} respectively, as indicated in the Equation (1):

$$I(e) = \frac{\sum_{i=features} \sum_{j=channels} C_{ij} F_{ij}(e)}{\sum_{i=features} \sum_{j=channels} K_{ij} F_{ij}(e)} \tag{1}$$

The purpose of the coefficients is to reduce or even eliminate the influence of certain channels of frequency-domain features, by setting the value in the range $[0, 1\rangle$, or increase the influence of certain channels of the frequency-domain features by setting the corresponding coefficient to a value in the range $[1, \infty\rangle$. There are 48 (6 channels and 8 features per channel) C coefficients and 48 K coefficients.

The ideal output of $I(e)$ should look like a step function (or an inverse step function), which would indicate a clear difference between the two stages: W and S1. Figure 2 illustrates the main features of the output. The output can be divided into two parts: the left one corresponds to stage W and the right one to S1. While the output in each part should be as smooth as possible, i.e., with minimal oscillations, it is expected that there will be a transition period between the phases, which may have significant oscillations. This transition period would ideally be the step function, but in realistic settings, it is expected that the transition between phases of brain activity will probably last several epochs and would not be considered as either stage W or S1.

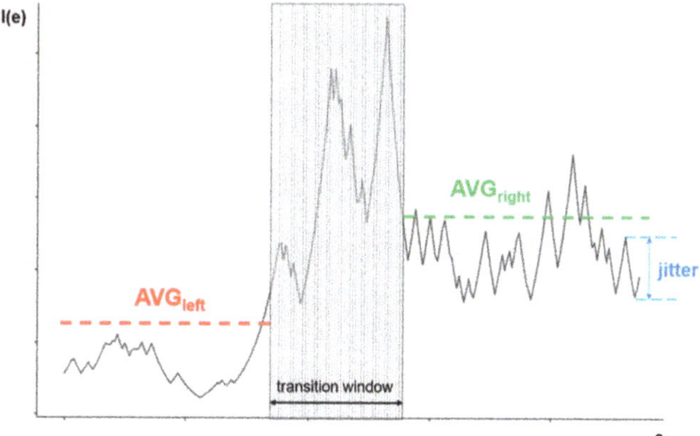

Figure 2. An illustration of all the elements needed for an evaluation of solutions of the multi-objective optimization in drowsiness detection.

In order to determine the appropriate value of the coefficients that would provide the output as close as possible to the ideal, at least two criteria must be taken into account: the absolute difference between the mean values left and right of the transition window and the quantification of the oscillations in each part. This can be defined as a multi-objective optimization problem that we want to solve using a metaheuristic multi-objective evolutionary optimization method, as described in the next section. To the best of our knowledge, this state transition problem has never been approached with evolutionary computation.

2.3. Multi-Objective Optimization

The optimization of a step function that is representative of the problem of flat surfaces is generally a challenge for any optimization algorithm because it does not provide information about which direction is favorable and an algorithm can get stuck on one of the flat plateaus [45]. To overcome this challenge, instead of optimizing the function according to one criterion, we define two objectives that we optimize simultaneously: (1) to maximize the absolute difference between the mean value of $I(e)$ output for the W and S1 stages, and (2) to minimize the oscillations of the output value around the mean value in each stage. According to Figure 2, the left part of the $I(e)$ output occurring before the transition phase corresponds to the W stage, and the right part, occurring after the transition phase, corresponds to the S1 stage. Since optimization problems are usually expressed as minimization problems, where the first objective function, $O1$, is defined as the inverse absolute difference between the mean value of $I(e)$ of the left part (avg_{left}) and the right part (avg_{right}), Equation (2) is established:

$$O1 = \frac{1}{\left| avg_{right} - avg_{left} \right|} \tag{2}$$

The second objective function, $O2$, expresses the oscillations in the function and is defined as the number of times the difference between the output values of $I(e)$ for two adjacent epochs was greater than a given limit. The exact value of this limit will be discussed later in this section as it is closely related to the specifics of the optimization method used. The main goal of the objective function $O2$ is to minimize the influence of the biggest flaw in the way that the objective $O1$ is calculated, i.e., to use the averaging function. For example, if a possible solution is a completely straight line, except for a large negative spike in the left part and a large positive spike in the right part, based only on the objective function $O1$, this would be a good solution, while the objective function $O2$ would penalize this solution.

As mentioned above, the transition between two stages will probably take several epochs and show significant oscillations of the function output values. According to the annotation made by clinical personnel, the transition phase should be approximately in the middle of the $I(e)$ output, but it cannot be determined exactly how long it will last. In our work, which is based on expert knowledge of human behavior in the case of drowsiness, we assume that it lasts about one minute, which corresponds to about 30 epochs. Within the transition window, neither one of the two objective functions is calculated, since it is assumed to belong neither to the W nor to the S1 stages. We also allow it to move around the center, shifting left and right, due to a possible error of the human observer who marked the data.

The multi-objective optimization problem can now be expressed as min{$O1,O2$}, where $O1$ and $O2$ are the conflicting objective functions, as defined above. The evolutionary meta-heuristic algorithm NSGA-II [46] was applied to solve this multi-objective optimization problem. The genetic algorithms (GAs) are normally used to solve complex optimization and search problems [47]. NSGA-II is one of the most popular evolutionary multicriteria optimization methods due to its versatility and ability to easily adapt to different types of optimization problems. The strong points of this MO algorithm are: (1) the fast non-dominated sorting ranking selection method used to emphasize Pareto-optimal solutions, (2) maintaining the population diversity by using the crowding distance and (3) the elitism approach, which ensures the preservation of best candidates through generations without the setting of any new parameters other than the normal genetic algorithm parameters, such as population size, termination parameter, crossover and mutation probabilities. Additionally, it was often used for the elimination of EEG channels with the similar purpose as in our case-dimensionality reduction [48]. This paper uses the implementation of NSGA-II provided by the MOEA framework [49] and is based on the guidelines defined in [46,50].

NSGA-II was used with the following configuration. The chromosome was divided into two parts: in the first part, genes represented the nominator coefficient values (C_{ij}), and in the second part, genes represented the denominator coefficient values (K_{ij}). In each part, the genes were grouped by frequency-domain features and channels, as illustrated in Figure 3. The genes were encoded as real values in the range [0.0, 10.0], and standard NSGA-II crossover and mutation operators were used to support operation on real values.

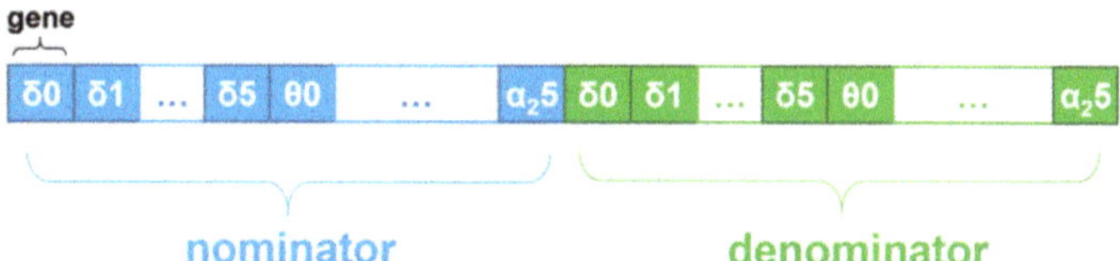

Figure 3. An illustration of a chromosome structure in the proposed optimization problem solution.

Each solution is evaluated based on the values of objectives $O1$ and $O2$, as described in the pseudocode in Algorithm 1. First, the chromosome is decoded (line 1). Then, for each test fragment, two values are calculated: (1) the inverse absolute difference (IAD) between the mean index value, $I(e)$, of the left part and the right part, represented by the invAbsDiff variable in the pseudocode, and (2) the oscillations in the function, represented by the oscillation variable in the pseudocode (lines 3–5). Finally, the value of each objective $O1$ for the given solution is defined as the average value of invAbsDiff for all test fragments, and the value of objective $O2$ is defined as the average value of oscillation for all test fragments (lines 7–8).

Algorithm 1. Evaluation.

1: decode chromosome to get coefficient values
2: **for** each fragment **do**
3:      indexVals[[] = calculate index value for each epoch
4:      invAbsDiff += IADCalc(indexVals[[], windowStart)
5:      oscillation += OscillationCalc(indexVals[[], windowStart, winSize)
6: **end for**
7: objective1 = invAbsDiff/number_of_fragments
8: objective2 = oscillation/number_of_fragments

The algorithm for the IAD calculation is provided in the pseudocode in Algorithm 2. The calculation of the IAD for each fragment was slightly modified compared to Equation (1) to allow a faster convergence of the search algorithm. The transition phase was not in the same position in each fragment but allowed to move more loosely away from the center because the annotation in the original dataset was performed manually and there was a possibility of human error in case the observer would register a transition from W to S1 a little too early or too late. The algorithm allows the transition phase to begin no earlier than 30 epochs from the fragment start, and end no later than 60 epochs before the fragment end (line 2). The algorithm assumes the transition phase by looking for a window of 30 epochs which has the maximum difference of index, $I(e)$, values between the left and the right part (lines 9–13).

The gradation of the absolute difference between the mean value of the left and the right parts is also introduced (lines 19–22) to allow easier and faster convergence of the algorithm. The optimization of the objective $O1$ can be considered as an optimization problem with soft constraints that are related to how much $O1$ deviates from the optimal value. However, it is quite difficult to determine the optimal value precisely a priori. As indicated in [51,52], constraints are often treated with penalties in optimization techniques. The basic idea is to transform a constrained optimization problem into an unconstrained one by introducing a penalty into the original objective function to penalize violations of constraints. According to a comprehensive overview in [51], the penalty should be based on the degree of constraint violation of an individual. In [53], it is also recommended that instead of having just one fixed penalty coefficient, the penalty coefficient should increase when higher levels of constraint violation are reached. The greatest challenge, however, is to determine the exact penalty values. If the penalty is too high or too low, evolutionary algorithms spend either too much or too little time exploring the infeasible region, so it is necessary to find the right trade-off between the objective function and the penalty function so that the search moves towards the optimum in the feasible space. As the authors have shown in [54], the choice of penalty boundaries is problem-dependent and difficult to generalize. Since we cannot strictly determine the optimal value of $O1$ in our case, we have chosen several thresholds for the absolute difference value, with the penalty increasing by a factor of 10 for each new threshold. The exact thresholds were selected based on the experience gained from the first few trial runs of the algorithm. Based on the observations from the trial runs, a third modification was also introduced: the difference is calculated with a relative, instead of absolute, value of $I(e)$. The relative value of $I(e)$ is calculated by using the lowest $I(e)$ value as a reference point, instead of zero, i.e., the zero is "moved", as shown in code lines 16–18 in Algorithm 2.

Algorithm 2. IAD Calculation.

```
 1: function IADCALC(indexVals[[], windowStart)
 2:    for j between 30 and (indexVals.size-60) do
 3:        maxAbsDiff = 0
 4:        left = 0
 5:        right = 0
 6:        avgLeft = average value of all Index values before j
 7:        avgRight = average value of all Index values after j+30
 8:        diff = ABS(avgRight–avgLeft)
 9:        if diff ≥ maxAbsDiff then
10:            maxAbsDiff 0 diff
11:            left = avgLeft
12:            right = avgRight
13:            windowStart = j
14:        end if
15:    end for
16:    lowestVal = GETLOWESTVAL(indexVals)
17:    movedZero = lowestVal–0.01*lowestVal
18:    absDiff = ABS(right–left)/MIN(left–movedZero, right–movedZero)
19:    if absDiff ≥ 5.0 then invAbsDiff = 1/absDiff
20:    else if absDiff ≥ 1.0 then invAbsDiff = 10/absDiff
21:    else if absDiff ≥ 0.5 then invAbsDiff = 100/absDiff
22:    else invAbsDiff = 1000
23:    end if
24:    return invAbsDiff
25: end function
```

The pseudocode for calculating the oscillations in the function as the second objective, $O2$, is provided in Algorithm 3. Again, the optimization of the oscillations can be considered a constrained optimization problem, so that, in the same way as in the case of the IAD calculation discussed previously, a gradation of the difference between the output values of $I(e)$ for two adjacent epochs is used to penalize the larger differences more severely (lines 7–10 and 15–18). The exact thresholds were chosen based on the experience gained from the first few trial runs of the algorithm. In order to make the algorithm converge more easily and quickly, the concept of "moved zero" was used again (lines 2, 3, 6 and 14).

Algorithm 3. Oscillation Calculation.

```
 1: function OSCILLATIONCALC(indexVals[[], windowStart, winSize)
 2:     lowestVal = GETLOWESTVAL(indexVals)
 3:     movedZero = lowestVal–0.01*lowestVal
 4:     oscillation = 0
 5:     for i between 1 and windowStart–1 do
 6:       absDiff = ABS((indexVals[i]-indexVals[i-1])/(indexVals[i-1] -movedZero))
 7:       if absDiff ≥ 5.0 then oscillation += 1000
 8:       else if absDiff ≥ 1.0 then oscillation += 100
 9:       else if absDiff ≥ 0.5 then oscillation += 10
10:         else if absDiff ≥ 0.25 then oscillation += 1
11:         end if
12:     end for
13:     for i between windowStart+winSize and indexVals.size()-1 do
14:       absDiff = ABS((indexVals[i]-indexVals[i-1])/(indexVals[i-1] -movedZero))
15:       if absDiff ≥ 5.0 then oscillation += 1000
16:         else if absDiff ≥ 1.0 then oscillation += 100
17:         else if absDiff ≥ 0.5 then oscillation += 10
18:         else if absDiff ≥ 0.25 then oscillation += 1
19:         end if
20:     end for
21:     return oscillation
22: end function
```

Finally, to further minimize the oscillations, and help the search algorithm converge more quickly, the maximum change in the $I(e)$ value between two adjacent epochs is set to 10% of the first of the two epochs. The mathematical formulation of this limit is provided in Equation (3):

$$Index(e) = \begin{cases} 1.1 * I(e-1), & if\ I(e) > 1.1 * I(e-1) \\ 0.9 * I(e-1), & if\ I(e) < 0.9 * I(e-1) \\ I(e),\ else \end{cases} \tag{3}$$

3. Results

The optimization algorithm was executed over 107 generations, using 100 randomly selected chromosomes as a starting point. Ideally, the optimization algorithm would have many C and K coefficients equal to zero and only a few non-zero coefficients in order to obtain a simple and easily understandable mathematical formulation of a novel multichannel ratio index. Unfortunately, even the best solutions of the optimization algorithm had only up to 20 C and K coefficients equal to zero. Although such a novel multichannel ratio index showed good behavior in detecting drowsiness, it is impractical to use a formula with 76 coefficients. We consider anything above 15 coefficients to be impractical.

In order to reduce the number of coefficients and to simplify the formulation of the novel multichannel ratio index, some coefficients were manually set to zero. In order to decide which coefficients have the least influence on the final solution, we counted how often a large value of the coefficient is fixed to a certain frequency-domain feature. By analyzing the coefficients of all solutions in the final population of the optimization algorithm, we concluded that the most frequently selected features were δ, α, $\alpha 1$ and $\alpha 2$. After manually fixing the coefficients of all other frequency-domain features to zero, the search range for the optimization algorithm was reduced to half.

Although 48 C and K coefficients remained in the solution at that time, the algorithm provided equally good results in terms of drowsiness detection, but with a much simpler mathematical formulation. In addition to the 48 coefficients that were manually set to zero, the algorithm often set many more coefficients to zero. A decision on the best solution in the final population was made based on the $O1$ and $O2$ values of the optimization algorithm in combination with the number of coefficients set to zero after using the floor operator on the coefficients. The floor operator was used to simplify the equation by removing the decimal numbers. Preferred solutions are those with a higher number of coefficients set to zero. Our choice was the solution with 13 non-zero coefficients, as shown in Equation (4):

$$I1(e) = \frac{\alpha_{F3} + 4\alpha_{O2} + 9\alpha 1_{F3} + 3\alpha 1_{C3} + 9\alpha 1_{C4} + \alpha 1_{O2} + 4\alpha 2_{O1} + 8\alpha 2_{O2}}{\delta_{F3} + 3\delta_{F4} + 3\delta_{C3} + 2\delta_{C4} + 9\delta_{O2}} \tag{4}$$

All C and K coefficients were rounded to a lower value (floor operator). Here, e represents the current epoch and all the features on the right side were from that same epoch.

The goal of the second condition of the optimization algorithm was to minimize the oscillations of the $I(e)$ function. The results were much better with this condition than without it, but the resulting function still oscillated strongly. In order to additionally minimize the oscillations, a limitation was performed. The maximum change between any two adjacent samples was set to 10% of the value of the first sample. Equation (5) shows the mathematical formulation of this limitation of the maximum change:

$$Index1(e) = \begin{cases} 1.1 * I1(e-1), & if \ I1(e) > 1.1 * I1(e-1) \\ 0.9 * I1(e-1), & if \ I1(e) < 0.9 * I1(e-1) \\ I1(e), & else \end{cases} \tag{5}$$

where $I1(e)$ is defined by Equation (4) and e is the current epoch. Limiting the maximum change of adjacent samples further improves the detection model, and therefore Equation (5) presents the first novel multichannel ratio index.

We have tried to further simplify the formulation of the multichannel ratio index. This time, brute force search for the best solution was applied with the following constraints: (1) encoding of all C and K coefficients was set to integer values of zero or one for the sake of simplicity, and (2) a maximum of five addends in the equation was allowed. With these constraints, we obtained Equation (6):

$$I2(e) = \frac{\delta_{F3} + \delta_{F4} + \delta_{O2}}{\alpha_{C3} + \alpha 2_{O2}} \tag{6}$$

Again, similar to the first index, the maximum change was limited, so that the final equation for the second ratio index was obtained as:

$$Index2(e) = \begin{cases} 1.1 * I2(e-1), & if \ I2(e) > 1.1 * I2(e-1) \\ 0.9 * I2(e-1), & if \ I2(e) < 0.9 * I2(e-1) \\ I2(e), & else \end{cases} \tag{7}$$

where $I2(e)$ is defined by Equation (6) and e is the current epoch. After obtaining the two novel indices, they were normalized to the range [0, 1] for each subject to eliminate interindividual differences between the subjects.

The two novel multichannel ratio indices defined by Equations (5) and (7) were compared with the seven existing indices θ/α and β/α [36], $(\theta + \alpha)/\beta$, θ/β and $(\theta + \alpha)/(\alpha + \beta)$ [37], and γ/δ and $(\gamma + \beta)/(\delta + \alpha)$ [10]. The indices γ/δ and $(\gamma + \beta)/(\delta + \alpha)$ were calculated based on the wavelet transform, i.e., in the same way as in the original paper. Figure 4 shows a comparison of our novel indices with the best and the worst channel for θ/α and $(\theta + \alpha)/\beta$ single-channel indices for subject tr08-0111. These two single-channel indices were selected because they are the best predictors of drowsiness for a given subject among all single-channel indices.

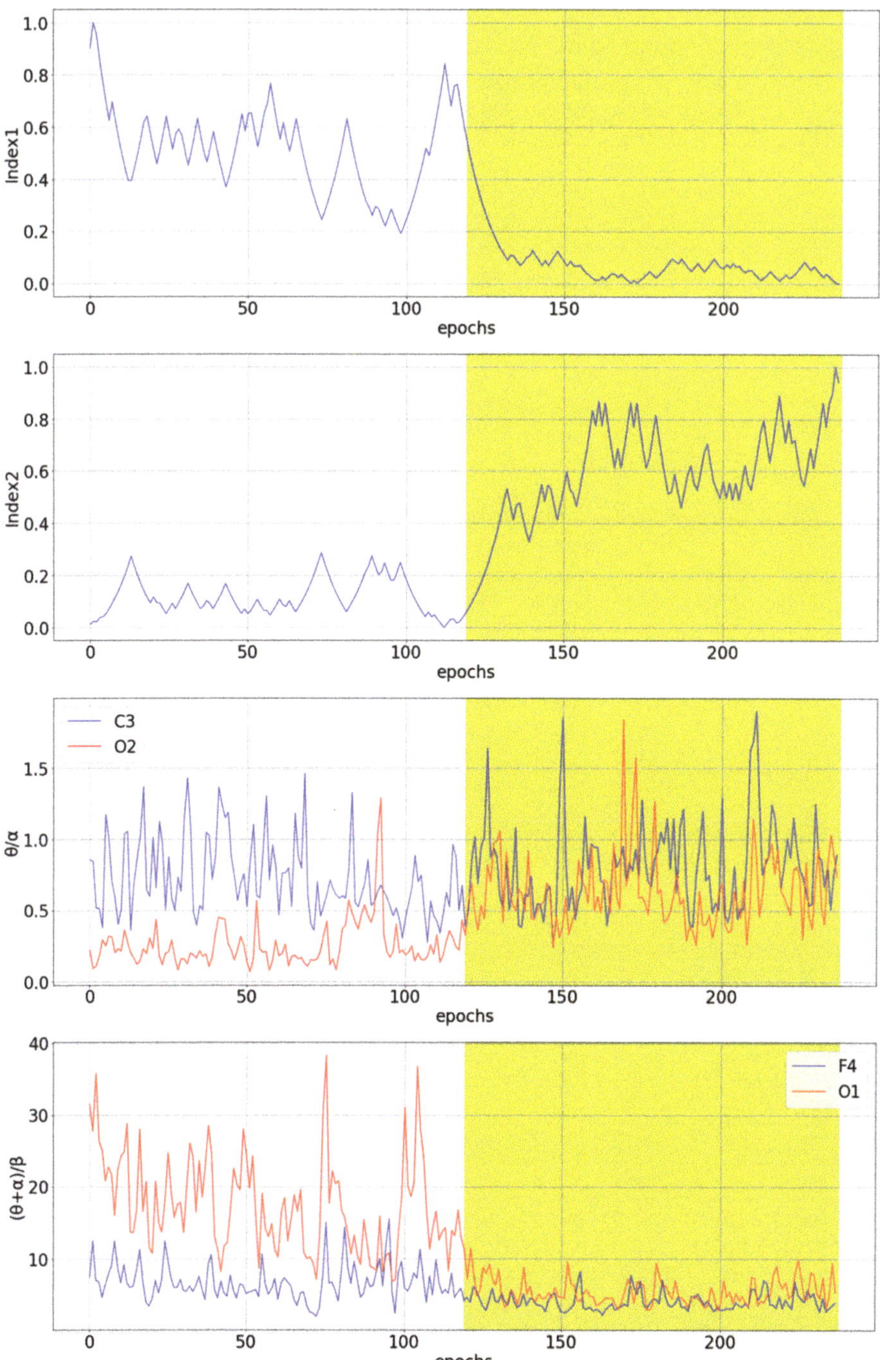

Figure 4. The comparison of the two novel multichannel indices with the best and the worst channel for θ/α and $(\theta + \alpha)/\beta$ single-channel indices for subject tr08-0111. The white part of the diagram represents an awake state, while the yellow part of the diagram represents stage 1 of sleep, i.e., a drowsiness state.

3.1. Statistical Analysis

The Wilcoxon signed-rank test [55] was used to analyze the statistical differences between the awake state and the S1 state. This test was chosen because it refers to data that do not necessarily follow the normal distribution. Table 2 shows p-values for each subject in the training set and each index. The significance level $\alpha_0 = 0.01$ was used together with the Bonferroni correction [55] to reduce the probability of false-positive results, as the test was repeated 144 times (16 subjects and 9 indices), giving us the final $\alpha_p = 6.9 \times 10^{-5}$. For the existing indices, the p-value was calculated for each channel, but only the p-values of the best channel (the lowest average of p-value for all subjects) are shown in Table 2.

The two novel indices show p-values lower than α_p for most subjects. From this, we can conclude that, for Index1, 14 of 16 subjects show two different distributions for the W stage and the S1 stage, while 13 of 16 subjects show significantly different distributions of the W stage and the S1 stage for Index2. There are only two existing indices where the p-value is lower than αp in more than ten cases. These are θ/β and $(\theta + \alpha)/(\alpha + \beta)$, both by Jap et al. [37].

Table 3 shows p-values for each subject in the test set and each index. Again, the two novel indices, together with the $(\gamma + \beta)/(\delta + \alpha)$ [10] index, show p-values lower than α_p for most subjects.

3.2. Drowsiness Prediction Analysis

An additional comparison of ratio indices was performed by analyzing the drowsiness detection accuracy and precision, as obtained with the XGBoost algorithm [56]. Default parameters were applied: learning rate eta equal to 0.3, gamma equal to 0 and a maximum depth of a tree equal to 6. For a detailed comparison of the indices, classification accuracy and precision were calculated for each subject. Namely, each subject has 238 epochs of the measured signal, with the first half representing the W state and the second half the S1 state. The algorithm classified the subject's state for each epoch (238 classifications per subject), and the accuracy for each subject was calculated based on these classifications. The leave-one-subject-out cross-validation method was applied on the training set, i.e., the algorithm was trained on the data of 15 subjects and tested on the subject excluded from the training set, and this was repeated 16 times to evaluate drowsiness detection on each subject from the training set. Table 4 shows the classification accuracy achieved on the training set.

Table 2. Statistical significance p-values were obtained by the Wilcoxon signed-rank test for distinguishing the awake state from the S1 state. The shaded green cells with bold text represent the lowest p-value for each subject in the training set. At the bottom, the index having p-values lower than α_p for most subjects is marked in the same way.

Subject	Index1	Index2	θ/α	β/α	$(\theta + \alpha)/\beta$	θ/β	$(\theta + \alpha)/(\alpha + \beta)$	γ/δ	$(\gamma + \beta)/(\delta + \alpha)$
	This Work		Eoh et al. [36]			Jap et al. [37]		da Silveira et al. [10]	
tr03-0092	1.12×10^{-6}	4.82×10^{-3}	1.23×10^{-4}	7.22×10^{-1}	4.97×10^{-2}	1.62×10^{-08}	**1.41×10^{-9}**	1.06×10^{-6}	3.09×10^{-4}
tr03-0256	3.92×10^{-8}	**1.43×10^{-13}**	2.42×10^{-6}	6.88×10^{-4}	1.11×10^{-3}	4.04×10^{-2}	$3.60 \times 10_{-1}$	9.08×10^{-3}	1.51×10^{-9}
tr03-0876	**1.37×10^{-20}**	3.87×10^{-17}	5.06×10^{-3}	1.63×10^{-7}	9.80×10^{-14}	2.94×10^{-6}	1.58×10^{-5}	1.09×10^{-1}	2.09×10^{-1}
tr03-1389	**9.77×10^{-11}**	1.35×10^{-8}	9.64×10^{-2}	2.06×10^{-1}	2.86×10^{-4}	1.27×10^{-1}	1.91×10^{-1}	1.82×10^{-1}	3.01×10^{-1}
tr04-0649	5.85×10^{-21}	**4.11×10^{-21}**	5.71×10^{-12}	2.38×10^{-9}	8.12×10^{-7}	1.89×10^{-1}	2.18×10^{-5}	3.33×10^{-7}	6.13×10^{-3}
tr04-0726	2.96×10^{-20}	3.19×10^{-20}	5.90×10^{-20}	2.31×10^{-16}	9.40×10^{-9}	2.78×10^{-14}	2.62×10^{-15}	2.36×10^{-19}	**2.19×10^{-20}**
tr05-1434	7.90×10^{-10}	9.79×10^{-13}	6.76×10^{-9}	3.70×10^{-10}	1.62×10^{-1}	3.96×10^{-17}	**2.00×10^{-19}**	5.42×10^{-17}	1.08×10^{-11}
tr05-1675	1.71×10^{-13}	1.85×10^{-11}	1.24×10^{-9}	7.82×10^{-3}	1.48×10^{-1}	1.10×10^{-14}	**4.47×10^{-16}**	1.58×10^{-2}	4.62×10^{-10}
tr07-0168	**2.88×10^{-21}**	5.15×10^{-21}	1.75×10^{-13}	4.73×10^{-11}	8.08×10^{-6}	1.05×10^{-8}	4.49×10^{-11}	8.87×10^{-16}	1.01×10^{-1}
tr07-0458	8.34×10^{-11}	**1.77×10^{-16}**	1.66×10^{-4}	1.77×10^{-4}	5.51×10^{-1}	1.32×10^{-2}	6.62×10^{-3}	3.68×10^{-4}	4.17×10^{-1}
tr07-0861	**2.88×10^{-21}**	3.11×10^{-21}	3.14×10^{-3}	3.55×10^{-7}	3.64×10^{-2}	9.96×10^{-8}	1.11×10^{-6}	1.49×10^{-17}	1.50×10^{-12}
tr08-0021	**2.88×10^{-21}**	**2.88×10^{-21}**	4.09×10^{-2}	2.54×10^{-9}	4.55×10^{-8}	3.34×10^{-10}	4.91×10^{-5}	4.19×10^{-13}	2.10×10^{-6}
tr08-0111	**2.88×10^{-21}**	**2.88×10^{-21}**	4.41×10^{-2}	7.54×10^{-5}	1.94×10^{-20}	2.04×10^{-20}	3.92×10^{-4}	4.50×10^{-15}	3.41×10^{-3}
tr09-0175	7.78×10^{-5}	7.92×10^{-2}	4.64×10^{-2}	3.10×10^{-4}	**5.35×10^{-14}**	2.23×10^{-5}	2.68×10^{-6}	7.18×10^{-2}	1.30×10^{-5}
tr $\times$ 10-0872	**2.62×10^{-15}**	1.96×10^{-14}	1.76×10^{-2}	3.89×10^{-2}	5.91×10^{-3}	5.09×10^{-6}	7.52×10^{-5}	2.14×10^{-5}	2.33×10^{-5}
tr13-0204	1.71×10^{-3}	6.62×10^{-1}	6.30×10^{-4}	4.91×10^{-5}	6.36×10^{-5}	2.91×10^{-10}	**2.63×10^{-10}**	5.59×10^{-1}	2.16×10^{-2}
No. subjects with $p < 6.9 \times 10^{-5}$	**14**	13	6	9	8	12	11	9	8

Table 3. Statistical significance *p*-values were obtained by the Wilcoxon signed-rank test for distinguishing the awake state from the S1 state. The shaded green cells with bold text represent the lowest *p*-value for each subject in the test set. At the bottom, the index having *p*-values lower than α_p for most subjects is marked in the same way.

Subject	Index1	Index2	θ/α	β/α	$(\theta+\alpha)/\beta$	θ/β	$(\theta+\alpha)/(\alpha+\beta)$	γ/δ	$(\gamma+\beta)/(\delta+\alpha)$
	This Work		Eoh et al. [36]			Jap et al. [37]		da Silveira et al. [10]	
tr04-0653	3.19×10^{-9}	2.71×10^{-9}	2.61×10^{-7}	1.80×10^{-5}	6.83×10^{-1}	3.63×10^{-9}	2.17×10^{-8}	2.69×10^{-6}	**3.33×10^{-10}**
tr05-0028	**1.26×10^{-9}**	2.36×10^{-7}	3.78×10^{-3}	2.99×10^{-1}	3.45×10^{-1}	3.56×10^{-1}	5.16×10^{-2}	1.70×10^{-1}	4.36×10^{-2}
tr05-0332	**2.88×10^{-21}**	**2.88×10^{-21}**	8.04×10^{-16}	2.10×10^{-5}	2.66×10^{-3}	7.39×10^{-14}	8.17×10^{-18}	1.84×10^{-16}	3.00×10^{-16}
tr07-0127	**3.71×10^{-18}**	1.27×10^{-15}	8.92×10^{-2}	1.44×10^{-2}	3.44×10^{-5}	1.38×10^{-2}	2.51×10^{-5}	9.07×10^{-16}	3.36×10^{-16}
tr08-0157	**2.88×10^{-21}**	**2.88×10^{-21}**	6.51×10^{-5}	1.26×10^{-7}	9.66×10^{-1}	6.67×10^{-3}	1.85×10^{-3}	2.82×10^{-11}	5.92×10^{-11}
tr09-0328	**1.80×10^{-10}**	1.14×10^{-2}	7.90×10^{-10}	9.56×10^{-7}	1.70×10^{-7}	2.01×10^{-1}	1.92×10^{-3}	3.54×10^{-2}	7.01×10^{-5}
tr09-0453	2.73×10^{-1}	7.80×10^{-4}	1.63×10^{-1}	2.96×10^{-1}	1.03×10^{-7}	3.89×10^{-2}	3.17×10^{-2}	**8.30×10^{-16}**	6.45×10^{-9}
tr12-0255	1.37×10^{-2}	5.55×10^{-10}	4.31×10^{-19}	2.17×10^{-18}	1.89×10^{-16}	**2.20×10^{-19}**	8.22×10^{-19}	1.11×10^{-7}	2.33×10^{-7}
tr12-0441	2.76×10^{-11}	9.26×10^{-9}	6.95×10^{-4}	6.89×10^{-1}	2.46×10^{-3}	**3.63×10^{-13}**	1.53×10^{-4}	2.13×10^{-3}	5.68×10^{-8}
tr13-0170	7.59×10^{-7}	3.60×10^{-5}	3.74×10^{-2}	4.73×10^{-2}	1.91×10^{-4}	9.71×10^{-4}	2.16×10^{-2}	**6.07×10^{-17}**	5.23×10^{-16}
tr13-0508	2.69×10^{-1}	7.61×10^{-2}	1.87×10^{-5}	1.99×10^{-2}	**1.10×10^{-9}**	9.17×10^{-2}	6.22×10^{-5}	7.26×10^{-5}	7.26×10^{-2}
tr13-0653	1.09×10^{-16}	**7.36×10^{-20}**	4.90×10^{-8}	4.16×10^{-4}	8.05×10^{-2}	1.66×10^{-2}	1.03×10^{-9}	3.47×10^{-2}	1.40×10^{-5}
No. subjects with $p < 6.9 \times 10^{-5}$	**9**	**9**	7	5	5	4	6	7	**9**

Table 4. The classification accuracy was obtained with the XGBoost algorithm for each subject in the training set. The shaded green cells with bold text show the highest accuracy obtained for each subject. At the bottom, the best mean accuracy for each ratio index is marked in the same way.

Subject	Index1	Index2	θ/α	β/α	$(\theta+\alpha)/\beta$	θ/β	$(\theta+\alpha)/(\alpha+\beta)$	γ/δ	$(\gamma+\beta)/(\delta+\alpha)$
	This Work		Eoh et al. [36]			Jap et al. [37]		da Silveira et al. [10]	
tr03-0092	0.5420	0.5252	**0.6387**	0.5168	0.5504	0.6092	0.6050	0.4684	0.5527
tr03-0256	0.5924	0.6303	0.5840	0.5504	0.5672	**0.6345**	**0.6345**	0.4473	0.4346
tr03-0876	0.6387	0.5672	0.6008	0.6218	0.4748	**0.6471**	0.6303	0.5781	0.5992
tr03-1389	0.3487	0.3908	0.4874	0.5588	0.5042	0.5378	0.4664	**0.5696**	0.5148
tr04-0649	0.6975	**0.8025**	0.5462	0.5462	0.5630	0.4832	0.5210	0.6118	0.5527
tr04-0726	**0.7983**	0.7605	0.6681	0.6176	0.5966	0.6134	0.6765	0.7637	0.7511
tr05-1434	0.3739	0.3697	0.4160	0.6218	**0.7059**	0.5882	0.6933	0.5485	0.5063
tr05-1675	0.6849	0.6513	0.6933	0.5546	0.6218	0.5630	**0.7227**	0.5781	0.5781
tr07-0168	0.7773	**0.8193**	0.6008	0.5504	0.5630	0.6092	0.6303	0.4473	0.5105
tr07-0458	0.3109	0.2689	**0.5378**	0.4748	0.5084	**0.5378**	0.5504	0.4684	0.4979
tr07-0861	0.6933	**0.7269**	0.5378	0.5420	0.5504	0.5168	0.5714	0.6540	0.6160
tr08-0021	**0.7857**	0.6387	0.5630	0.4874	0.4664	0.3866	0.4748	0.6329	0.6245
tr08-0111	0.6891	**0.8025**	0.7143	0.7101	0.7227	0.5462	0.4748	0.6287	0.6118
tr09-0175	0.6050	0.5252	0.5966	0.4748	0.5420	0.6218	0.6008	0.5401	**0.6498**
tr10-0872	**0.6134**	0.6008	0.5084	0.5168	0.4874	0.5252	0.4832	0.5274	0.4810
tr13-0204	0.5042	0.4538	**0.6597**	0.4790	0.6387	0.6471	0.6303	0.5570	0.5570
Average	**0.6035**	0.5959	0.5846	0.5515	0.5664	0.5667	0.5853	0.5638	0.5649

Index1 has the highest average accuracy and the highest classification accuracy for 3 of 16 subjects. Index2 has the second-highest average accuracy and the highest classification accuracy for 4 of 16 subjects, which is the most of all indices. θ/α [36] and $(\theta+\alpha)/(\alpha+\beta)$ [37] are the only other indices with an average classification accuracy above 0.58, while θ/α [36] and θ/β [37] are the only other indices with the highest accuracy for 3 of 16 subjects. The β/α [36] index has the lowest average classification accuracy on the training set (0.5515).

Table 5 shows the classification accuracy on the test set. Index1 has the highest average accuracy and the highest classification accuracy for 3 of 12 subjects. Index2 has the third-highest average accuracy and the highest classification accuracy for 4 of 12 subjects, which is the most of all indices. The only other index with comparable accuracy is θ/α [36], with the second-highest average accuracy. All other indices have at least 2.5% lower accuracy than the two novel indices.

Table 6 shows the degree of precision of drowsiness detection on the training set. Index2 has the highest average precision of drowsiness detection and the highest precision of drowsiness detection for five subjects, which is the highest of all indices. Index1 has the second-best average precision of drowsiness detection. $(\theta + \alpha)/(\alpha + \beta)$ [37] and γ/δ [10] have a precision of drowsiness detection comparable to Index1 and Index2, while all other ratio indices have lower precision.

Table 5. The classification accuracy was obtained with the XGBoost algorithm for each subject in the test set. The shaded green cells with bold text show the highest accuracy obtained for each subject. At the bottom, the best mean accuracy for each ratio index is marked in the same way.

Subject	Index1	Index2	θ/α	β/α	$(\theta + \alpha)/\beta$	θ/β	$(\theta + \alpha)/(\alpha + \beta)$	γ/δ	$(\gamma + \beta)/(\delta + \alpha)$
	This Work		Eoh et al. [36]			Jap et al. [37]		da Silveira et al. [10]	
tr04-0653	0.5672	**0.6345**	0.6303	0.5042	0.5168	0.5840	0.5630	0.5612	0.5738
tr05-0028	0.4454	0.4202	0.5294	**0.5588**	0.4664	0.4076	0.5462	0.5021	0.5443
tr05-0332	0.8067	**0.8277**	0.7563	0.5630	0.5294	0.6555	0.6134	0.5654	0.6118
tr07-0127	0.5462	**0.6092**	0.4916	0.5294	0.5252	0.5000	0.4118	0.4304	0.4051
tr08-0157	0.6303	**0.6681**	0.5294	0.5294	0.5000	0.5084	0.5294	0.5232	0.4810
tr09-0328	**0.6050**	0.5084	0.5966	0.5168	0.5000	0.5042	0.5672	0.5738	0.5401
tr09-0453	0.5588	0.5252	0.5714	0.5420	0.5294	0.5504	**0.5840**	0.5105	0.4557
tr12-0255	0.5420	0.5546	**0.6639**	0.5462	0.4748	0.5630	0.5924	0.6329	0.5654
tr12-0441	**0.6891**	0.5756	0.5840	0.5168	0.5378	0.5630	0.5798	0.6498	0.5274
tr13-0170	0.6008	0.5630	0.5924	0.5546	0.6261	**0.6471**	0.5336	0.5612	0.4430
tr13-0508	0.4538	0.5084	**0.6555**	0.6008	0.5588	0.6261	0.6092	0.5654	0.5443
tr13-0653	**0.6807**	0.6303	0.5210	0.5462	0.5630	0.5336	0.6008	0.5359	0.5063
Average	**0.5938**	0.5854	0.5935	0.5424	0.5273	0.5536	0.5609	0.5510	0.5165

Table 6. The precision of drowsiness detection was obtained with the XGBoost algorithm for each subject in the training set. The shaded green cells with bold text show the highest precision obtained for each subject. At the bottom, the best mean precision for each ratio index is marked in the same way.

Subject	Index1	Index2	θ/α	β/α	$(\theta + \alpha)/\beta$	θ/β	$(\theta + \alpha)/(\alpha + \beta)$	γ/δ	$(\gamma + \beta)/(\delta + \alpha)$
	This Work		Eoh et al. [36]			Jap et al. [37]		da Silveira et al. [10]	
tr03-0092	0.5439	0.5439	0.5439	0.5439	0.5439	0.5439	0.5439	0.5439	0.5439
tr03-0256	0.6222	**0.6348**	0.5676	0.5349	0.5571	0.6127	0.6096	0.4270	0.3974
tr03-0876	0.6514	0.5800	0.6765	0.5973	0.4808	**0.7397**	0.7123	0.5804	0.6264
tr03-1389	0.2273	0.3774	0.4906	0.5455	0.5034	0.5391	0.4556	**0.5519**	0.5120
tr04-0649	0.7582	0.8273	0.5733	0.7895	0.7778	0.4878	0.5294	**0.9063**	0.6000
tr04-0726	0.8318	**1.0000**	0.7128	0.6373	0.6055	0.6627	0.7333	0.8370	0.7706
tr05-1434	0.4051	0.4083	0.1429	0.6028	0.7168	0.7692	**0.7805**	0.6571	0.5027
tr05-1675	0.6642	0.6011	0.6885	0.5607	0.6559	0.5862	**0.7265**	0.5425	0.5433
tr07-0168	0.7500	**0.7923**	0.5759	0.5375	0.5397	0.5730	0.5963	0.4488	0.5156
tr07-0458	0.2816	0.0492	0.5437	0.4789	0.5088	0.5446	**0.5732**	0.4500	0.4930
tr07-0861	0.6264	**0.6688**	0.5338	0.5342	0.5349	0.5105	0.5521	0.6216	0.5780
tr08-0021	**0.7464**	0.6854	0.6119	0.4717	0.4535	0.4000	0.4688	0.6325	0.6355
tr08-0111	0.6692	**0.8214**	0.7297	0.8205	0.7912	0.5314	0.4840	0.6500	0.5985
tr09-0175	**0.6404**	0.5263	0.5742	0.4762	0.5379	0.6142	0.6053	0.5273	0.6207
tr10-0872	**0.6174**	0.5909	0.5045	0.5130	0.4892	0.5254	0.4882	0.5349	0.4627
tr13-0204	0.5054	0.4545	0.6357	0.4820	0.6170	**0.6636**	0.6348	0.6032	0.5970
Average	0.5963	**0.5976**	0.5691	0.5704	0.5821	0.5815	0.5934	0.5946	0.5623

Table 7 shows the degree of precision of drowsiness detection achieved on the test set. Index1 has the highest average precision. θ/α [36] and γ/δ [10] have 1% lower precision than Index1, while all other indices have at least 4% lower precision. Index2 has the second-highest average precision.

Table 7. The precision of drowsiness detection was obtained with the XGBoost algorithm for each subject in the test set. The shaded green cells with bold text show the highest precision obtained for each subject. At the bottom, the best mean precision for each ratio index is marked in the same way.

Subject	Index1	Index2	θ/α	β/α	$(\theta+\alpha)/\beta$	θ/β	$(\theta+\alpha)/(\alpha+\beta)$	γ/δ	$(\gamma+\beta)/(\delta+\alpha)$
	This Work		Eoh et al. [36]			Jap et al. [37]		da Silveira et al. [10]	
tr04-0653	0.5769	0.6569	**0.6742**	0.5030	0.5133	0.6000	0.5862	0.5795	0.5914
tr05-0028	0.3818	0.4296	0.5178	0.5385	0.4762	0.4214	0.5364	0.5000	**0.5862**
tr05-0332	0.7483	**0.9333**	0.7905	0.5547	0.5321	0.7846	0.6709	0.8571	0.8824
tr07-0127	0.5401	**0.6512**	0.4722	0.5294	0.5231	0.5000	0.3600	0.4309	0.4207
tr08-0157	0.5812	**0.6087**	0.5158	0.5574	0.5000	0.5048	0.5153	0.5122	0.4886
tr09-0328	**0.6147**	0.5078	0.5650	0.5137	0.5000	0.5041	0.5678	0.5620	0.5372
tr09-0453	0.5398	0.5176	0.5436	0.5301	0.5248	0.5306	**0.5538**	0.5049	0.4721
tr12-0255	0.5316	0.5351	**0.6054**	0.5342	0.4826	0.5393	0.5545	0.5886	0.5389
tr12-0441	**0.8000**	0.5789	0.5633	0.5093	0.5249	0.5478	0.5785	0.6496	0.5231
tr13-0170	0.6333	0.5466	0.5724	0.5355	0.6056	0.6296	0.5313	**0.6944**	0.3478
tr13-0508	0.4337	0.5091	**0.6331**	0.5674	0.5443	0.6154	0.6140	0.5478	0.5352
tr13-0653	**0.7048**	0.6000	0.5177	0.5314	0.5419	0.5213	0.5845	0.5392	0.5037
Average	**0.5905**	0.5896	0.5809	0.5337	0.5224	0.5582	0.5544	0.5805	0.5356

3.3. Computational Analysis

With regard to the classification and the use of machine learning algorithms, an advantage of using the novel multichannel indices compared to the existing single-channel indices is also the saving of memory and time, due to the reduction of dimensionality. The accuracies of Index1 and Index2 from Table 4 were achieved with the model constructed from the single feature only, while all other indices had six features since the dataset contains six EEG channels. For this reason, storing the novel indices consumes six times less memory. The time consumption was measured as an average of 100 executions. The measured time included classifier initialization, classifier training, classifications on the test subject and calculation of classification accuracy. Table 8 shows the results of time consumption measurements. The use of the novel multichannel indices saves about 30% of time compared to all other traditionally used single-channel ratio indices.

Table 8. The average time of 100 executions of the XGBoost classifier's initialization, training, classifications on the test subject and calculation of classification accuracy, expressed in milliseconds. The shaded green cells with bold text represent the best values for each subject and the best average value.

Subject	Index1	Index2	θ/α	β/α	$(\theta+\alpha)/\beta$	θ/β	$(\theta+\alpha)/(\alpha+\beta)$	γ/δ	$(\gamma+\beta)/(\delta+\alpha)$
	This Work		Eoh et al. [36]			Jap et al. [37]		da Silveira et al. [10]	
tr03-0092	**86.3772**	86.9689	122.7764	124.0313	129.9221	129.5543	130.6956	128.6490	128.9000
tr03-0256	**86.7446**	87.1034	123.3161	123.2911	130.3844	130.2867	130.6147	128.6415	128.7518
tr03-0876	**86.3508**	87.0188	122.6970	123.5249	130.2485	130.7789	130.6456	128.7160	128.3419
tr03-1389	**85.9344**	86.8811	122.1586	123.8243	130.4414	130.1170	131.3382	129.2281	128.8390
tr04-0649	**86.9833**	87.5527	123.6650	124.0832	130.2565	130.0574	130.2316	129.2234	128.8357
tr04-0726	**86.5498**	87.5921	123.7690	123.6945	129.6285	129.6750	130.6256	128.9002	128.6363
tr05-1434	**86.5450**	86.6549	123.0505	130.6853	131.9267	130.6205	131.3138	130.6215	131.3438
tr05-1675	**87.0534**	87.4627	123.3135	130.0399	130.4660	129.1552	129.5943	130.5365	128.9256
tr07-0168	**86.9143**	87.2251	122.9559	129.6185	130.5158	130.1070	129.7780	129.6381	128.7795
tr07-0458	**86.5651**	86.8690	122.6074	129.9533	130.1468	130.2667	130.4915	128.8906	129.5788
tr07-0861	**86.8634**	87.2801	122.7545	130.2319	130.0104	128.1135	129.9124	129.3949	128.7760
tr08-0021	**86.9239**	88.9566	123.0910	130.0868	129.1948	130.0221	130.1670	129.3241	129.1652
tr08-0111	**86.6697**	87.4626	122.7216	130.3879	130.5019	130.4011	129.8419	128.8413	129.1940
tr09-0175	**87.3240**	87.3827	123.4803	129.6729	130.9953	130.1271	131.1785	128.7062	128.9786
tr10-0872	**86.9690**	87.5381	124.0918	130.5091	129.4638	130.6490	130.2964	129.4843	129.3599
tr13-0204	87.2509	**87.1135**	123.2010	131.7928	130.4062	131.4087	130.3568	128.9199	128.1530
Average	**86.7512**	87.3164	123.1031	127.8392	130.2818	130.0838	130.4426	129.2322	129.0350

4. Discussion

The main idea of our research was to combine frequency-domain features from different brain regions into a multichannel ratio index to improve frequency-domain features for the detection of drowsiness and to gain new insights into drowsiness. The results in Tables 2–8 suggest that two novel multichannel ratio indices improve the detection of drowsiness based on the frequency-domain features and reduce the time required for detection.

We must note that the main idea of this research was not to create the best possible model for drowsiness detection but only to bring improvement into frequency-domain features that are often used for drowsiness detection. Our focus was on developing the method for obtaining these novel indices, which is explained in Section 2.3 "Multi-Objective Optimization". In order to confirm that our conclusions also hold for other classifiers besides XGBoost, Table 9 shows the average accuracy on the test set obtained with Naïve Bayes, k nearest neighbors, logistic regression, decision tree, random forest and support vector machine classifiers (using the scikit-learn library at default settings). The average accuracies of two novel indices vary from 56% to 65% among the algorithms. All the algorithms show that our novel multichannel indices are better than existing single-channel indices.

Table 9. The average accuracy was obtained on the test set with different classification algorithms. Each row is colored with a pallet of colors ranging from dark green for the highest number in the row to dark red for the lowest number in the row. The algorithms are: NB—Naïve Bayes, KNN—k nearest neighbors, Logistic—logistic regression, DT—decision tree, RF—random forest and SVM—support vector machine.

Algorithm	Index1	Index2	θ/α	β/α	$(\theta + \alpha)/\beta$	θ/β	$(\theta + \alpha)/(\alpha + \beta)$	γ/δ	$(\gamma + \beta)/(\delta + \alpha)$
	This Work		Eoh et al. [36]		Jap et al. [37]			da Silveira et al. [10]	
NB	0.6399	0.6535	0.5947	0.5462	0.5432	0.5308	0.5663	0.5316	0.5277
KNN	0.5785	0.5840	0.5588	0.5387	0.5399	0.5452	0.5525	0.5378	0.5277
Logistic	0.6396	0.6543	0.6029	0.5131	0.5383	0.5626	0.5735	0.5793	0.5613
DT	0.5717	0.5629	0.5456	0.5050	0.5074	0.5420	0.5267	0.5356	0.5223
RF	0.5719	0.5659	0.5762	0.5380	0.5360	0.5549	0.5501	0.5321	0.5222
SVM	0.6325	0.6526	0.6200	0.5695	0.5714	0.5731	0.5801	0.5541	0.5478

Our results were compared with the seven existing single-channel ratio indices that are currently state-of-the-art frequency-domain features. The newest one was introduced in 2016 [10], but all of these single-channel ratio indices are often used in the more recent drowsiness detection papers [57–59].

The authors in the aforementioned research report 92% accuracy as the best-obtained accuracy [57]. This accuracy was obtained based on the epoch-level validation. Epoch-level validation is a cross-validation procedure on the epoch level, which means that there is a very high probability that all subjects will have epochs in the training set and in the test set at the same time. On the other hand, subject-level validation is validation where it is ensured that subjects in the test set are not contained in the training set. An example of a subject-level validation is the leave-one-subject-out cross-validation that we used in this research. The only proper way for model validation is subject-level validation, as it represents the real-life setting in which the data from a new subject are used only for testing the model. Empirical tests conducted in related research showed a large difference in the accuracies between epoch-level validation and subject-level validation [60].

In a study from Mehreen et al. [57], the authors also provide subject-level validation, and the accuracy achieved was 71.5% based on 15 frequency-domain features. The highest accuracy achieved in our research is shown in Table 9, and it was 65.45%, achieved by logistic regression. This 65.45% accuracy is relatively close to 71.5%, and it must be noted that it was obtained based only on the Index2 feature, with a simple algorithm and without any parameter optimization. Due to this, we are confident that the addition of our two multichannel ratio indices would lead to an improvement in all state-of-the-art drowsiness detection systems that use frequency-domain features. Again, our aim was not to create the best possible drowsiness detection model but to prove that the novel multichannel indices are better than the existing single-channel frequency-domain features.

The Equations (4) and (6) for these multichannel ratio indices, obtained after optimizing the parameters with the optimization algorithm, suggest that alpha and delta are two of the most important frequency power bands for drowsiness detection. Equation (6) suggests that delta power in the frontal region describes drowsiness better than in the central region, while alpha power in the occipital and central regions describes drowsiness better than in the frontal region.

These results are consistent with several previous research papers on drowsiness detection that reported the importance of increasing alpha power [22,35,61,62]. Delta power is usually only present in deep sleep stages [36], so some researchers studying drowsiness do not include delta in their research [63]. However, there is still much research that includes delta power. The increase in delta power is considered to be an indicator of drowsiness [4]. Our research found that theta and beta powers are not as good drowsiness indicators as alpha and delta powers, while many other research studies disagree. A decrease in beta power was found to be an indicator of drowsiness in [4,36,64,65] and an increase in theta power was found to be an indicator of drowsiness in [27,34,61,62,65]. Wang et al. [38], in their study of microsleep events, found that alpha and delta rhythms characterize microsleep events. As mentioned earlier, there is an inconsistency in terminology, and some researchers consider sleep stage S1 as drowsiness [5–9], while Johns [11] considers it equivalent to microsleep events in the driving scenario. We used the data from sleep stage S1 and referred to it in this research as drowsiness. Since our results suggest that delta and alpha are the most significant for the detection of drowsiness, as in the work of Wang et al. [38] on microsleep events, our work suggests that sleep stage S1 may be more similar to microsleep events than to drowsiness, but further research is needed to support this as a fact.

Apart from the indication that drowsiness is closely related to microsleep events, it may also be closely linked to driver fatigue. Some researchers even use the term fatigue as a synonym for drowsiness [66]. Fatigue is a consequence of prolonged physical or mental activity [67] and can lead to drowsiness [68]. Normally, rest and inactivity relieve fatigue, however, they exacerbate drowsiness [69]. Lal and Craig [70] found that delta and theta band activities increase significantly during fatigue. Craig et al. [71] reported significant changes in the alpha 1, alpha 2, theta and beta bands, while they did not find any significant changes in the delta band when observing driver fatigue. Simon et al. [68] report that alpha band power and alpha spindles correlate with fatigue.

These three research papers [68,70,71] all use visual inspection to define the ground truth of fatigue. This approach to defining the ground truth is prone to subjectivity. A similar problem occurs when drowsiness is defined by using subjective drowsiness ratings, such as the Karolinska sleepiness scale [72].

Driver drowsiness, driver fatigue and microsleep events are defined as different internal states of the brain, but show similar behavior when observing the features obtained from the EEG. Possible explanations could be that fatigue, drowsiness and microsleep have a similar effect on brain functions and cause the driver's inability to function at the desired level. Most researchers of these three driver states only use frequency-domain features, while there are a number of other features (nonlinear features [30], spatiotemporal features [31] and entropies [32]) that could be used. Further studies with these features could find some features of the EEG signal that distinguish drowsiness, fatigue and microsleep. Distinguishing features of these three brain states could lead to the exact definitions of these terms. Precise and standardized definitions of fatigue, drowsiness and microsleep would help researchers to compare their work more easily.

Figure 4 shows that the proposed procedure for creating the novel ratio indices has succeeded in creating step-like indices for a given subject. In addition to Index1 and Index2, which show desirable behavior, the indices $(\theta + \alpha)/\beta$ and θ/β show similar, favorable behavior for a few channels. Figure 5 shows a comparison of novel ratio indices with the best and the worst channel for γ/δ and $(\gamma + \beta)/(\delta + \alpha)$ single-channel indices for subject tr04-0726. Index1, index2, θ/α [36], $(\theta + \alpha)/\beta$, $(\theta + \alpha)/(\alpha + \beta)$ [37] and γ/δ [10] show similar behavior. These indices seem to detect drowsiness well, but with about a 50 epochs delay. Since several different single-channel indices that were previously shown to correlate with drowsiness together with two novel multichannel indices show the same delay in detecting drowsiness, this suggests that there may be shortcomings in the labeling of the initial signals. The manual for scoring sleep [42] provides guidelines for labeling, and it may be possible that the professionals who labeled the sleep signals labeled an approximate time of transition from the W state to the S1 state, as it is known that labeling any kind of several-hour-long EEG signal is a very tedious, hard and time-consuming job [73]. For this reason, the loose transition window is applied in the optimization algorithm, as described in Section 2.3.

The main shortcoming in applying our approach is the need to place six EEG electrodes on the driver's scalp while driving. Apart from being intrusive, there is also a problem with noise in real-world applications that cannot be neutralized with the current state-of-the-art filter technology. All electrophysiological signals measured with wearable devices have a similar problem with intrusiveness and noise. ECG measurements, for example, are somewhat less susceptible to noise than EEG. Several recent works have shown that ECG can be used as a good predictor of sleep stages based on deep learning classifiers. Sun et al. [74] combined ECG with abdominal respiration and obtained a kappa value of 0.585, while Sridhar et al. [75] obtained a kappa value of 0.66. Combining EEG and ECG measurements has also been proposed in the context of driver drowsiness detection under simulator-based laboratory conditions [76]. Despite the problems of intrusiveness and noise susceptibility, research based on the electrophysiological signals brings a shift towards a precise definition of drowsiness. Once there is an exact definition of drowsiness or at least guidelines and manuals that accurately describe drowsiness (similar to the manuals for evaluating sleep stages), a big step will be taken to solve the problem of early detection of drowsiness [77]. It is doubtful that a wearable system based on electrophysiological signals will ever be widely used in real-world driving, but they still need to be developed. In our opinion, such wearable electrophysiological devices are more likely to be used for calibration/validation of non-intrusive systems (such as the driving performance-based or video-based systems) in controlled/simulated driving scenarios. In such scenarios, it is possible to control ambient noise, leading to a reduction in the effects of noise sensitivity.

An additional limitation of this work is that we were able to download data from 393 of 992 subjects completely, and only 28 of these 393 subjects were included in our study due to the inclusion condition that we described in Section 2.1 "Dataset, Preprocessing and Feature Extraction". Although it is a small subset of data, with the use of 12 subjects as a test set, we showed that the dataset is large enough to provide a good generalization (as seen in Tables 3, 5 and 7). In a recent review paper about state-of-the-art drowsiness detection [33], the authors reviewed 39 papers, and the average number of subjects in the included works is 23.5, which also indicates that our number of subjects included in the current study (28) is acceptable.

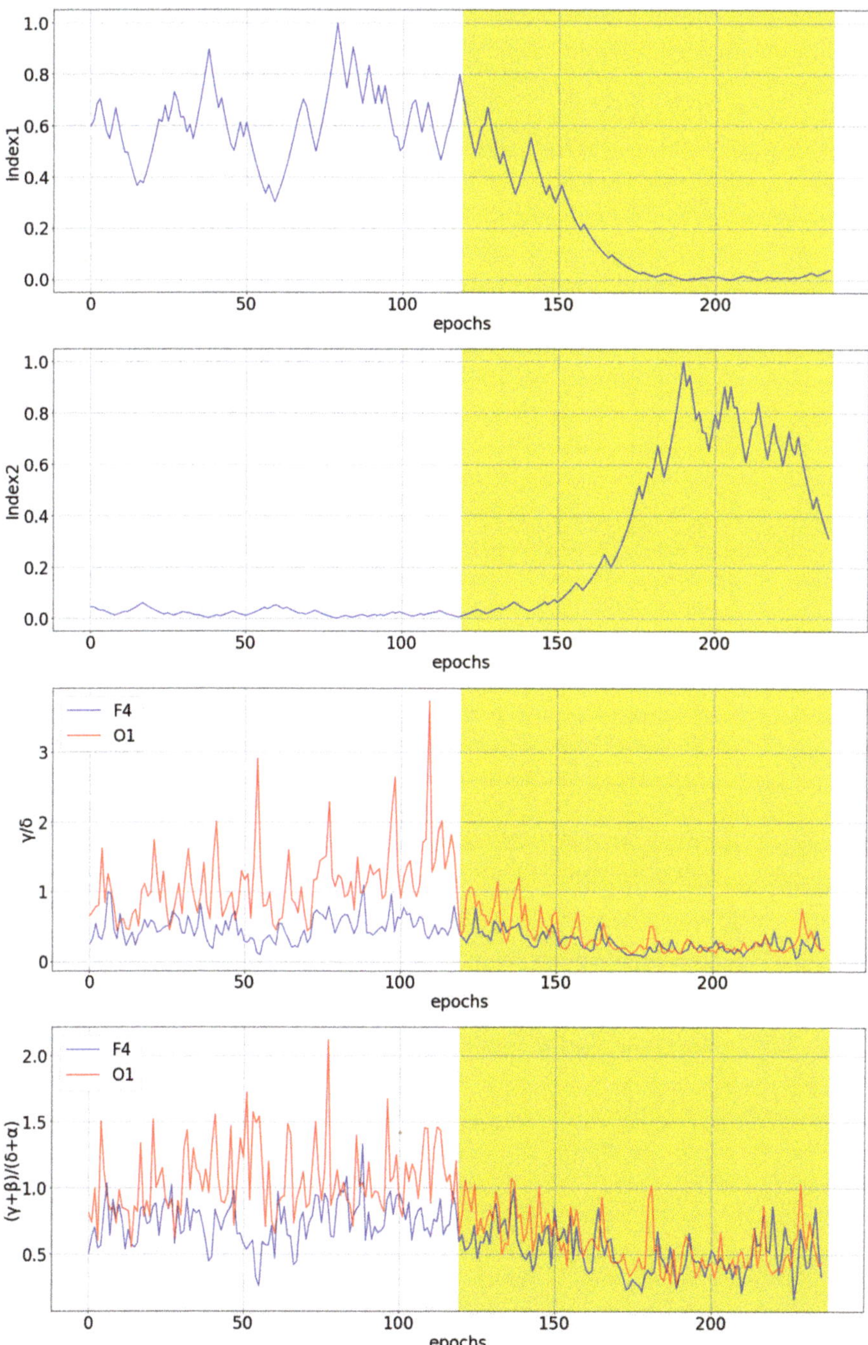

Figure 5. The comparison of the two novel multichannel indices with the best and the worst channel for γ/δ and $(\gamma + \beta)/(\delta + \alpha)$ single-channel indices for subject tr04-0726. The white part of the diagram represents the awake state, while the yellow part of the diagram represents the stage 1 of sleep, i.e., the drowsiness state.

5. Conclusions

This paper presented two novel multichannel ratio indices for the detection of drowsiness obtained by multi-objective optimization based on evolutionary computation. The results suggested that alpha and delta powers are good drowsiness indicators. The novel multichannel ratio indices were compared with seven existing single-channel ratio indices and showed better results in detecting drowsiness measured with precision and in the overall classification accuracy of both states using several machine learning algorithms. Our work suggests that a more precise definition of drowsiness is needed, and that accurate early detection of drowsiness should be based on multichannel frequency-domain ratio indices. The multichannel features also reduced the time needed for classification. The process of obtaining these indices by using a multi-objective optimization algorithm can also be applied to other areas of EEG signal analysis.

Research such as this, together with research on small hardware for physiology-based drowsiness detection, can eventually lead to an easy-to-use, non-intrusive device that reliably detects drowsiness. In addition, research on a reliable and standardized definition of drowsiness is needed and it would lead to improvements in the field of drowsiness detection.

Author Contributions: Conceptualization, I.S., N.F. and A.J.; methodology, I.S., N.F. and A.J.; software, I.S. and N.F.; validation, I.S., N.F., M.C. and A.J.; formal analysis, I.S., N.F. and A.J.; investigation, I.S.; resources, I.S. and N.F.; data curation, I.S.; writing—original draft preparation, I.S. and N.F.; writing—review and editing, I.S., N.F., M.C. and A.J.; visualization, I.S.; supervision, A.J.; project administration, M.C.; funding acquisition, M.C. All authors have read and agreed to the published version of the manuscript.

Funding: This work has been carried out within the project "Research and development of the system for driver drowsiness and distraction identification—DFDM" (KK.01.2.1.01.0136), funded by the European Regional Development Fund in the Republic of Croatia under the Operational Programme Competitiveness and Cohesion 2014–2020.

Institutional Review Board Statement: The study was conducted according to the guidelines of the Declaration of Helsinki, and approved by the Ethics Committee of the University of Zagreb, Faculty of Electrical Engineering and Computing (on 26 September 2020).

Informed Consent Statement: All the available information regarding patients is available from the PhysioNet portal [47], from the 2018 PhysioNet computing in cardiology challenge [48], at: https://physionet.org/content/challenge-2018/1.0.0/ (accessed on 24 September 2021).

Data Availability Statement: The data used in this paper were obtained from the PhysioNet portal [47], from the 2018 PhysioNet computing in cardiology challenge [48], at: https://physionet.org/content/challenge-2018/1.0.0/ (accessed on 24 September 2021).

Conflicts of Interest: The authors declare no conflict of interest.

References

1. Jackson, M.L.; Kennedy, G.A.; Clarke, C.; Gullo, M.; Swann, P.; Downey, L.A.; Hayley, A.C.; Pierce, R.J.; Howard, M.E. The utility of automated measures of ocular metrics for detecting driver drowsiness during extended wakefulness. *Accid. Anal. Prev.* **2016**, *127*, 127–133. [CrossRef]
2. Kamran, M.A.; Mannan, M.M.N.; Jeong, M.Y. Drowsiness, Fatigue and Poor Sleep's Causes and Detection: A Comprehensive Study. *IEEE Access* **2019**, *7*, 167172–167186. [CrossRef]
3. Lal, S.K.L.; Craig, A. A critical review of the psychophysiology of driver fatigue. *Biol. Psychol.* **2001**, *173*, 173–194. [CrossRef]
4. Papadelis, C.; Chen, Z.; Kourtidou-Papadeli, C.; Bamidis, P.; Chouvarda, I.; Bekiaris, E.; Maglaveras, N. Monitoring sleepiness with on-board electrophysiological recordings for preventing sleep-deprived traffic accidents. *Clin. Neurophysiol.* **2007**, *1906*, 1906–1922. [CrossRef]
5. Chowdhury, A.; Shankaran, R.; Kavakli, M.; Haque, M.M. Sensor Applications and Physiological Features in Drivers' Drowsiness Detection: A Review. *IEEE Sens. J.* **2018**, *3055*, 3055–3067. [CrossRef]
6. Oken, B.S.; Salinsky, M.C.; Elsas, S.M. Vigilance, alertness, or sustained attention: Physiological basis and measurement. *Clin. Neurophysiol.* **2006**, *1885*, 1885–1901. [CrossRef]

7. Majumder, S.; Guragain, B.; Wang, C.; Wilson, N. On-board Drowsiness Detection using EEG: Current Status and Future Prospects. In Proceedings of the 2019 IEEE International Conference on Electro Information Technology (EIT), Brookings, SD, USA, 16–18 May 2019; pp. 483–490.

8. Sriraam, N.; Shri, T.P.; Maheshwari, U. Recognition of wake-sleep stage 1 multichannel eeg patterns using spectral entropy features for drowsiness detection. *Australas. Phys. Eng. Sci. Med.* **2016**, *797*, 797–806. [CrossRef]

9. Budak, U.; Bajaj, V.; Akbulut, Y.; Atila, O.; Sengur, A. An Effective Hybrid Model for EEG-Based Drowsiness Detection. *IEEE Sens. J.* **2019**, *7624*, 7624–7631. [CrossRef]

10. da Silveira, T.L.; Kozakevicius, A.J.; Rodrigues, C.R. Automated drowsiness detection through wavelet packet analysis of a single EEG channel. *Expert Syst. Appl.* **2016**, *559*, 559–565. [CrossRef]

11. Johns, M.W. A new perspective on sleepiness. *Sleep Biol. Rhythm.* **2010**, *170*, 170–179. [CrossRef]

12. Moller, H.J.; Kayumov, L.; Bulmash, E.L.; Nhan, J.; Shapiro, C.M. Simulator performance, microsleep episodes, and subjective sleepiness: Normative data using convergent methodologies to assess driver drowsiness. *Psychosom J. Res.* **2006**, *335*, 335–342. [CrossRef] [PubMed]

13. Martensson, H.; Keelan, O.; Ahlstrom, C. Driver Sleepiness Classification Based on Physiological Data and Driving Performance From Real Road Driving. *IEEE Trans. Intell. Transp. Syst.* **2019**, *421*, 421–430. [CrossRef]

14. Phillips, R.O. A review of definitions of fatigue—And a step towards a whole definition. *Transp. Res. Part F Traffic Psychol. Behav.* **2015**, *48*, 48–56. [CrossRef]

15. Slater, J.D. A definition of drowsiness: One purpose for sleep? *Med. Hypotheses* **2008**, *641*, 641–644. [CrossRef]

16. Subasi, A. Automatic recognition of alertness level from EEG by using neural network and wavelet coefficients. *Expert Syst. Appl.* **2005**, *701*, 701–711. [CrossRef]

17. Orasanu, J.; Parke, B.; Kraft, N.; Tada, Y.; Hobbs, A.; Anderson, B.; Dulchinos, V. *Evaluating the Effectiveness of Schedule Changes for Air Traffic Service (ATS) Providers: Controller Alertness and Fatigue Monitoring Study*; Technical Report; Federal Aviation Administration, Human Factors Division: Washington, DC, USA, 2012.

18. Hart, C.A.; Dinh-Zarr, T.B.; Sumwalt, R.; Weener, E. *Most Wanted List of Transportation Safety Improvements: Reduce Fatigue-Related Accidents*; National Transportation Safety Board: Washington, DC, USA, 2018.

19. Gonçalves, M.; Amici, R.; Lucas, R.; Åkerstedt, T.; Cirignotta, F.; Horne, J.; Léger, D.; McNicholas, W.T.; Partinen, M.; Téran-Santos, J.; et al. Sleepiness at the wheel across Europe: A survey of 19 countries. *Sleep J. Res.* **2015**, *24*, 242–253. [CrossRef]

20. Balandong, R.P.; Ahmad, R.F.; Saad, M.N.M.; Malik, A.S. A Review on EEG-Based Automatic Sleepiness Detection Systems for Driver. *IEEE Access* **2018**, *6*, 22908–22919. [CrossRef]

21. Kundinger, T.; Sofra, N.; Riener, A. Assessment of the Potential of Wrist-Worn Wearable Sensors for Driver Drowsiness Detection. *Sensors* **2020**, *20*, 1029. [CrossRef]

22. Zheng, W.-L.; Gao, K.; Li, G.; Liu, W.; Liu, C.; Liu, J.-Q.; Wang, G.; Lu, B.-L. Vigilance Estimation Using a Wearable EOG Device in Real Driving Environment. *IEEE Trans. Intell. Transp. Syst.* **2020**, *170*, 170–184. [CrossRef]

23. Fu, R.; Wang, H.; Zhao, W. Dynamic driver fatigue detection using hidden Markov model in real driving condition. *Expert Syst. Appl.* **2016**, *397*, 397–411. [CrossRef]

24. Lin, C.-T.; Chuang, C.-H.; Tsai, S.-F.; Lu, S.-W.; Chen, Y.-H.; Ko, L.-W. Wireless and Wearable EEG System for Evaluating Driver Vigilance. *IEEE Trans. Biomed. Circuits Syst.* **2014**, *165*, 165–176.

25. Cassani, R.; Falk, T.H.; Horai, A.; Gheorghe, L.A. Evaluating the Measurement of Driver Heart and Breathing Rates from a Sensor-Equipped Steering Wheel using Spectrotemporal Signal Processing. In Proceedings of the 2019 IEEE Intelligent Transportation Systems Conference (ITSC), Auckland, New Zealand, 27–30 October 2019; pp. 2843–2847.

26. Li, Z.; Li, S.; Li, R.; Cheng, B.; Shi, J. Online Detection of Driver Fatigue Using Steering Wheel Angles for Real Driving Conditions. *Sensors* **2017**, *17*, 495. [CrossRef] [PubMed]

27. Aeschbach, D.; Matthews, J.R.; Postolache, T.T.; Jackson, M.A.; Giesen, H.A.; Wehr, T.A. Dynamics, of the human EEG during prolonged wakefulness: Evidence for frequency-specific circadian and homeostatic influences. *Neurosci. Lett.* **1997**, *239*, 121–124. [CrossRef]

28. Barua, S.; Ahmed, M.U.; Ahlström, C.; Begum, S. Automatic driver sleepiness detection using EEG, EOG and contextual information. *Expert Syst. Appl.* **2019**, *121*, 121–135. [CrossRef]

29. Wang, L.; Li, J.; Wang, Y. Modeling and Recognition of Driving Fatigue State Based on R-R Intervals of ECG Data. *IEEE Access* **2019**, *7*, 175584–175593. [CrossRef]

30. Stam, C.J. Nonlinear dynamical analysis of EEG and MEG: Review of an emerging field. *Clin. Neurophysiol.* **2005**, *2266*, 2266–2301. [CrossRef]

31. Bastos, A.M.; Schoffelen, M.J. A Tutorial Review of Functional Connectivity Analysis Methods and Their Interpretational Pitfalls. *Front. Syst. Neurosci.* **2016**, *9*, 175. [CrossRef] [PubMed]

32. Acharya, U.R.; Hagiwara, Y.; Deshpande, S.N.; Suren, S.; Koh, J.E.W.; Oh, S.L.; Arunkumar, N.; Ciaccio, E.J.; Lim, C.M. Characterization of focal EEG signals: A review. *Future Gener. Comput. Syst.* **2019**, *290*, 290–299. [CrossRef]

33. Stancin, I.; Cifrek, M.; Jovic, A. A Review of EEG Signal Features and Their Application in Driver Drowsiness Detection Systems. *Sensors* **2021**, *21*, 3786. [CrossRef]

34. Cajochen, C.; Brunner, D.P.; Krauchi, K.; Graw, P.; Wirz-Justice, A. Power Density in Theta/Alpha Frequencies of the Waking EEG Progressively Increases During Sustained Wakefulness. *Sleep* **1995**, *890*, 890–894. [CrossRef]

35. Astolfi, L.; Mattia, D.; Vecchiato, G.; Babiloni, F.; Borghini, G. Measuring neurophysiological signals in aircraft pilots and car drivers for the assessment of mental workload, fatigue and drowsiness. *Neurosci. Biobehav. Rev.* **2012**, *58*, 58–75.
36. Eoh, H.J.; Chung, M.K.; Kim, S.H. Electroencephalographic study of drowsiness in simulated driving with sleep deprivation. *Int. Ind. J. Ergon.* **2005**, *307*, 307–320. [CrossRef]
37. Jap, B.T.; Lal, S.; Fischer, P.; Bekiaris, E. Using EEG spectral components to assess algorithms for detecting fatigue. *Expert Syst. Appl.* **2009**, *36*, 2352–2359. [CrossRef]
38. Wang, C.; Guragain, B.; Verma, A.K.; Archer, L.; Majumder, S.; Mohamud, A.; Flaherty-Woods, E.; Shapiro, G.; Almashor, M.; Lenné, M.; et al. Spectral Analysis of EEG During Microsleep Events Annotated via Driver Monitoring System to Characterize Drowsiness. *IEEE Trans. Aerosp. Electron. Syst.* **2020**, *1346*, 1346–1356. [CrossRef]
39. Goldberger, A.L.; Amaral, L.A.N.; Glass, L.; Hausdorff, J.M.; Ivanov, P.C.; Mark, R.G.; Mietus, J.E.; Moody, G.B.; Peng, C.-K.; Stanley, H.E. PhysioBank, PhysioToolkit, and PhysioNet: Components of a new research resource for complex physiologic signals. *Circulation* **2000**, *101*, e215–e220. [CrossRef] [PubMed]
40. Ghassemi, M.; Moody, B.; Lehman, L.-W.; Song, C.; Li, Q.; Sun, H.; Westover, B.; Clifford, G. You Snooze, You Win: The PhysioNet/Computing in Cardiology Challenge 2018. In Proceedings of the 2018 Computing in Cardiology Conference (CinC), Maastricht, The Nethelands, 23–26 September 2018.
41. Institute of Medicine (US) Committee on Sleep Medicine and Research; Colten, H.; Altevogt, B. Sleep Physiology. In *Sleep Disorders and Sleep Deprivation: An Unmet Public Health Problem*; National Academies Press (US): Washington, DC, USA, 2006.
42. Berry, R.B.; Quan, S.F.; Abreu, A.R.; Bibbs, M.L.; Del Rosso, L.; Harding, S.M.; Mao, M.; Plante, D.T.; Pressman, M.R.; Troester, M.M.; et al. *The AASM Manual for the Scoring of Sleep and Associated Events: Rules, Terminology and Technical Specifications*; Version 2.6; American Academy of Sleep Medicine: Darien, IL, USA, 2020.
43. Welch, P. The use of fast Fourier transform for the estimation of power spectra: A method based on time averaging over short, modified periodograms. *IEEE Trans. Audio Electroacoust.* **1967**, *15*, 70–73. [CrossRef]
44. Basha, A.J.; Balaji, B.S.; Poornima, S.; Prathilothamai, M.; Venkatachalam, K. Support vector machine and simple recurrent network based automatic sleep stage classification of fuzzy kernel. *J. Ambient Intell. Humaniz. Comput.* **2020**, *12*, 6189–6197. [CrossRef]
45. Feoktistov, V. *Differential Evolution 5*; Springer US: Boston, MA, USA, 2006.
46. Deb, K.; Pratap, A.; Agarwal, S.; Meyarivan, T. A fast and elitist multiobjective genetic algorithm: NSGA-II. *IEEE Trans. Evol. Comput.* **2002**, *182*, 182–197. [CrossRef]
47. Chugh, T.; Sindhya, K.; Hakanen, J.; Miettinen, K. A survey on handling computationally expensive multiobjective optimization problems with evolutionary algorithms. *Soft Comput.* **2019**, *23*, 3137–3166. [CrossRef]
48. Moctezuma, L.A.; Molinas, M. Towards a minimal EEG channel array for a biometric system using resting-state and a genetic algorithm for channel selection. *Sci. Rep.* **2020**, *10*, 14917. [CrossRef]
49. Hadka, D. MOEA Framework. Available online: http://moeaframework.org/ (accessed on 20 September 2021).
50. Coello Coello, C.; Lamont, G.B.; van Veldhuizen, D.A. *Evolutionary Algorithms for Solving Multi-Objective Problems*; Springer: Boston, MA, USA, 2007.
51. Coello Coello, C.A. Theoretical and numerical constraint-handling techniques used with evolutionary algorithms: A survey of the state of the art. *Comput. Methods Appl. Mech. Eng.* **2002**, *191*, 1245–1287. [CrossRef]
52. Wang, Y.; Cai, Z.; Zhou, Y.; Zeng, W. An Adaptive Tradeoff Model for Constrained Evolutionary Optimization. *IEEE Trans. Evol. Comput.* **2008**, *80*, 80–92. [CrossRef]
53. Homaifar, A.; Qi, C.X.; Lai, S.H. Constrained Optimization Via Genetic Algorithms. *Simulation* **1994**, *242*, 242–253. [CrossRef]
54. Runarsson, T.P.; Yao, X. Stochastic ranking for constrained evolutionary optimization. *IEEE Trans. Evol. Comput.* **2000**, *284*, 284–294. [CrossRef]
55. McDonald, J.H. *Handbook of Biological Statistics*, 3rd ed.; Sparky House Publishing: Baltimore, MD, USA, 2014.
56. Chen, T.; Guestrin, C. XGBoost. In Proceedings of the 22nd ACM SIGKDD International Conference on Knowledge Discovery and Data Mining, San Francisco, CA, USA, 13–17 August 2016; pp. 785–794.
57. Mehreen, A.; Anwar, S.M.; Haseeb, M.; Majid, M.; Ullah, M.O. A Hybrid Scheme for Drowsiness Detection Using Wearable Sensors. *IEEE Sens. J.* **2019**, *5119*, 5119–5126. [CrossRef]
58. Seok, W.; Yeo, M.; You, J.; Lee, H.; Cho, T.; Hwang, B.; Park, C. Optimal Feature Search for Vigilance Estimation Using Deep Reinforcement Learning. *Electronics* **2020**, *9*, 142. [CrossRef]
59. Wu, E.Q.; Peng, X.Y.; Zhang, C.Z.; Lin, J.X.; Sheng, R.S.F. Pilots' Fatigue Status Recognition Using Deep Contractive Autoencoder Network. *IEEE Trans. Instrum. Meas.* **2019**, *3907*, 3907–3919. [CrossRef]
60. Kamrud, A.; Borghetti, B.; Schubert Kabban, C. The Effects of Individual Differences, Non-Stationarity, and the Importance of Data Partitioning Decisions for Training and Testing of EEG Cross-Participant Models. *Sensors* **2021**, *21*, 3225. [CrossRef]
61. Lin, F.-C.; Ko, L.-W.; Chuang, C.-H.; Su, T.-P.; Lin, T.-C. Generalized EEG-Based Drowsiness Prediction System by Using a Self-Organizing Neural Fuzzy System. *IEEE Trans. Circuits Syst. I Regul. Pap.* **2012**, *2044*, 2044–2055. [CrossRef]
62. Jap, B.T.; Lal, S.; Fischer, P. Comparing combinations of EEG activity in train drivers during monotonous driving. *Expert Syst. Appl.* **2011**, *996*, 996–1003. [CrossRef]

63. Akbar, I.A.; Rumagit, A.M.; Utsunomiya, M.; Morie, T.; Igasaki, T. Three drowsiness categories assessment by electroencephalogram in driving simulator environment. In Proceedings of the 2017 39th Annual International Conference of the IEEE Engineering in Medicine and Biology Society (EMBC), Jeju Island, Korea, 11–15 July 2017; pp. 2904–2907.

64. Keckluno, G.; Åkersteot, T. Sleepiness in long distance truck driving: An ambulatory EEG study of night driving. *Ergonomics* **1993**, *1007*, 1007–1017. [CrossRef]

65. Dussault, C.; Jouanin, J.-C.; Philippe, M.; Guezennec, Y.C. EEG and ECG changes during simulator operation reflect mental workload and vigilance. *Aviat. Space Environ. Med.* **2005**, *344*, 344–351.

66. Khushaba, R.N.; Kodagoda, S.; Lal, S.; Dissanayake, G. Driver drowsiness classification using fuzzy wavelet-packet-based feature-extraction algorithm. *IEEE Trans. Biomed. Eng.* **2011**, *121*, 121–131. [CrossRef]

67. Brown, I.D. Driver Fatigue. *Hum. Factors* **1994**, *36*, 298–314. [CrossRef] [PubMed]

68. Simon, M.; Schmidt, E.A.; Kincses, W.E.; Fritzsche, M.; Bruns, A.; Aufmuth, C.; Bogdan, M.; Rosenstiel, W.; Schrauf, M. EEG alpha spindle measures as indicators of driver fatigue under real traffic conditions. *Clin. Neurophysiol.* **2011**, *122*, 1168–1178. [CrossRef]

69. Johns, M.W.; Chapman, R.; Crowley, K.; Tucker, A. A new method for assessing the risks of drowsiness while driving. *Somnologie—Schlafforsch. Schlafmed.* **2008**, *66*, 66–74. [CrossRef]

70. Lal, S.K.L.; Craig, A. Driver fatigue: Electroencephalography and psychological assessment. *Psychophysiology* **2002**, *313*, 313–321. [CrossRef]

71. Craig, A.; Tran, Y.; Wijesuriya, N.; Nguyen, H. Regional brain wave activity changes associated with fatigue. *Psychophysiology* **2012**, *574*, 574–582. [CrossRef]

72. Kaida, K.; Takahashi, M.; Åkerstedt, T.; Nakata, A.; Otsuka, Y.; Haratani, T.; Fukasawa, K. Validation of the Karolinska sleepiness scale against performance and EEG variables. *Clin. Neurophysiol.* **2006**, *1574*, 1574–1581. [CrossRef]

73. Gajic, D.; Djurovic, Z.; Di Gennaro, S.; Gustafsson, F. Classification of EEG signals for detection of epileptic seizures based on wavelets and statistical pattern recognition. *Biomed. Eng. Appl. Basis Commun.* **2014**, *26*, 1450021. [CrossRef]

74. Sun, H.; Ganglberger, W.; Panneerselvam, E.; Leone, M.J.; Quadri, S.A.; Goparaju, B.; Tesh, R.A.; Akeju, O.; Thomas, R.J.; Westover, M.B. Sleep staging from electrocardiography and respiration with deep learning. *Sleep* **2020**, *43*, zsz306. [CrossRef] [PubMed]

75. Sridhar, N.; Shoeb, A.; Stephens, P.; Kharbouch, A.; Shimol, D.B.; Burkart, J.; Ghoreyshi, A.; Myers, L. Deep learning for automated sleep staging using instantaneous heart rate. *NPJ Digit. Med.* **2020**, *3*, 106. [CrossRef]

76. Awais, M.; Badruddin, N.; Drieberg, M. A Hybrid Approach to Detect Driver Drowsiness Utilizing Physiological Signals to Improve System Performance and Wearability. *Sensors* **2017**, *17*, 1991. [CrossRef] [PubMed]

77. Dong, Y.; Hu, Z.; Uchimura, K.; Murayama, N. Driver Inattention Monitoring System for Intelligent Vehicles: A Review. *IEEE Trans. Intell. Transp. Syst.* **2011**, *12*, 596–614. [CrossRef]

Article

Fusion Method to Estimate Heart Rate from Facial Videos Based on RPPG and RBCG

Hyunwoo Lee [1], Ayoung Cho [1] and Mincheol Whang [2,*]

[1] Department of Emotion Engineering, Sangmyung University, Seoul 03016, Korea; lhw4846@naver.com (H.L.); joa6391@gmail.com (A.C.)
[2] Department of Intelligence Informatics Engineering, Sangmyung University, Seoul 03016, Korea
* Correspondence: whang@smu.ac.kr; Tel.: +82-2-2287-5293

Abstract: Remote sensing of vital signs has been developed to improve the measurement environment by using a camera without a skin-contact sensor. The camera-based method is based on two concepts, namely color and motion. The color-based method, remote photoplethysmography (RPPG), measures the color variation of the face generated by reflectance of blood, whereas the motion-based method, remote ballistocardiography (RBCG), measures the subtle motion of the head generated by heartbeat. The main challenge of remote sensing is overcoming the noise of illumination variance and motion artifacts. The studies on remote sensing have focused on the blind source separation (BSS) method for RGB colors or motions of multiple facial points to overcome the noise. However, they have still been limited in their real-world applications. This study hypothesized that BSS-based combining of colors and the motions can improve the accuracy and feasibility of remote sensing in daily life. Thus, this study proposed a fusion method to estimate heart rate based on RPPG and RBCG by the BSS methods such as ensemble averaging (EA), principal component analysis (PCA), and independent component analysis (ICA). The proposed method was verified by comparing it with previous RPPG and RBCG from three datasets according to illumination variance and motion artifacts. The three main contributions of this study are as follows: (1) the proposed method based on RPPG and RBCG improved the remote sensing with the benefits of each measurement; (2) the proposed method was demonstrated by comparing it with previous methods; and (3) the proposed method was tested in various measurement conditions for more practical applications.

Keywords: heart rate measurement; remote HR; remote PPG; remote BCG; blind source separation

Citation: Lee, H.; Cho, A.; Whang, M. Fusion Method to Estimate Heart Rate from Facial Videos Based on RPPG and RBCG. *Sensors* **2021**, *21*, 6764. https://doi.org/10.3390/s21206764

Academic Editor: Yvonne Tran

Received: 17 August 2021
Accepted: 7 October 2021
Published: 12 October 2021

Publisher's Note: MDPI stays neutral with regard to jurisdictional claims in published maps and institutional affiliations.

1. Introduction

Remote sensing of vital signs has been studied to improve the measurement burden. In particular, camera-based methods allow one to measure heart rate using a smartphone both anywhere and anytime. Despite the potential of camera-based methods, they have still been limited in their applications in daily life due to the noise generated by illumination variance and motion artifacts [1].

The camera-based method is based on two concepts, i.e., color and motion. First, the color-based method uses the principle of photoplethysmography (PPG), and thus it is called remote PPG (RPPG). RPPG measures the color variation generated by reflectance of blood due to the cardiac cycle from the heart and the head through the carotid arteries [2]. The color variance is most clearly measured at the green wavelength (i.e., 510~560 nm) since the pulsations of the arteries are able to be monitored through elastic-mechanical interaction of deep arteries with the superficial dermis [3]. However, the illumination variance of various frequency ranges that occur in daily life distorts the color variance caused by the heartbeat, so it is difficult to remove the noise using only green color. The studies on RPPG have focused on the blind source separation (BSS) method from the RGB colors to overcome the noise of illumination variance. Poh et al. [4] first proposed RPPG, which extracts the color variation from the RGB colors and estimates the plethysmographic signal by using the BSS method based on independent component analysis (ICA). Xu et al. [5] reduced the noise of illumination variance using the partial least squares (PLS) method and multivariate empirical mode decomposition (MEMD). They evaluated their method in illumination changing conditions and showed more improved accuracy than the ICA-based method. Zhang [6] applied the separation of the luminance on RPPG by converting from RGB color space to LAB color space. They demonstrated that the luminance separation is resistant to natural, motion-induced, and artificial illumination variances. Although RPPG has been improved to overcome the noise of illumination variance, it is still difficult to apply in daily life since all RGB colors are affected by lighting at the same time and most proposed noise removal methods were validated in limited scenarios.

Second, the motion-based method uses the principle of ballistocardiography (BCG), so it is called remote BCG (RBCG). The RBCG measures the subtle motion generated by the contraction of the heart and the ejection of the blood from the ventricles into the vasculature [7]. The subtle motion is sensitive to measurement since its displacement is small at about 0.5 mm [8]. Thus, the major motion such as head movement and facial expression makes the measurement of subtle motion difficult. The studies on RBCG have developed by applying the BSS method to the subtle motions extracted from as many feature points as possible. Balakrishnan et al. [8] first proposed RBCG which extracts the subtle motions from feature points of the forehead and nose regions and applied the BSS method based on principal component analysis (PCA). Shan et al. [9] extracted the subtle motion from one feature point of the forehead region and applied the ICA on the displacement calculated in both the x-axis and y-axis. Haque et al. [10] increased the number of feature points by combining good features to track (GFTT) with facial landmark detection based on the supervised descent method (SDM). They demonstrated that increasing the number of feature points overcomes the tracking noise for extracting the subtle motion. Hassan et al. [11] improved the subtle motion extraction by the skin color-based foreground segmentation and the motion artifact removal by the BSS method based on singular value decomposition (SVD). Despite the improvement of RBCG, it is difficult to apply this method in daily life due to the limitation of computational performance in real time on the measurement device and of noise removal on irregular motion artifacts.

Recent studies on remote sensing have focused on the combining of RPPG with RBCG. Shao et al. [12] developed the simultaneous monitoring of RPPG and RBCG. They demonstrated that it potentially provides a low-cost solution based on a single camera in daily life, but did not consider combining them together. Liu et al. [13] combined RPPG with BCG measured using an additional motion sensor, not a camera. They corrected the tracking noise caused by motion artifacts in RPPG by referring to BCG. They only used BCG to remove motion artifacts in RPPG and did not consider their combination. Thus, it is still necessary to develop the fusion method of RPPG and RBCG by considering their interaction effect.

In summary, the BSS method has been developed to overcome the noise of illumination variance and motion artifacts in both RPPG and RBCG. Also, the fusion method of RPPG and RBCG need to be further developed by considering their interaction effect. Thus, this study hypothesized that the BSS-based combining of RPPG and RBCG can improve the accuracy and feasibility of remote sensing in daily life. This study examined the BSS methods using ensemble averaging (EA), PCA, and ICA to estimate heart rate based on combining RPPG with RBCG. The proposed method was compared with the previous methods, which only used the color variation (i.e., RPPG) or the subtle motion (i.e., RBCG). The contributions of this study can be summarized as follows: (1) the proposed method based on RPPG and RBCG improved the remote sensing with the benefits of each measurement; (2) the proposed method was demonstrated by comparing it with previous methods; and (3) the proposed method was tested in various measurement conditions for a more practical application.

2. Proposed Method

This study proposed a fusion method to estimate heart rate based on RPPG and RBCG by their ensemble averaging. Figure 1 depicts the procedure of the proposed method. First, the face was detected and tracked from the consequence frame of facial video. Then, photoplethysmographic and ballistocardiographic signals were extracted from the face by RPPG and RBCG, respectively. These signals were combined with each other to minimize noise and to maximize cardiac components. Finally, the heart rate was estimated from the combined signal in the frequency domain.

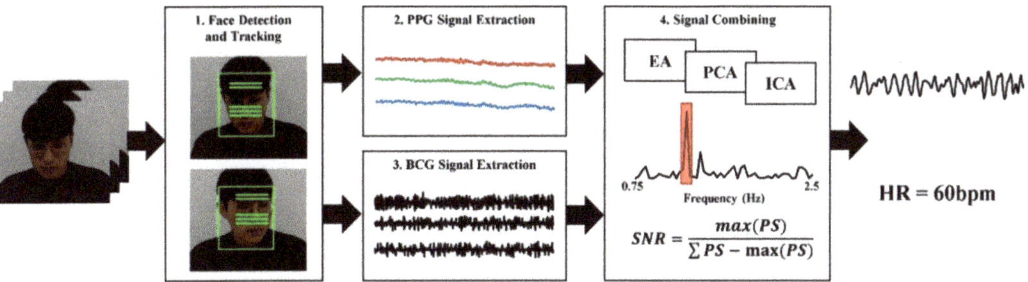

Figure 1. Overview of proposed method.

2.1. Face Detection and Tracking

To extract color variation and subtle motion, the face was detected and tracked from facial video. This study focused on low misdetection and fast inference time for practical application. The Viola-Jones algorithm [14] is a basic and widely known face detection using AdaBoost with Haar features. AdaBoost selected the candidates from an image using simple Haar features and detected the face from the candidates using complex Haar features to improve inference time. It is easy to use since it is implemented in the OpenCV library [15], but it frequently mis-detected the face. Histogram of oriented gradients (HOG) algorithm [16] is also a representative face detection algorithm implemented in the DLIB library [17]. It computed the spatial gradients from an image and detected the face using histogram of gradient orientation. Although it has less misdetection than the Viola-Jones algorithm, it can detect only the frontal face and has a slow inference time.

Recently, face detection methods using deep learning have been proposed and have shown more enhanced performance than the Viola-Jones and HOG algorithms. Although their algorithm was complex and difficult to implement, they are easier to use than before by development of open-source framework such as Tensorflow [18]. Thus, this study employed the single shot detector (SSD) [19] with ResNet [20] trained by WIDER FACE dataset [21]. The ResNet extracted a high dimensional feature map which has facial features such as contrast or facial contour from an image. The SSD detected the face from the feature map by extracting image pyramids of various sizes. It has been implemented and available in OpenCV library [22] recently.

In addition, face tracking is important to extract the same facial region from successive frames. The facial region was divided into sub-regions by cropping the middle 50% of the width and top 20% of the height (i.e., forehead) and the middle 50% of the width and middle 25% of height (i.e., nose). Then, 80 facial points were determined within the sub-regions by dividing the forehead region into 32 cells and the nose region into 48 cells, respectively. The facial points were tracked on xy-coordinates by the Kanade-Lucas-Tomasi (KLT) tracker [23]. By empirically considering tracking accuracy and computing cost, the window size of the small subarea cells was determined as the detected face size divided by 10. If the tracked facial points suddenly moves more than 10 pixels, the facial points were re-defined in the forehead and nose regions as described above. Then, the similarity transformation matrix was extracted from locations of the facial points on the previous and next frames to estimate the transformation of the facial region. The similarity transformation matrix hypothesized three characteristics (i.e., translation, rotation, and scaling). Finally, the facial region on next frame was tracked from previous frame by the similarity transformation matrix. Figure 2 shows the procedure of face detection and tracking.

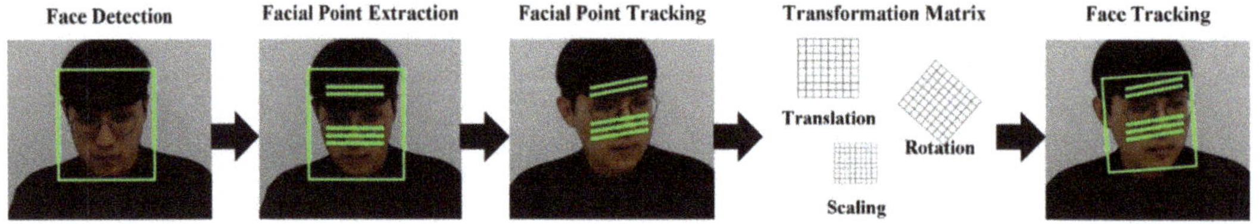

Figure 2. Procedure of face detection and tracking.

2.2. Photoplethysmographic Signal Extraction

Color variation caused by the heartbeat was prominent in the cheek area [24], so that the photoplethysmographic signal was extracted by the RGB spectrums on the middle 50% of the width and middle 25% of height (i.e., nose), on the left 20–35% of width and top 45–70% of height (i.e., left cheek), 220–35% of width and top 45–50% of height (i.e., right cheek). Each RGB signal was normalized by subtracting its mean since the mean indicates melanin components (i.e., skin color) [25]. Then, RBG signals were combined with each other based on pulse blood-volume vector (PBV) modeling [26] for each face area. The combined signals were filtered by a second order Butterworth bandpass filter with a cut-off of 0.75–2.5 Hz corresponding to 45–150 bpm. Finally, the photoplethysmographic signal was extracted by applying the ICA on the filtered signals and selecting the signal with the highest signal to noise ratio (SNR) from three components. The SNR was calculated from the frequency domain of each component as:

$$SNR = max(PS)/(\sum PS - max(PS)), \tag{1}$$

where *SNR* is a signal to noise ratio and *PS* is a power spectrum of the signal. Figure 3 depicts the procedure of photoplethysmographic signal extraction.

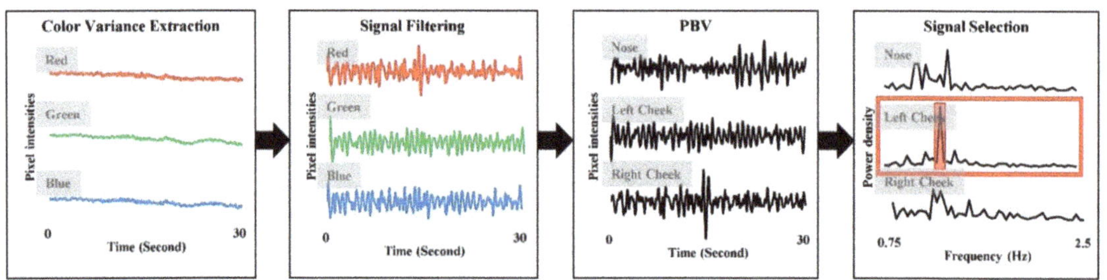

Figure 3. Procedure of photoplethysmographic signal extraction.

2.3. Ballistocardiographic Signal Extraction

Ballistocardiographic head movements were generated up and down by the heartbeat so that the ballistocardiographic signal was extracted by the y-coordinate of each facial point on successive frames. In this study, the 80 ballistocardiographic signals were extracted from the 80 facial points and were normalized by subtracting its mean to make the unit the same as RPPG. The normalized signals were filtered by a second order Butterworth bandpass filter with a cut-off of 0.75–2.5 Hz. Voluntary head movements distorted the signals to have a large amplitude, so that the signals were corrected as the mean if the amplitude is larger than twice the standard deviation. Also, facial expressions distorted the signals extracted from specific muscles (e.g., smile moves the cheek and eye muscles). Thus, the SNR was calculated from each signal and the signals which have lower SNR than mean SNR were removed in the next step. Finally, the ballistocardiographic signal was extracted by applying the PCA on the filtered signals and selecting the signal with the highest SNR from five components. Figure 4 shows the procedure of ballistocardiographic signal extraction.

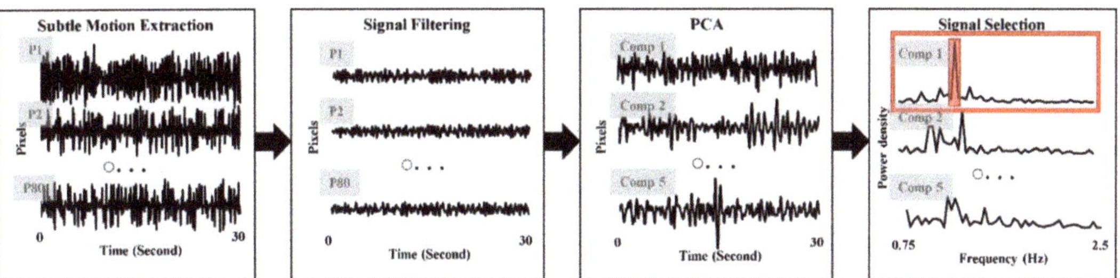

Figure 4. Procedure of ballistocardiographic signal extraction.

2.4. Signal Combining

This study hypothesized that RPPG are robust to the motion artifacts and sensitive to the illumination variance, whereas RBCG are robust to the illumination variance and sensitive to the motion artifacts. Thus, the photoplethysmographic signal and the ballistocardiographic signal had combined with each other to improve robustness to both the illumination variance and the motion artifacts. This study tested three BSS methods (i.e., EA, PCA, and ICA) to combine RPPG and RBCG. EA was hypothesized to reduce the random noise by averaging repetitive signals and has verified its enhancement as a fusion method in a previous BCG study [27]. EA was calculated as:

$$EA = (RPPG + RBCG)/2, \tag{2}$$

where *RPPG* is a signal extracted by *RPPG* and *RBCG* is a signal extracted by *RBCG*. In addition, PCA and ICA are the representative BSS methods to reduce the various noise. PCA and ICA were applied on PPG signal and BCG signal extracted from Sections 2.2 and 2.3 and two components were extracted because the BSS applied on two signals. Then, the combined signal was selected from two components by highest SNR, respectively. Figure 5 depicts the procedure of signal combining.

2.5. Heart Rate Estimation

The heart rate estimated from the frequency domain of the signal. The power spectrum was converted from the combined signal by fast Fourier transform (FFT). This study extracted the band of power spectrum between the ranges of 0.75 and 2.5 Hz corresponding to the ranges of 45 and 150 bpm. The dominant frequency was identified with the highest power from the band of power spectrum. The heart rate was calculated by multiplying by the dominant frequency and 60 as:

$$HR = 60 \times freq, \tag{3}$$

where *HR* is a heart rate and *freq* is a dominant frequency. The heart rate for RPPG (i.e., HR_{RPPG}) was estimated from the PPG signal extracted by Section 2.2, whereas the heart rate for RBCG (i.e., HR_{RBCG}) was estimated from the BCG signal extracted by Section 2.3. The heart rates for proposed methods (i.e., HR_{EA}, HR_{PCA} and HR_{ICA}) were estimated from the combined signals based on RPPG and RBCG extracted Section 2.4.

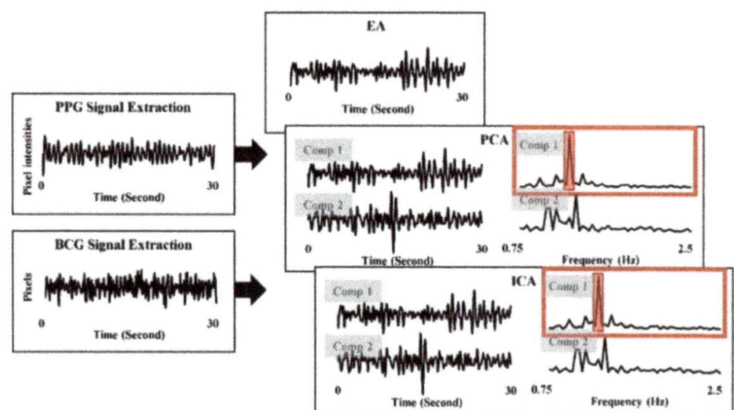

Figure 5. Procedure of signal combining.

3. Experiments

The experiments were conducted to evaluate the proposed method according to three measurement conditions in this study. The measurement conditions were determined by illumination and motion artifacts (i.e., normal, facial expressions, and human computer interactions). The experimental procedure was approved by the Institutional Review Board of the Sangmyung University, Seoul, Korea (BE2018-35).

3.1. Experiment 1: Normal

This experiment is to collect the normal dataset without illumination variance and motion artifacts. The participants consisted of 20 persons (12 males) and were asked to sit 1 m away from a camera for 3 min with a stationary state. The facial video was recorded by an RGB webcam (Logitech Webcam C270) with 640 × 360 resolution at 30 fps. Also, the ECG signal was simultaneously measured by an ECG measurement system with Lead-I (BIOPAC Systems Inc., Goleta, CA, USA) at a sampling rate of 500 Hz. It was employed as a ground-truth for the evaluation of the proposed methods.

3.2. Experiment 2: Facial Expressions

This experiment is to collect the dataset with facial expressions. The 20 persons who participated in experiment 1 were also asked to sit in front of a camera for 3 min with a stationary state. They then followed the six basic facial expressions (i.e., happiness, sadness, surprise, anger, disgust, and fear), which were displayed on the monitor for 30 s in random order to minimize the ordering effect. The facial video and the ECG signal were recorded and measured at the same manner in experiment 1.

3.3. Experiment 3: Human Computer Interactions

This experiment is to collect the dataset including illumination variance and motion artifacts occurred in human computer interactions. The 17 persons (8 males) were participated and asked to watch the four videos to cause to some emotions. Then, they wrote their emotions on self-report by two-dimensional model [28]. While watching the videos and writing the self-report, they asked to freely express the facial expressions and move their head. Also, the illumination on the face was changed by reflecting the light from the video. The facial video was recorded by an RGB webcam (Logitech Webcam C270) with a 1920 × 1080 resolution at 30 fps. The ECG signal was also measured at the same manner in experiment 1.

4. Results

This study demonstrated the improvement of proposed fusion methods by comparing them with previous RPPG and RBCG methods. The heart rate of ECG was calculated by the QRS detection algorithm [29] and determined as the ground-truth for evaluation of the proposed methods. The proposed method was verified by calculating the four metrics and by representing the one plot as follows: mean absolute error (MAE), standard deviation of absolute error (SDAE), root mean squared error (RMSE), Pearson's correlation coefficient (CC), and the Bland-Altman plot. MAE, SDAE, and RMSE describes the difference of mean heart rate and variation, respectively. CC determines the statistical similarity of heart rates over time, so that the coefficient value indicates a strong positive similarity if it is approaching to the one. The Bland-Altman plot represents graphically the statistical differences by assigning the mean (x-axis) and difference (y-axis) between the two measurements. The line on the plot is indicated the 95% limits of an agreement based on mean difference and the ± 1.96 standard deviation of the differences. The statistical parameters were calculated from the heart rates for RPPG (i.e., HR_{RPPG}), RBCG (i.e., HR_{RPPG}), and the proposed methods (i.e., HR_{EA}, HR_{PCA} and HR_{ICA}) by comparing them with ECG.

4.1. Experiment 1: Normal

Table 1 shows the estimation of heart rates from the normal dataset using RPPG, RBCG, and the proposed fusion methods (i.e., EA, PCA, and ICA). All fusion methods were more accurate than RPPG and RBCG. In addition, the ICA-based fusion method showed the lowest errors in the normal dataset (MAE = 1.04, SDAE = 0.91, RMSE = 1.39, CC = 0.999).

Table 1. Estimation of heart rates from the normal dataset without illumination variance and motion artifacts.

Methods	MAE (bpm)	SDAE (bpm)	RMSE (bpm)	CC (r)
RPPG	1.84	1.66	2.49	0.981 **
RBCG	2.56	2.26	3.48	0.927 **
Fusion (EA)	1.05	0.93	1.42	0.996 **
Fusion (PCA)	1.08	0.99	1.49	0.996 **
Fusion (ICA)	**1.04**	**0.91**	**1.39**	**0.999 ****

MAE, mean absolute error; SDAE, standard deviation of absolute error; RMSE, root mean square error; CC, Pearson's correlation coefficient. Two asterisk represents significant correlation levels at p-value < 0.01. The lowest error and highest correlation values are bolded.

The Bland-Altman plots of estimated heart rates from the normal dataset without illumination variance and motion artifacts using RPPG, RBCG, and the proposed fusion methods are shown in Figure 6. The mean errors were 2.19 with 95% limits of agreement (LOA) in -3.43 to 7.81 (RPPG), -1.93 with 95% LOA in -8.35 to 4.49 (RBCG), 0.37 with 95% LOA in -1.19 to 1.93 (EA), 0.34 with 95% LOA in -1.26 to 1.94 (PCA), and 0.18 with 95% LOA in -0.63 to 0.98 (ICA). The heart rates estimated using the ICA-based fusion method showed the lowest mean difference and variances.

4.2. Experiment 2: Facial Expressions

The heart rates were estimated from the dataset with facial expressions using RPPG, RBCG, and the proposed fusion methods as shown in Table 2. All fusion methods showed lower errors than RPPG and RBCG. In addition, the errors of the PCA-based fusion method were lower than one of other methods (MAE = 2.76, SDAE = 2.34, RMSE = 3.23, CC = 0.968).

Figure 7 shows the Bland-Altman plots of the heart rates estimated from the dataset with facial expressions using RPPG, RBCG, and the proposed fusion methods. The mean errors were 5.54 with 95% limits of agreement (LOA) in -6.39 to 17.48 (RPPG), -2.48 with 95% LOA in -8.56 to 3.59 (RBCG), 2.48 with 95% LOA in -4.59 to 9.55 (EA), 2.20 with 95% LOA in -4.04 to 8.45 (PCA), and 1.87 with 95% LOA in -4.12 to 7.86 (ICA). The heart rates estimated using the PCA-based fusion method showed the lowest mean difference and variances.

Table 2. Estimation of heart rates from the dataset with facial expressions.

Methods	MAE (bpm)	SDAE (bpm)	RMSE (bpm)	CC (r)
RPPG	3.13	2.59	4.09	0.947 **
RBCG	3.94	4.07	5.71	0.920 **
Fusion (EA)	3.06	2.56	4.53	0.955 **
Fusion (PCA)	**2.76**	**2.34**	**3.23**	0.968 **
Fusion (ICA)	2.99	2.53	4.17	**0.972 **

MAE, mean absolute error; SDAE, standard deviation of absolute error; RMSE, root mean square error; CC, Pearson's correlation coefficient. Two asterisk represents significant correlation levels at p-value < 0.01. The lowest error and highest correlation values are bolded.

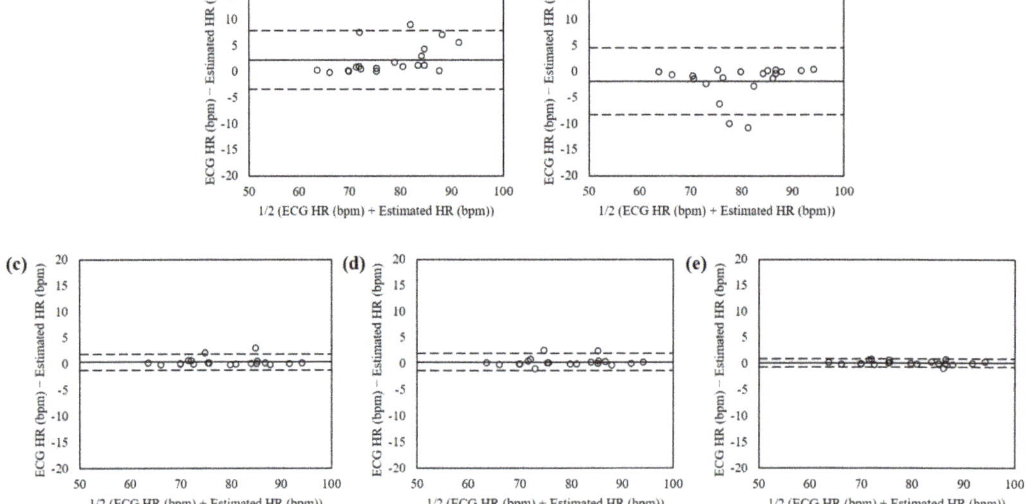

Figure 6. Bland-Altman plots of heart rates estimated from the normal dataset without illumination variance and motion artifacts using interactions using (**a**) RPPG, (**b**) RBCG, (**c**) EA, (**d**) PCA, and (**e**) ICA. The lines are the mean errors and 95% LOA.

4.3. Experiment 3: Human Computer Interactions

Table 3 shows the estimation of heart rates from the dataset in human computer interactions using RPPG, RBCG, and the proposed fusion methods. All fusion methods showed lower errors than RPPG and RBCG. Unlike the other datasets, the EA-based fusion method showed lowest errors in this dataset (MAE = 4.79, SDAE = 2.13, RMSE = 5.181, CC = 0.629).

Figure 8 shows the Bland-Altman plots of the heart rates estimated from the dataset in human computer interactions using RPPG, RBCG, and the proposed fusion methods. The mean errors were 6.05 with 95% limits of agreement (LOA) in −9.71 to 21.81 (RPPG), −12.35 with 95% LOA in −32.88 to 8.18 (RBCG), −0.09 with 95% LOA in −15.49 to 15.32 (EA), −0.49 with 95% LOA in −15.14 to 15.16 (PCA), and 0.25 with 95% LOA in −16.15 to 16.65 (ICA). The heart rates estimated using the EA-based fusion method showed the lowest mean difference and variances.

Table 3. Estimation of heart rates from the dataset in human computer interactions.

Methods	MAE (bpm)	SDAE (bpm)	RMSE (bpm)	CC (r)
RPPG	5.68	2.93	6.53	**0.713 ****
RBCG	14.06	6.75	15.86	0.051
Fusion (EA)	**4.79**	**2.13**	**5.81**	0.629 **
Fusion (PCA)	5.42	4.2.84	6.13	0.622 **
Fusion (ICA)	5.66	3.59	6.48	0.617 **

MAE, mean absolute error; SDAE, standard deviation of absolute error; RMSE, root mean square error; CC, Pearson's correlation coefficient. Two asterisk represents significant correlation levels at p-value < 0.01. The lowest error and highest correlation values are bolded.

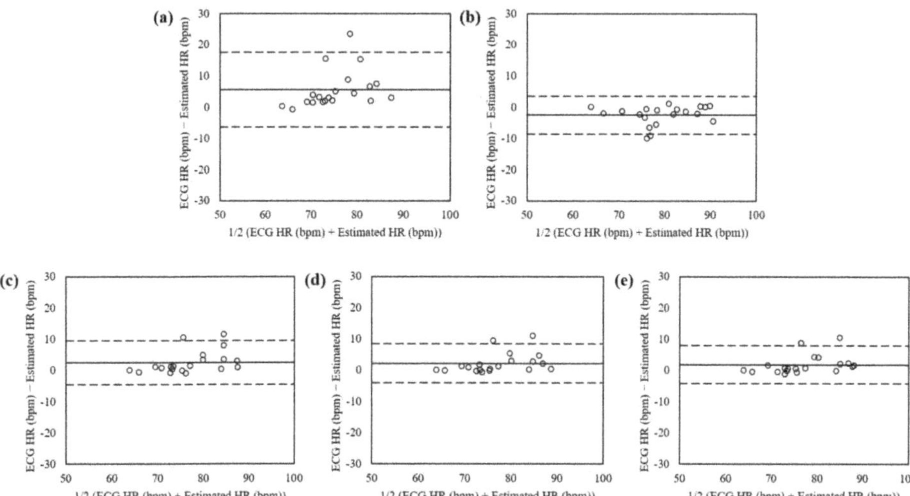

Figure 7. Bland-Altman plots of heart rates estimated from the dataset with facial ex-pressions interactions using (**a**) RPPG, (**b**) RBCG, (**c**) EA, (**d**) PCA, and (**e**) ICA. The lines are the mean errors and 95% LOA.

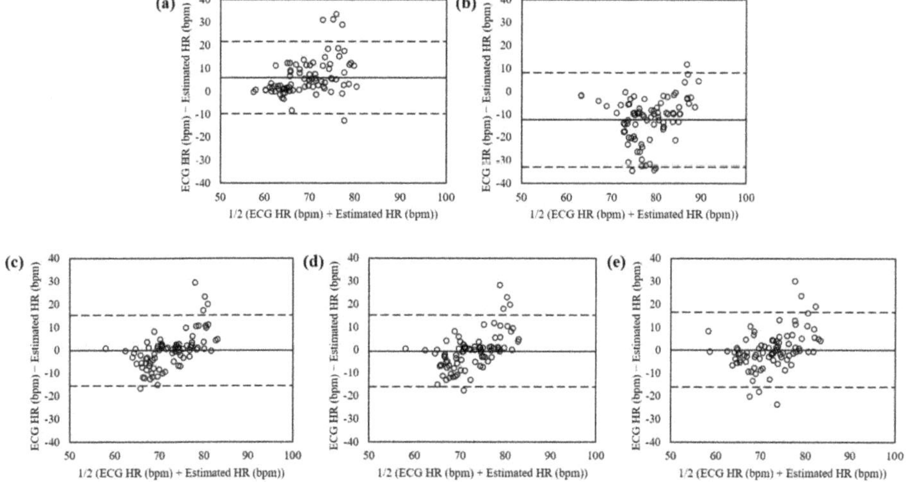

Figure 8. Bland-Altman plots of heart rates estimated from the dataset in human computer interactions using (**a**) RPPG, (**b**) RBCG, (**c**) EA, (**d**) PCA, and (**e**) ICA. The lines are the mean errors and 95% LOA.

5. Discussion

In this study, the fusion method based on RPPG and RBCG was developed to enhance the heart rate estimation using EA, PCA, and ICA. This study evaluated the proposed method on three datasets according to illumination variance and motion artifacts as follows: (1) normal, (2) facial expressions, and (3) human computer interactions. The proposed method was more accurate than the previous RPPG and RBCG in all datasets. This result indicated that with the advancement of fusion methods based on RPPG and RBCG, ECG could eventually be replaced by remote sensing in daily life. Thus, this study strongly encourages the fusion method based on RPPG and RBCG as a requirement to estimate heart rate using a camera more accurately.

Overall, this study has drawn four significant findings. First, PCA and ICA were better than EA in the datasets including less noise such as normal and facial expressions. It indicated that the BSS algorithms can be improved if RPPG and RBCG are enhanced by reducing the noise. On the other hand, EA was better than PCA and ICA in daily life not yet.

Second, experiment 3 evaluated the proposed method and the previous RPPG and RBCG in human computer interactions environment which is similar to daily life. As shown in the Bland-Altman plots (Figure 7), the previous RPPG and RBCG estimated the heart rate higher (LOA in −9.71 to 21.81) and lower (LOA in −32.88 to 8.18) than the ground-truth (i.e., ECG), respectively. It indicated that RPPG has high frequency noise whereas RBCG has low frequency noise. The illumination variance has high frequency because it appears at a high rate that humans cannot perceive. On the other hand, the motion artifacts have low frequency because they occur instantaneously and briefly. RPPG and RBCG are sensitive to the illumination variance and motion artifacts, respectively, so that the result is acceptable. Note that the proposed method reduced both illumination variance and motion artifacts (LOA in −15.49 to 15.32) since it has low frequency and high frequency noise evenly and little.

Third, this study showed significant results compared to other fusion methods. Particularly, Liu et al. [13] developed a fusion method, combined RPPG with BCG measured using an additional motion sensor, and showed MAE of 6.20 bpm in motion state. Although the measurement condition is different, our proposed method showed MAE of 4.79 bpm in experiment 3 including illumination variance and motion artifacts. Note that it indicates possibility of combining PPG and BCG with only a camera without an additional sensor.

Finally, this study presented the fusion method based on RPPG and RBCG to enhance the heart rate estimation using a camera. Although it should be more improved, a novel approach was developed for the possibility of practical use of remote sensing in daily life. Many studies have been presented before, but it has not been developed as a product for real users due to restrictions on the use environment. The product for real users should be developed and tested to minimize the measurement burden and to improve the use environment, so that it can be used in practical domains such as self-driving cars or non-face-to-face communication.

6. Conclusions

This study developed a fusion method to estimate heart rate from facial videos based on RPPG and RBCG. The proposed methods using EA, PCA, and ICA had compared them with the previous RPPG and RBCG. As a result, the proposed methods showed enhanced accuracy from three datasets according to illumination variance and motion artifacts. The findings are a significant step toward ensuring the enhanced development of RPPG and RBCG. This study is expected to contribute to enhanced heart rate measurement by overcoming noise of illumination variance and motion artifacts and consequently improve the possibility of applications of remote sensing in daily life.

Author Contributions: H.L. and M.W. conceived and designed the experiments; H.L. and A.C. performed the experiments; H.L. analyzed the data; H.L. and M.W. wrote the paper. All authors have read and agreed to the published version of the manuscript.

Funding: This work was supported by Basic Science Research Program through the National Research Foundation of Korea (NRF) funded by the Ministry of Education (2021R1I1A1A01052752) and by Electronics and Telecommunications Research Institute (ETRI) grant funded by the Korean government. [21ZS1100, Core Technology Research for Self-Improving Integrated Artificial Intelligence System].

Institutional Review Board Statement: This study was conducted according to the guidelines of the Declaration of Helsinki, and approved by the Institutional Review Board of the Sangmyung University, Seoul, Korea (BE2018-35).

Informed Consent Statement: Written informed consent has been obtained from the patients to publish this paper.

Data Availability Statement: The data presented in this study are available on request from the corresponding author. The data are not publicly available due to privacy.

Conflicts of Interest: The authors declare no conflict of interest.

References

1. Rouast, P.V.; Adam, M.T.P.; Chiong, R.; Cornforth, D.; Lux, E. Remote heart rate measurement using low-cost RGB face video: A technical literature review. *Front. Comput. Sci.* **2018**, *12*, 858–872. [CrossRef]
2. Hertzman, A.B.; Dillon, J.B. Applications of photoelectric plethysmography in peripheral vascular disease. *Am. Heart J.* **1940**, *20*, 750–761. [CrossRef]
3. Kamshilin, A.A.; Nippolainen, E.; Sidorov, I.S.; Vasilev, P.V.; Erofeev, N.P.; Podolian, N.P.; Romashko, R.V. A new look at the essence of the imaging photoplethysmography. *Sci. Rep.* **2015**, *5*, 1–9. [CrossRef] [PubMed]
4. Poh, M.Z.; McDuff, D.J.; Picard, R.W. Non-contact, automated cardiac pulse measurements using video imaging and blind source separation. *Opt. Express* **2010**, *18*, 10762–10774. [CrossRef] [PubMed]
5. Xu, L.; Cheng, J.; Chen, X. Illumination variation interference suppression in remote PPG using PLS and MEMD. *Electron. Lett.* **2017**, *53*, 216–218. [CrossRef]
6. Zhang, Y.; Dong, Z.; Zhang, K.; Shu, S.; Lu, F.; Chen, J. Illumination variation-resistant video-based heart rate monitoring using LAB color space. *Opt. Lasers Eng.* **2021**, *136*, 106328. [CrossRef]
7. Starr, I.; Rawson, A.J.; Schroeder, H.A.; Joseph, N.R. Studies on the Estimation of Cardiac Output in Man, and of Abnormalities in Cardiac Function, from the heart's Recoil and the blood's Impacts; the Ballistocardiogram. *Am. J. Physiol. Leg. Content* **1939**, *127*, 1–28. [CrossRef]
8. Balakrishnan, G.; Durand, F.; Guttag, J. Detecting pulse from head motions in video. In Proceedings of the IEEE Conference on Computer Vision and Pattern Recognition (CVPR), Portland, OR, USA, 23–28 June 2013; pp. 3430–3437.
9. Shan, L.; Yu, M. Video-based heart rate measurement using head motion tracking and ICA. In Proceedings of the 2013 6th International Congress on Image and Signal Processing (CISP), Hangzhou, China, 16–18 December 2013; pp. 160–164.
10. Haque, M.A.; Nasrollahi, K.; Moeslund, T.B.; Irani, R. Facial video-based detection of physical fatigue for maximal muscle activity. *IET Comput. Vis.* **2016**, *10*, 323–330. [CrossRef]
11. Hassan, M.A.; Malik, A.S.; Fofi, D.; Saad, N.M.; Ali, Y.S.; Meriaudeau, F. Video-Based Heartbeat Rate Measuring Method Using Ballistocardiography. *IEEE Sens. J.* **2017**, *17*, 4544–4557. [CrossRef]
12. Shao, D.; Tsow, F.; Liu, C.; Yang, Y.; Tao, N. Simultaneous monitoring of ballistocardiogram and photoplethysmogram using a camera. *IEEE Trans. Biomed. Eng.* **2016**, *64*, 1003–1010. [CrossRef] [PubMed]
13. Liu, Y.; Qin, B.; Li, R.; Li, X.; Huang, A.; Liu, H.; Liu, M. Motion-Robust Multimodal Heart Rate Estimation Using BCG Fused Remote-PPG With Deep Facial ROI Tracker and Pose Constrained Kalman Filter. *IEEE Trans. Instrum. Meas.* **2021**, *70*, 1–15.
14. Viola, P.; Jones, M. Rapid object detection using a boosted cascade of simple features. In Proceedings of the IEEE Conference on Computer Vision and Pattern Recognition (CVPR), Kauai, HI, USA, 8–14 December 2001; pp. 511–518.
15. OpenCV: Cascade Classifier. Available online: https://docs.opencv.org/4.2.0/db/d28/tutorial_cascade_classifier.html (accessed on 28 May 2020).
16. Dalal, N.; Triggs, B. Histograms of oriented gradients for human detection. In Proceedings of the IEEE Computer Society Conference on Computer Vision and Pattern Recognition (CVPR), San Diego, CA, USA, 20–26 June 2005; pp. 886–893.
17. Dlib: Face Detector. Available online: http://dlib.net/face_detector.py.html (accessed on 28 May 2020).
18. Abadi, M.; Barham, P.; Chen, J.; Chen, Z.; Davis, J.; Kudlur, M. Tensorflow: A system for large-scale machine learning. In Proceedings of the USENIX Symposium on Operating Systems Design and Implementation (OSDI), Savannah, GA, USA, 2–4 November 2016; pp. 265–283.
19. Liu, W.; Anguelov, D.; Erhan, D.; Szegedy, C.; Reed, S.; Fu, C.Y.; Berg, A.C. Ssd: Single shot multibox detector. In Proceedings of the European Conference on Computer Vision (ECCV), Amsterdam, The Netherlands, 8–16 October 2016; pp. 21–37.

20. He, K.; Zhang, X.; Ren, S.; Sun, J. Deep residual learning for image recognition. In Proceedings of the IEEE Conference on Computer Vision and Pattern Recognition (CVPR), Las Vegas, NV, USA, 27–30 June 2016; pp. 770–778.
21. Yang, S.; Luo, P.; Loy, C.C.; Tang, X. Wider face: A face detection benchmark. In Proceedings of the IEEE Conference on Computer Vision and Pattern Recognition (CVPR), Las Vegas, NV, USA, 27–30 June 2016; pp. 5525–5533.
22. OpenCV: Face Detector by SSD in DNN Module. Available online: https://github.com/opencv/opencv/tree/master/samples/dnn/face_detector (accessed on 28 May 2020).
23. Bouguet, J.Y. Pyramidal implementation of the affine lucas kanade feature tracker description of the algorithm. *Intel Corp.* **2001**, *5*, 4.
24. Lempe, G.; Zaunseder, S.; Wirthgen, T.; Zipser, S.; Malberg, H. ROI selection for remote photoplethysmography. In Proceedings of the Bildverarbeitung für die Medizin, Heidelberg, Germany, 3–5 March 2013; pp. 99–103.
25. De Hann, G.; Jeanne, V. Robust pulse rate from chrominance-based rPPG. *IEEE Trans. Biomed. Eng.* **2013**, *60*, 2878–2886. [CrossRef] [PubMed]
26. De Haan, G.; Van Leest, A. Improved motion robustness of remote-PPG by using the blood volume pulse signature. *Physiol. Meas.* **2014**, *35*, 1913. [CrossRef] [PubMed]
27. Lee, H.; Lee, H.; Whang, M. An enhanced method to estimate heart rate from seismocardiography via ensemble averaging of boy movements at six degrees of freedom. *Sensors* **2018**, *18*, 238. [CrossRef] [PubMed]
28. Russell, J.A. A circumplex model of affect. *J. Personal. Soc. Psychol.* **1980**, *39*, 1161. [CrossRef]
29. Pan, J.; Tompkins, W.J. A real-time QRS detection algorithm. *IEEE Trans. Biomed. Eng.* **1985**, *3*, 230–236. [CrossRef] [PubMed]

 *sensors*

Article

Non-Invasive Hemodynamics Monitoring System Based on Electrocardiography via Deep Convolutional Autoencoder

Muammar Sadrawi [1], Yin-Tsong Lin [2], Chien-Hung Lin [2], Bhekumuzi Mathunjwa [1], Ho-Tsung Hsin [1,3], Shou-Zen Fan [4], Maysam F. Abbod [5] and Jiann-Shing Shieh [1,*

[1] Department of Mechanical Engineering, Yuan Ze University, Taoyuan 32003, Taiwan; muammarsadrawi@yahoo.com (M.S.); mathunjwabhekie@gmail.com (B.M.); hsinht@gmail.com (H.-T.H.)
[2] AI R&D Department, New Era AI Robotic Inc., Taipei 105, Taiwan; lotusytlin@neweraai.com (Y.-T.L.); lance_lin@neweraai.com (C.-H.L.)
[3] Cardiovascular Intensive Care Unit, Far-Eastern Memorial Hospital, New Taipei City 220, Taiwan
[4] Department of Anesthesiology, College of Medicine, National Taiwan University, Taipei 100, Taiwan; shouzen@gmail.com
[5] Department of Electronic and Electrical Engineering, Brunel University London, Uxbridge UB8 3PH, UK; Maysam.Abbod@brunel.ac.uk
* Correspondence: jsshieh@saturn.yzu.edu.tw

Abstract: This study evaluates cardiovascular and cerebral hemodynamics systems by only using non-invasive electrocardiography (ECG) signals. The Massachusetts General Hospital/Marquette Foundation (MGH/MF) and Cerebral Hemodynamic Autoregulatory Information System Database (CHARIS DB) from the PhysioNet database are used for cardiovascular and cerebral hemodynamics, respectively. For cardiovascular hemodynamics, the ECG is used for generating the arterial blood pressure (ABP), central venous pressure (CVP), and pulmonary arterial pressure (PAP). Meanwhile, for cerebral hemodynamics, the ECG is utilized for the intracranial pressure (ICP) generator. A deep convolutional autoencoder system is applied for this study. The cross-validation method with Pearson's linear correlation (R), root mean squared error (RMSE), and mean absolute error (MAE) are measured for the evaluations. Initially, the ECG is used to generate the cardiovascular waveform. For the ABP system—the systolic blood pressure (SBP) and diastolic blood pressures (DBP)—the R evaluations are 0.894 ± 0.004 and 0.881 ± 0.005, respectively. The MAE evaluations for SBP and DBP are, respectively, 6.645 ± 0.353 mmHg and 3.210 ± 0.104 mmHg. Furthermore, for the PAP system—the systolic and diastolic pressures—the R evaluations are 0.864 ± 0.003 mmHg and 0.817 ± 0.006 mmHg, respectively. The MAE evaluations for systolic and diastolic pressures are, respectively, 3.847 ± 0.136 mmHg and 2.964 ± 0.181 mmHg. Meanwhile, the mean CVP evaluations are 0.916 ± 0.001, 2.220 ± 0.039 mmHg, and 1.329 ± 0.036 mmHg, respectively, for R, RMSE, and MAE. For the mean ICP evaluation in cerebral hemodynamics, the R and MAE evaluations are 0.914 ± 0.003 and 2.404 ± 0.043 mmHg, respectively. This study, as a proof of concept, concludes that the non-invasive cardiovascular and cerebral hemodynamics systems can be potentially investigated by only using the ECG signal.

Keywords: non-invasive system; hemodynamics; electrocardiography; arterial blood pressure; central venous pressure; pulmonary arterial pressure; intracranial pressure; deep convolutional autoencoder

Citation: Sadrawi, M.; Lin, Y.-T.; Lin, C.-H.; Mathunjwa, B.; Hsin, H.-T.; Fan, S.-Z.; Abbod, M.F.; Shieh, J.-S. Non-Invasive Hemodynamics Monitoring System Based on Electrocardiography via Deep Convolutional Autoencoder. *Sensors* **2021**, *21*, 6264. https://doi.org/10.3390/s21186264

Academic Editor: Biswanath Samanta

Received: 10 August 2021
Accepted: 15 September 2021
Published: 18 September 2021

Publisher's Note: MDPI stays neutral with regard to jurisdictional claims in published maps and institutional affiliations.

1. Introduction

In the intensive care unit (ICU), the most precise health monitoring system is utilized to thoroughly observe the critically ill patients. Hemodynamics is the blood physical phenomena in the circulatory system and a fundamental standard utilized in the ICU. Arterial blood pressure (ABP), pulmonary arterial pressure (PAP), and central venous pressure (CVP) are hemodynamics measures related to the cardiovascular system. Meanwhile, the intracranial pressure (ICP) is applied pressure due to the fluid inside the skull, measured for the cerebral hemodynamics system evaluation. These measures are essential for the patients in the ICU.

For the cardiovascular hemodynamics, CVP evaluation is fundamental for heart failure patients. It is significant in measuring the cardiac preload and blood volume information. However, the measurement requires the central venous catheter, which is highly invasive. Meanwhile, another important investigation of cardiovascular hemodynamics is PAP, which is the driven force given by the heart to pump the blood from the heart to the lung. Elevated PAP is detected for pulmonary hypertension. This evaluation, similar to CVP, is highly invasive through right heart catheterization. Further, ABP is the least invasive system compared to CVP and PAP in the cardiovascular hemodynamics system. For the cerebral hemodynamics, ICP is critical to the brain condition. The elevated ICP can have a significant indication for traumatic brain injury (TBI), and hemorrhagic stroke patients [1–3]. However, besides their benchmark accuracy, these evaluations are highly invasive procedures and potentially lead to infection [4]. Therefore, these measurements are not suitable for a monitoring system.

Recently, non-invasive technologies have been developed for the monitoring of the cardiovascular and cerebral hemodynamics systems. Specifically, for the cardiovascular-based hemodynamics, the utilization of artificial intelligence (AI) with photoplethysmography (PPG) has been widely used for ABP evaluations non-invasively. A study was applied to a single PPG signal with artificial neural networks (ANN) to predict ABP [5]. Furthermore, electrocardiography (ECG) and PPG, with ANN and long short-term memory (LSTM) methods were applied for the evaluations [6]. Backpropagation with a genetic algorithm (GA) was also used for hemodynamics evaluation [7]. Meanwhile, another study utilized ballistocardiography (BCG) and ECG alongside the PPG with convolutional neural networks (CNN) and a gated recurrent units (GRU)-based technique [8]. Additionally, other studies investigated the continuous arterial blood pressure evaluation based on the single PPG signal using the LSTM network [9] and hybrid GA-based optimization with a convolutional autoencoder [10]. However, even though the PPG provides a well-classified result of the blood pressure estimation, the sensor is very sensitive to motion artifacts for vertical movement and several typical activities [11,12].

Several previous studies have also been conducted for non-invasive CVP and PAP measurements. An ultrasound-based system was developed for the non-invasive CVP without central venous access [13]. In addition, a linear regression method by utilizing ICU patient data was implemented for evaluating the CVP signal using echocardiography with comparison to right heart catheterization [14]. Another study also performed a linear regression method from heart failure patients. The patients were under right heart catheterization for the CVP evaluation. The inferior vena cava utilizing echocardiography was applied for the input signal. For the non-invasive PAP system, a study utilized electrical impedance tomography of the input of the system [15]. This previous study was initiated for healthy subjects. However, a single-lead ECG system is more suitable compared to these methods for an intensive monitoring system.

For the cerebral hemodynamics, several previous studies were investigated. Most of them used the ABP and cerebral blood flow velocity (CBFV). A study by Jaishankar et al. [2] utilized ABP and CBFV as inputs for interpreting the ICP and the frequency-based method was selected for the evaluation. Another study applying a Bayesian-based approach was conducted by Imaduddin et al. [3] from the patients with TBI, hydrocephalus, and hemorrhagic stroke, using ABP and CBFV signals in order to evaluate the ICP. The utilization of ABP makes these previous studies less invasive, but not fully non-invasive.

In general, ECG is the more regularly used measurement to evaluate the cardiovascular activities for a longer period. Most importantly, ECG is a non-invasive system. Furthermore, it has been fundamentally applied for the physiological signal evaluation of arrhythmia [16], anesthesia [17,18], and sleep-related evaluations [19–21].

Prior studies have investigated the interconnection between ECG alterations and cardiovascular hemodynamics. It was an indication of increased P wave incidence from the ECG that correlates to hypertension [22]. The P wave also appeared later from the ECG, especially when the diastolic blood pressure was greater than 120 mmHg in hypertensive patients compared to the normal subjects. Another P wave phenomenon was also revealed. A study depicted that the hypertensive patients have a wider P wave compared to the normal subjects [23]. Furthermore, a higher T wave with extended PR intervals was also seen on hypertensive patients [24]. In addition, ST segment depression [25] and higher QRS amplitude also exhibited the information of hypertension [26]. Meanwhile, the ECG can be potentially used for pulmonary hypertension evaluation [27,28]. Recently, a single ECG signal has been utilized to investigate the hypertension system [29,30]. As it can be seen, some previously conducted studies have morphologically enlightened some relationship between cardiac electricity and blood pressure.

On the other hand, an ECG also shows several promising results when evaluating the cerebrovascular hemodynamics. The brain–heart interaction studies revealed several relationships between these associating signals. ECG abnormalities—increased corrected QT (QTc) interval, much higher P wave amplitude, higher QRS amplitude, and longer ST segment—appeared in patients with head injury compared to healthy subjects [31,32]. Furthermore, another study also revealed that the more obvious the abnormalities, the more deteriorated the consciousness level of the patients. Meanwhile, subarachnoid hemorrhage condition (SAH) can be seen morphologically in T and R wave abnormalities [33]. Specifically, in head trauma, abnormalities are discovered from prolonged QTc [34]. Furthermore, this previous study also depicted that the more severe SAH, the more prolonged QTc. For intracranial investigation, morphologically, some changes appeared in the ECG—U wave, T wave, ST-T segment, QT interval, J wave [35,36]. In addition, ECG is considered as the secondary effect of traumatic brain injury (TBI) [37].

As previously mentioned, the evaluation of hemodynamics is significant, especially for cardiovascular- and cerebral-related conditions. However, most of the precise procedures are measured invasively. Therefore, the aim of this study is to generate cardiovascular and cerebral hemodynamics monitoring systems non-invasively using the ECG signal and a deep convolutional autoencoder system.

2. Materials and Methods

This study used two databases from PhysioNet [38]: Massachusetts General Hospital/Marquette Foundation (MGH/MF) Waveform Database [39] and Cerebral Hemodynamic Autoregulatory Information System Database (CHARIS DB) [40]. For the hemodynamics system, the shorter window size was evaluated, and the better prediction was obtained. This compensates for the rapid change in some circumstances. However, the ECG is unpractical to be analyzed in very short period. This is due to, for normal subjects, the QRS cycle being sometimes not fully formed within a too short period. Therefore, the 2 s window of signal was selected for the input and output system. It was then combined and randomly separated from all the patients for training and testing with no overlapping selection.

A deep convolutional autoencoder (DCAE) system was used for the signal generation model. The DCAE model is a modified model from the original U-Net model used in a biomedical image generator segmentation system [41]. This modified model deploys a multi-atrous U-Net based deep convolutional autoencoder (MA-UDCAE) system, which was originally applied by [10]. Figures 1 and 2 show the MA-UDCAE model structures for both cardiovascular and cerebral hemodynamics systems.

Figure 1 shows the convolutional autoencoder system applied to the cardiovascular hemodynamics system. For the first half, the encoder system decreases the shape of the layer. Meanwhile, the decoder increases the shape of the layer. This structure consists of several layers: convolution layer (CV), down pooling layer (DP), up sampling layer (UP), and merging layer (ME). Initially, the input of this system is the ECG. The next layer, the first convolution layer (CV_01) with multi dilation is applied. The next layer of this multi-dilation layer is merged into ME_01 followed by the down pooling (DP_01) layer. For the down pooling, this study utilizes the max pooling system. The second convolution layer (CV_02) is used and merged into the ME_02 layer. Later, the ME_02 layer is down pooled into DP_02. This block—the convolution layer, concatenation, and down pooling—is replicated until the 4th layer. For the decoder, this system is identical to the encoder system. However, the down sampling system is switched into the up sampling. Furthermore, the U-Net based system is also applied into this system. A layer in encoder is merged with the decoder layer. There are four merging layers: DP_03-ME_07, DP_02-ME_09, DP_01-ME_11, and input layer ME_14. Finally, several convolution layers are applied before the output layer, which has three channels: ABP, CVP, and PAP signals.

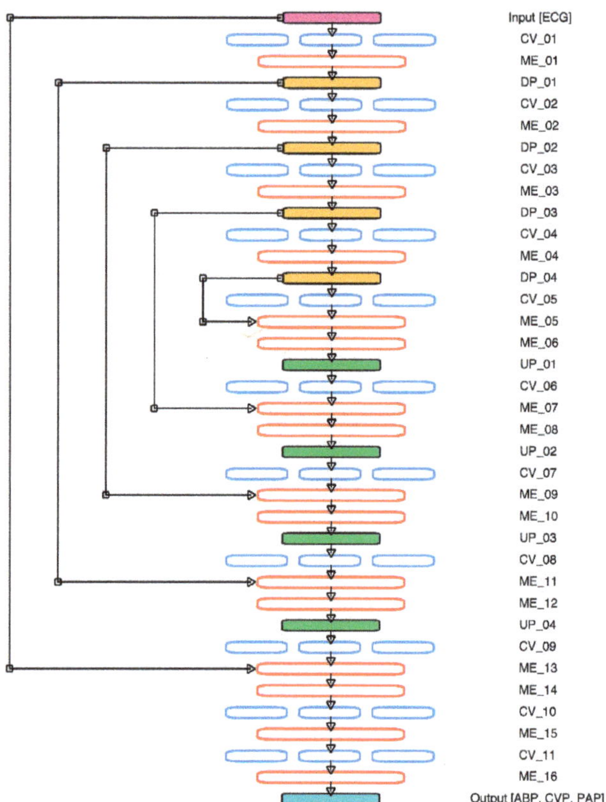

Figure 1. Cardiovascular hemodynamics MA-UDCAE model structure. Note: CV: convolution layer; ME: merge layer; DP: down pooling layer; UP: up sampling layer.

Figure 2 shows the cerebral hemodynamics system, which is identical to Figure 1. However, this figure has two merged layers. Furthermore, this system only has a single output channel—intracranial pressure signal (ICP). Initially for the encoder, the ECG input processed by the multi-altrous convolution system (CV_01), followed by the merging layer (ME_01), and a down pooling layer (DP_01). The next block is identical—the convolutional layer is followed by the concatenating layer and down pooling layer. For the decoder block, the first up sampling layer (UP_01) is also processed by the convolutional layer (CV_04) and concatenating layer (ME_04). Moreover, in this decoder system, the ME_02 is merged with ME_04 to form the ME_05 layer, similar to ME_07 by combining between ME_01 and ME_06 layers. Finally, the last layer is the ICP waveform.

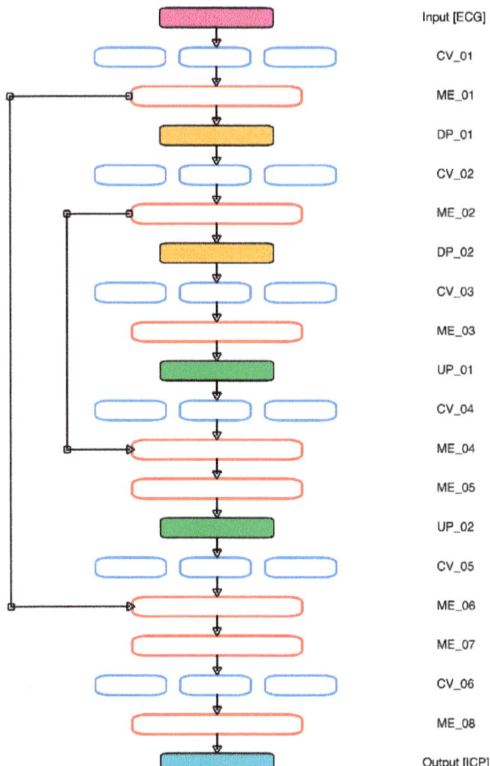

Figure 2. Cerebral hemodynamics MA-UDCAE model structure. Note: CV: convolution layer; ME: merge layer; DP: down pooling layer; UP: up sampling layer.

Initially, the Massachusetts General Hospital/Marquette Foundation (MGH/MF) Waveform Database was used for the cardiovascular hemodynamics system. It contains several physiological signals: ECG, ABP, CVP, and PAP. This system was uniformly sampled for 360 Hz. From this database, this study uses only the lead II ECG signal, as the input signal, to initially generate the waveforms of ABP, CVP, and PAP, as the outputs. The utilization of the lead II ECG signal is due to this lead being mostly utilized in ECG record consideration [42]. Furthermore, the systolic and diastolic pressures of ABP and PAP were calculated by taking the maximum and minimum values, respectively, from the signal. Meanwhile, the mean CVP value was taken from averaging the 2 s CVP waveform. There were 59,401 and 14,850 of 2 sec segments, respectively, for training and testing.

Furthermore, the Cerebral Hemodynamic Autoregulatory Information System Database (CHARIS DB) was used for the intracranial pressure (ICP) evaluation. This database was sampled for 50 Hz. The input signal was the ECG signal, meanwhile the output signal was the ICP signal. A 2 s window was also selected, both for the input and output systems. Furthermore, there were 1,451,281 and 362,560 of 2 sec segments, respectively, for training and testing. Even though this database contains smaller number of patients, the data collection was much longer compared to the dataset utilized for the cardiovascular hemodynamics system.

For pre-processing the data and post-processing the result, MATLAB R2014b (Math-Works, Inc., Natick, MA, USA) was utilized. The data were initially filtered manually by visualization. The input and output signal must be appropriate for the training and testing. If either the input or output signal is noisy, both will be deleted. For training the deep convolutional autoencoder system, Python 3.6 was utilized with TensorFlow (Ver. 1.15.2) [43] and Keras (Ver. 2.3.1) under Google Colaboratory (Google Inc., 1600 Amphitheatre Parkway Mountain View, CA, USA). In more detail, the training of both cardiovascular and cerebral hemodynamics systems were conducted with the checkpoint system.

For the evaluation, a 5-fold cross-validation was performed with shuffled training data. Generally, this strategy is performed to evaluate the regularity of the data to the model. Furthermore, Pearson's linear correlation (R), root mean squared error (RMSE), and mean absolute error (MAE) evaluations were conducted. Furthermore, the Bland–Altman plot was conducted for the further evaluation. R, RMSE, and MAE evaluations are shown in Equations (1)–(3).

$$R_{x,y} = \frac{\sum_{i=1}^{n}(x_i - \overline{x})(y_i - \overline{y})}{\sqrt{\left[\sum_{i=1}^{n}(x_i - \overline{x})^2 \sum_{i=1}^{n}(y_i - \overline{y})^2\right]}} \tag{1}$$

$$\text{MAE} = \frac{1}{n}\sum_{i=1}^{n}|x_i - y_i| \tag{2}$$

$$\text{RMSE} = \sqrt{\frac{1}{n}\sum_{i=1}^{n}(x_i - y_i)^2} \tag{3}$$

where x_i is the reference, y_i is the predicted result, n is the number of samples, $\overline{x}$ is the mean of the reference, and $\overline{y}$ is the mean of the predicted result.

3. Results

The multi-atrous deep convolutional autoencoder models were applied to both cardiovascular and cerebral hemodynamics systems using a single ECG signal. Pearson's linear correlation, root mean squared error, and mean absolute error evaluations were deployed as performance indicators. Finally, a Bland-Altman plot was conducted to investigate the performance of the deep learning for both cardiovascular and cerebral hemodynamics models. The models initially generated the hemodynamics waveform. This waveform generating system is fundamental due to it accommodating the morphological cardiovascular conditions [44]. Finally, from the generated waveform, the systolic, diastolic, or the mean values were extracted.

3.1. Cardiovascular Hemodynamics

The Massachusetts General Hospital/Marquette Foundation (MGH/MF) Waveform Database was used for this cardiovascular hemodynamics system. The ECG was used to generate the ABP, CVP, and PAP signals. The model was trained for 500 epochs, as shown in Figure 3. As it can be seen, the model starts to saturate at 400 epochs. The training and validation curves are relatively stable. The 5-fold cross-validation results show 0.942 ± 0.001, 7.833 ± 0.061 mmHg, and 4.959 ± 0.044 mmHg, respectively, for the R, RMSE, and MAE of ABP waveform evaluations. Furthermore, the CVP evaluation records are: 0.852 ± 0.002, 3.155 ± 0.018 mmHg, 2.024 ± 0.020 mmHg for R, RMSE, and MAE, respectively. Finally, for the PAP waveform evaluation, the evaluations of R, RMSE, and MAE are 0.873 ± 0.002, 4.853 ± 0.031 mmHg, and 3.263 ± 0.034 mmHg. The detail about the waveform evaluations is given in Table 1.

Furthermore, the systolic and diastolic pressures were calculated for the ABP and PAP. Meanwhile, the mean CVP value was also predicted from the waveform. For the ABP system, the systolic and diastolic pressures, the R evaluations are 0.894 ± 0.004 and 0.881 ± 0.005, respectively. The RMSE evaluations are 8.997 ± 0.401 mmHg and 4.726 ± 0.129 mmHg, respectively. The MSE evaluations are, respectively, 6.645 ± 0.353 mmHg and 3.210 ± 0.104 mmHg. Furthermore, for the PAP system, the systolic and diastolic pressures, the R evaluations are 0.864 ± 0.003 mmHg and 0.817 ± 0.006 mmHg, respectively. The RMSE evaluations are 5.833 ± 0.075 mmHg and 4.360 ± 0.179 mmHg, respectively. The MSE evaluations are 3.847 ± 0.136 mmHg and 2.964 ± 0.181 mmHg, respectively.

Meanwhile, the mean CVP evaluations are 0.916 ± 0.001, 2.220 ± 0.039 mmHg, and 1.329 ± 0.036 mmHg for R, RMSE, and MSE, respectively. The details about these evaluations are shown in Tables 2–4.

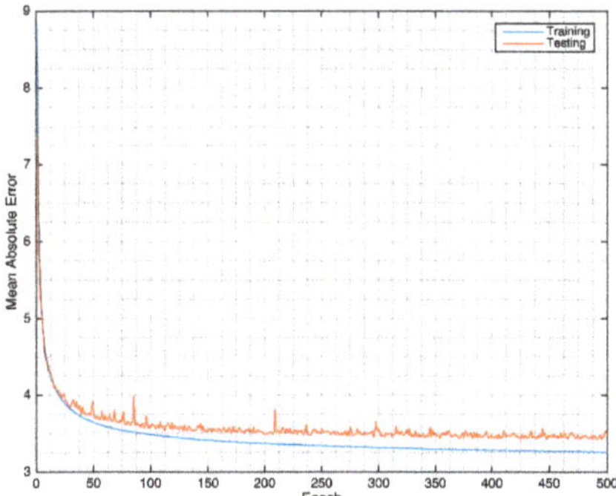

Figure 3. MA-UDCAE model convergence for cardiovascular hemodynamics.

Table 1. Waveform evaluations of cardiovascular hemodynamics.

CV	ABP			CVP			PAP		
	R	RMSE (mmHg)	MAE (mmHg)	R	RMSE (mmHg)	MAE (mmHg)	R	RMSE (mmHg)	MAE (mmHg)
1	0.941	7.874	4.950	0.850	3.168	2.032	0.871	4.863	3.270
2	0.942	7.915	5.036	0.854	3.124	1.994	0.874	4.805	3.207
3	0.941	7.819	4.927	0.851	3.156	2.015	0.872	4.863	3.262
4	0.942	7.797	4.938	0.852	3.165	2.045	0.872	4.888	3.295
5	0.943	7.761	4.945	0.855	3.162	2.032	0.875	4.847	3.281
Mean	0.942	7.833	4.959	0.852	3.155	2.024	0.873	4.853	3.263
STD	0.001	0.061	0.044	0.002	0.018	0.020	0.002	0.031	0.034

Table 2. Systolic and diastolic arterial blood pressure evaluations.

CV	Arterial Blood Pressure					
	R		RMSE (mmHg)		MAE (mmHg)	
	SBP	DBP	SBP	DBP	SBP	DBP
1	0.890	0.875	8.980	4.906	6.517	3.344
2	0.894	0.884	9.640	4.575	7.260	3.093
3	0.892	0.878	8.926	4.782	6.593	3.217
4	0.895	0.881	8.905	4.638	6.482	3.123
5	0.900	0.888	8.534	4.730	6.371	3.272
Mean	0.894	0.881	8.997	4.726	6.645	3.210
STD	0.004	0.005	0.401	0.129	0.353	0.104

Table 3. Mean central venous pressures.

CV	Central Venous Pressure		
	R	RMSE (mmHg)	MAE (mmHg)
1	0.916	2.232	1.332
2	0.918	2.156	1.277
3	0.915	2.219	1.312
4	0.916	2.228	1.369
5	0.917	2.264	1.353
Mean	0.916	2.220	1.329
STD	0.001	0.039	0.036

Table 4. Systolic and diastolic pulmonary arterial pressure evaluations.

CV	Pulmonary Arterial Pressure					
	R		RMSE (mmHg)		MAE (mmHg)	
	SBP	DBP	SBP	DBP	SBP	DBP
1	0.859	0.818	5.864	4.295	3.863	2.892
2	0.864	0.813	5.769	4.457	3.766	3.042
3	0.863	0.810	5.819	4.329	3.817	2.938
4	0.864	0.817	5.766	4.598	3.717	3.218
5	0.868	0.827	5.947	4.122	4.070	2.730
Mean	0.864	0.817	5.833	4.360	3.847	2.964
STD	0.003	0.006	0.075	0.179	0.136	0.181

The waveform evaluation result is shown in Figure 4. This figure also investigates the information of systolic, diastolic, and mean pressures from ABP, CVP, and PAP. Most importantly, the ability of MA-UDCAE deals with several conditions of the ECG, ABP, CVP, and PAP signals within two seconds and is also given in this figure. From Figure 4a, it can be seen how a relatively normal heartbeat generates normal ABP, CVP, and PAP signals. From this figure, the model predicts accurate systolic, diastolic, and mean values. In Figure 4b, ECG has a higher heart rate. In this case, the subject has very low ABP information, high CVP, and high PAP measures. A high error is given from the systolic ABP in the last period of the CVP waveform. However, the PAP is relatively good. Normal heart rate ECG is given in Figure 4c. However, this subject has relatively high systolic ABP, and relatively high CVP and PAP. Furthermore, Figure 4d shows relatively higher ECG heartbeat for generating normal ABP, normal CVP, and high PAP. It can be seen that the systolic PAP has a relatively high error.

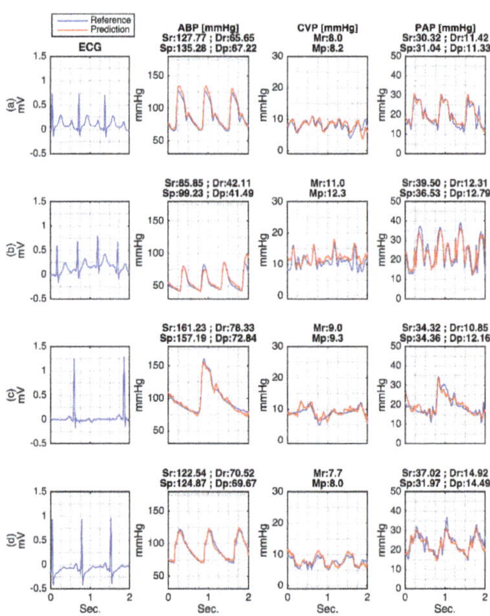

Figure 4. MA-UDCAE-generated waveforms of ABP, CVP, and PAP: (**a**) Normal ECG with normal hemodynamics; (**b**) Fast ECG with abnormal hemodynamics; (**c**) Slow ECG with abnormal hemodynamics; (**d**) Normal ECG with abnormal hemodynamics. Note: Systolic reference (Sr); diastolic reference (Dr); systolic prediction (Sp); diastolic prediction (Dp); mean reference (Mr); mean prediction (Mp).

Finally, the Pearson's linear correlation and Bland–Altman plot are given in Figures 5 and 6. Figure 5 shows the Pearson's linear correlation result. It can be seen that the correlation coefficient of systolic ABP and diastolic ABP are 0.89 and 0.86, respectively. Meanwhile, the SPAP and DPAP are, respectively, 0.86 and 0.81. Finally, the mean CVP is 0.92. Figure 6 shows the Bland–Altman plot for the cardiovascular hemodynamics system. For systolic ABP, the reference and prediction have the mean difference of -4.182 mmHg, -1.96 STD of -21.268 mmHg, and $+1.96$ STD of 12.904. For the diastolic ABP, it has a mean difference of 0.202 mmHg, -1.96 STD of -9.987 mmHg, and $+1.96$ STD of 10.391. Furthermore, for the systolic PAP, the mean difference is 0.668 mmHg, -1.96 STD of -10.563 mmHg, and $+1.96$ STD of 11.899 mmHg. For diastolic PAP, the mean difference is -1.827 mmHg, -1.96 STD of 6.140 mmHg, and $+1.96$ STD of -9.794 mmHg. Finally, the mean CVP has mean difference of -0.121 mmHg, -1.96 STD of 4.099, and $+1.96$ STD of -4.341 mmHg. The graphs show some negative values appearing in the prediction. This situation likely happens due to the complexity of patient monitoring, and the method for selecting the systolic and diastolic peaks using maximum and minimum values.

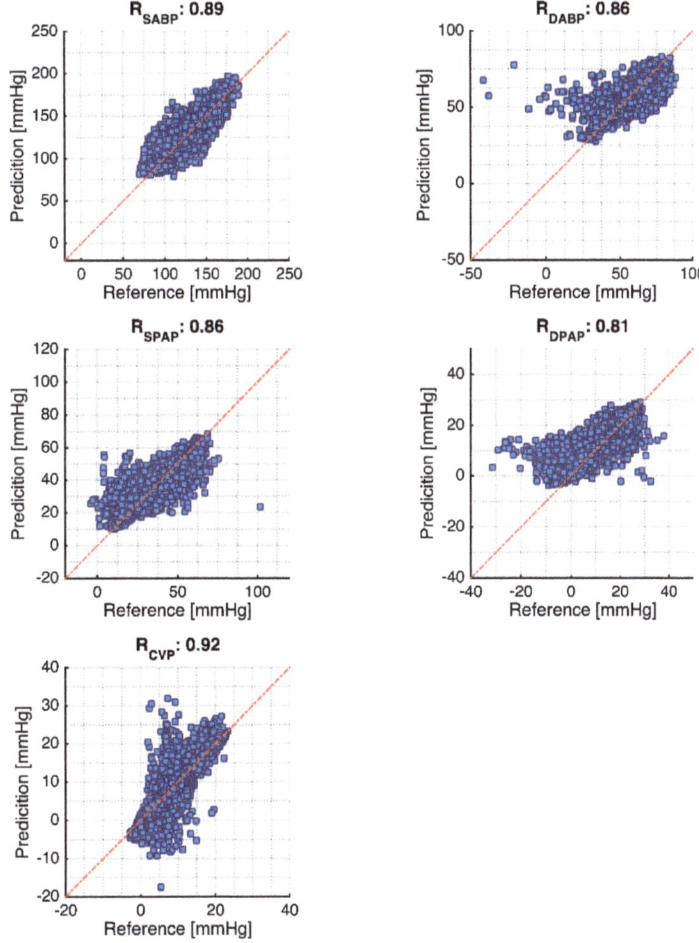

Figure 5. Pearson's linear correlation coefficient results. Note: Red dotted line is the diagonal line.

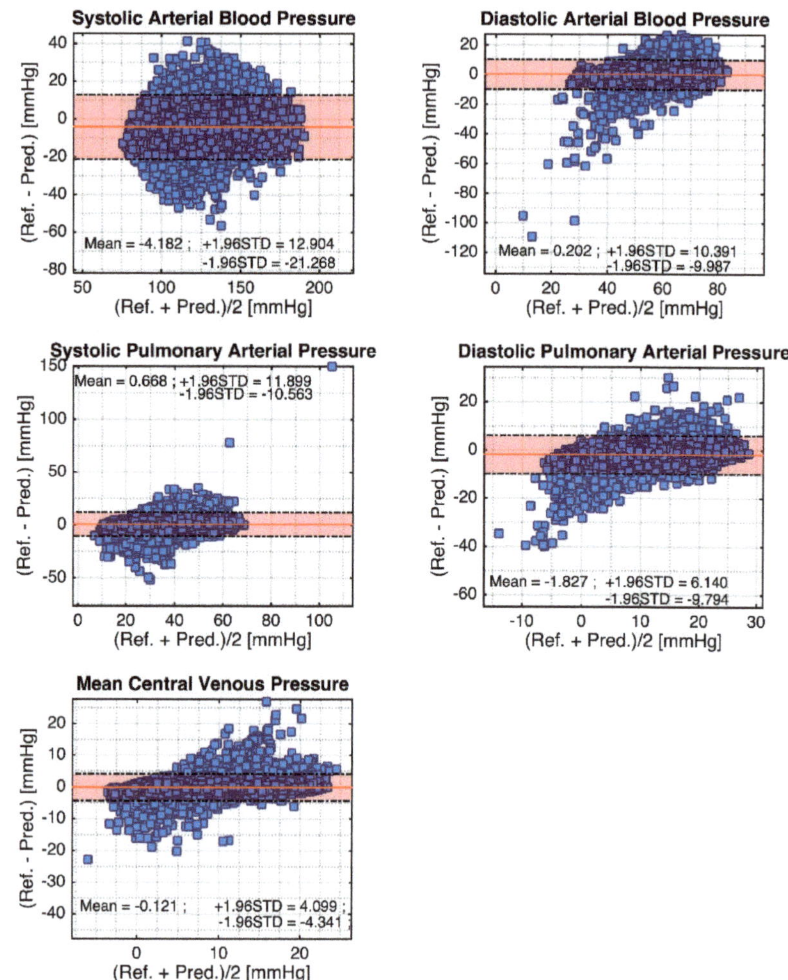

Figure 6. Bland–Altmann plots for cardiovascular hemodynamics system. Note: Red line is the mean, and dash-dot lines are ±1.96 standard deviations.

3.2. Intracranial Pressure

For the intracranial pressure evaluation, the Cerebral Hemodynamic Autoregulatory Information System Database (CHARIS DB) was utilized. The MA-UDCAE model was also applied for the ECG signal to understand the ICP pattern. For the evaluation of the deep learning system from the ECG to ICP, the MA-UDCAE model was applied. This system was conducted for 20 epochs. The mean absolute error was used for the evaluation. The ICP training convergence is shown in Figure 7.

After the training season, the testing data were subsequently evaluated into the trained model. Some of the ECG-generated ICP waveforms can be seen in Figure 8. These results are selected based on the variation of the reference ICP. From this figure, it can be seen that the model is not only reasonably robust in handling the data that has either a relatively low or high ICP index, but also gives some information on how the model decodes the ICP waveform.

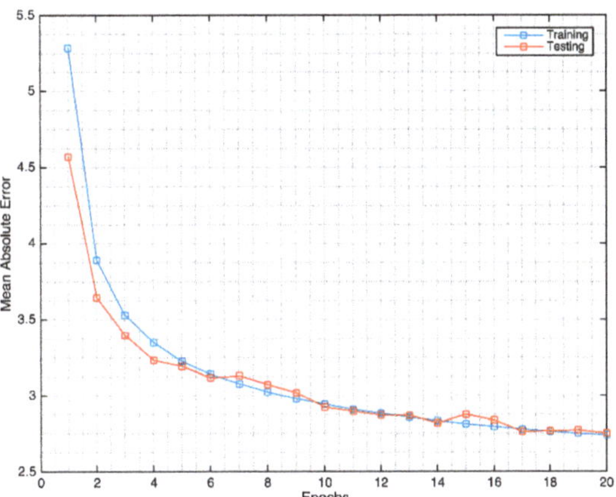

Figure 7. The MA-UDCAE model convergence for intracranial pressure.

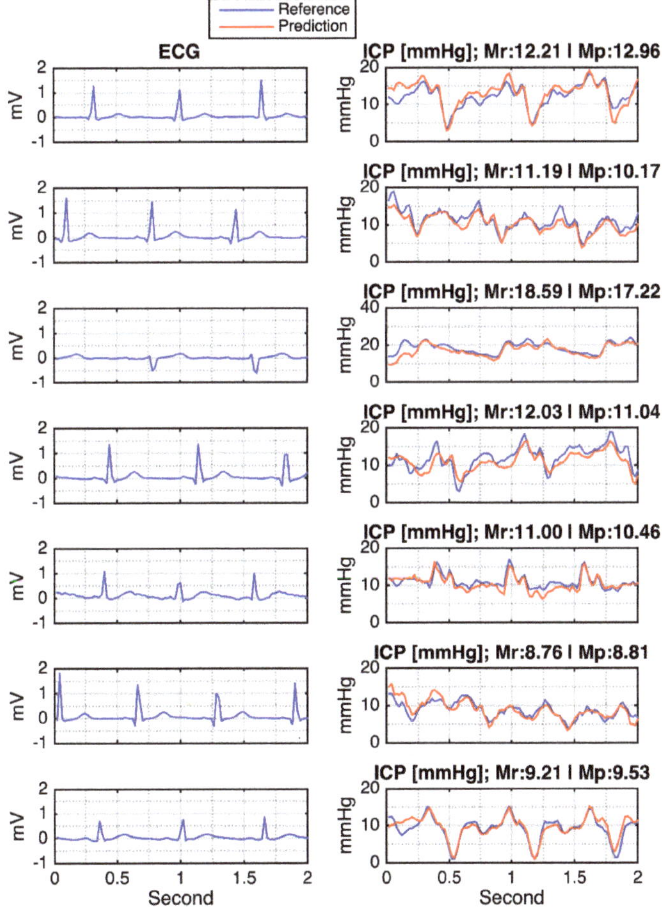

Figure 8. MA-UDCAE-generated waveform of ICP. Note: Mean reference (Mr); mean prediction (Mp).

Furthermore, in more detail, for the intracranial pressure waveform evaluations, the cross-validation results generate 0.887 ± 0.003, 5.306 ± 0.041 mmHg, and 2.765 ± 0.026 mmHg, respectively, for R, RMSE, and MAE. For the mean ICP evaluation, the R, RMSE and MAE evaluations are 0.914 ± 0.003, 4.582 ± 0.044 mmHg, and 2.404 ± 0.043 mmHg, as shown in Table 5. The intracranial evaluations for Pearson's linear correlation and Bland–Altman are shown in Figure 9. Pearson's linear correlation for the ICP is 0.89, as given in Figure 9a. From this figure, it can be seen that the model likely starts to generate a bigger error when dealing with an ICP higher than 30 mmHg. In addition, Figure 9b shows the Bland–Altman plot for ICP evaluations. Results in Figure 9b support Pearson's linear correlation. As it can be seen, the mean difference is 0.492 mmHg, -1.96 STD of -8.32 mmHg, and $+1.96$ STD of 9.310 mmHg. Even though it has relatively low prediction error, the model gives lower accuracy on higher ICP measures. The entire ICP evaluation is shown in Table 5, indicating that all evaluations of R, MAE, and RMSE have fairly low standard deviation values. However, some negative values can be seen appearing in the prediction. This situation likely happens due to the complexity of the patient data.

Table 5. Intracranial pressure evaluations.

CV	R		RMSE (mmHg)		MAE (mmHg)	
	Waveform	Mean	Waveform	Mean	Waveform	Mean
1	0.890	0.917	5.330	4.592	2.786	2.435
2	0.884	0.912	5.342	4.603	2.762	2.398
3	0.885	0.910	5.333	4.637	2.758	2.393
4	0.888	0.914	5.254	4.550	2.792	2.453
5	0.889	0.915	5.269	4.526	2.726	2.343
Mean	0.887	0.914	5.306	4.582	2.765	2.404
STD	0.003	0.003	0.041	0.044	0.026	0.043

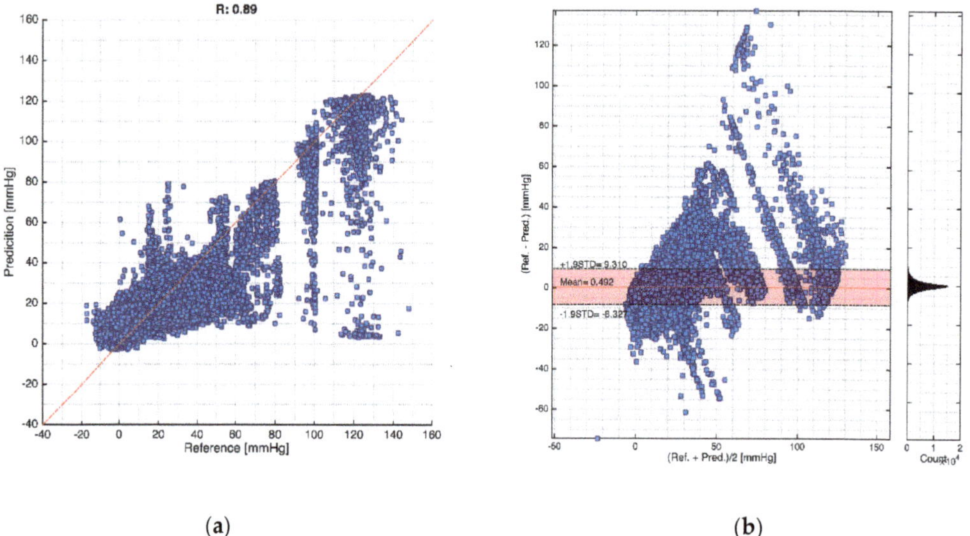

Figure 9. Intracranial pressure evaluations: (**a**) Pearson's linear correlation result for intracranial pressure evaluation. Note: Red dotted line is the diagonal line; (**b**) Bland–Altmann plot for intracranial pressure evaluation. Note: Red line is the mean, and dash-dot lines are for ± 1.96 standard deviations.

4. Discussion

The novelty in this study is the utilization of a non-invasive ECG signal to investigate cardiovascular and cerebral hemodynamics. Initially, the models generated the cardiovascular hemodynamics ABP, PAP, and CVP from MGHDB. This study used two databases from PhysioNet: Massachusetts General Hospital/Marquette Foundation (MGH/MF) Waveform Database and Cerebral Hemodynamic Autoregulatory Information System Database (CHARIS DB). The MA-UDCAE deep learning model was deployed for modeling these systems.

Most of the previously conducted studies as shown in Table 6 utilized PPG signal as an input of the model. Slapničar et al. [45] used PPG signal of 510 subjects of MIMIC III from a PhysioNet and ResNet-based model to investigate hypertension. This previous study had a MAE of 9.43 mmHg and 6.88 mmHg, respectively, for SBP and DBP. Chowdhury et al. administered the PPG signal of 126 subjects and a regression model. Their results achieved a correlation coefficient of 0.95 and 0.96, respectively, for SBP and DBP, and MAE of SBP and DBP are, respectively, 3.02 mmHg and 1.74 mmHg [46]. Meanwhile, Zadi et al. [47] applied the ARMA algorithm to PPG signals from 15 subjects. They achieved an RMSE of 7.21 mmHg and 5.12 mmHg for SBP and DBP, respectively. However, none of these studies provides information about the continuous waveform evaluation.

Table 6. Cardiovascular hemodynamics evaluations. Note: RMSE and MAE are in mmHg.

Studies	Dataset	Input Signal	Cont. ABP	Method	Perf. Eval.	Waveform	SBP	DBP
Tanveer et al. [6]	39 subjects, MIMIC, PhysioNet	ECG + PPG	No	ANN + LSTM	RMSE MAE R	N/A N/A N/A	1.26 0.93 0.999	0.73 0.52 0.998
Wu et al. [7]	27 subjects	ECG + PPG	No		RMSE	N/A	3.404	3.289
Eom et al. [8]	15 subjects	ECG + PPG + BCG	No	CNN + Bi-GRU + Attention	MAE R^2	N/A N/A	4.06 ± 4.04 0.52	3.33 ± 3.42 0.49
Sideris et al. [9]	42 subjects, MIMIC, PhysioNet	PPG	Yes	LSTM	RMSE R	6.04 ± 3.26 0.95 ± 0.05	2.58 ± 1.23 N/A	1.98 ± 1.06 N/A
Sadrawi et al. [10]	18 Patients, NTUH, Taiwan	PPG	Yes	GDCAE	RMSE MAE R	3.46 2.33 0.984	3.41 2.54 0.981	2.14 1.48 0.979
Fan et al. [30]	MIMIC II, PhysioNet	ECG	No	BiLSTM + FCN	RMSE MAE	N/A N/A	12.3 7.69	6.88 4.36
Slapničar et al. [45]	510 subjects, MIMIC III, PhysioNet	PPG	No	Spectro temporal ResNet	MAE	N/A	9.43	6.88
Chowdhury et al. [46]	222 records, 126 subjects	PPG	No	Gaussian process regression	RMSE MAE R MSE	N/A N/A N/A N/A	6.74 3.02 0.95 45.49	3.59 1.74 0.96 12.89
Aguirre et al. [48]	1131 subjects, MIMIC, PhysioNet	PPG	Yes	Seq2seq + Attention	RMSE MAE R R^2	8.67 7.39 0.98 N/A	15.96 12.08 N/A 0.39	7.4 5.56 N/A 0.41
Zadi et al. [47]	15 subjects	PPG	No	ARMA	RMSE	N/A	7.21	5.12
Proposed	250 subjects, MGH/MF, PhysioNet	ECG	Yes	MA-UDCAE	RMSE MAE R	7.83 ± 0.06 4.95 ± 0.04 0.94 ± 0.00	8.99 ± 0.40 6.64 ± 0.35 0.89 ± 0.00	4.73 ± 0.13 3.21 ± 0.10 0.88 ± 0.01

Several previous studies, shown in Table 6, have utilized PPG signal as their input, and investigated the waveform evaluation. Sideris et al. [9] performed evaluation of continuous blood pressure using PPG signals with the LSTM method, delivering an RMSE of 6.04 ± 3.26 mmHg, 2.58 ± 1.23 mmHg, and 1.98 ± 1.06 mmHg, respectively, on SBP and DBP evaluations, and a waveform correlation coefficient of 0.95 ± 0.045. Meanwhile, Sadrawi et al. [10] utilized PPG for arterial blood pressure estimation, reporting 0.984 linear correlation and MAEs of 2.54 mmHg and 1.48 mmHg for the systolic and diastolic, respectively. Furthermore, Aguirre et al. [48] utilized PPG signal of 1131 subjects from the PhysioNet database using the Seq2seq algorithm to evaluate the hemodynamics. This study also investigated the arterial blood pressure waveform. The generated waveform had a correlation coefficient of 0.98. Meanwhile, the MAE of systolic and diastolic were 12.08 mmHg and 5.56 mmHg, respectively [48].

On the other hand, prior studies as shown in Table 6 also correlated ECG signal with blood pressure estimation. Tanveer et al. [6] used a combination of ECG and PPG, utilizing ANN and LSTM. This study reported 0.93 mmHg and 0.52 mmHg for SBP and DBP, respectively, on the MAE estimations. This earlier study has very high correlation coefficients, 0.999 and 0.998, respectively, for the SBP and DBP. Another study by Wu et al. [7] also used ECG and PPG with the hybrid ANN method, and produced 3.404 mmHg and 3.289 mmHg, respectively, for SBP and DBP. Meanwhile, a study by Eom et al. [8] conducted an investigation of multiple input signals, ECG, PPG, and BCG, combined with CNN, Bi-GRU, and attention methods. This previous study produced an MAE of 4.06 ± 4.04 and 3.33 ± 3.42, respectively, for SBP and DBP evaluations. Meanwhile, it had 0.52 and 0.49 for the R^2 evaluations for the SBP and DBP, respectively. The study that only used the ECG signal to investigate hypertension was conducted by Fan et al. [23]. This previously conducted study reported the MAE for the systolic and diastolic as 7.69 and 4.36 mmHg, respectively.

For a comparison of intracranial pressure evaluation, this study is compared to several sub-studies, given in Table 7. These previously conducted works were undertaken by utilizing the ABP and cerebral blood flow velocity (CBFV). Imaduddin et al. [3] used these signals to estimate the ICP from 13 patients with the Bayesian system. This previous study provided an RMSE of 3.7 mmHg. Another study conducted by Jaishankar et al. [4] had RMSEs of 5.1 mmHg and 4.5 mmHg, respectively, for 13 pediatric and 5 adult subjects. This prior study utilized the spectral method. Even though the CBFV is a non-invasive system, the ABP is classified as an invasive technique. This study reported 4.582 ± 0.044 mmHg RMSE from five cross-validation systems. This result is slightly inferior compared to the previously compared study. However, the novelty of this study is the utilization of the non-invasive ECG signal as the input signal to generate the ICP waveform with further evaluation of the mean ICP.

Table 7. Intracranial pressure evaluations. Note: RMSE and MAE are in mmHg.

Studies	Dataset	Input Signal	Method	Performance Evaluation	Mean ICP (mmHg)
Imaduddin et al. [3]	13 subjects	ABP + CBFV	Bayesian model	RMSE	3.7
Jaishankar et al. [4]	13 pediatric subjects	ABP + CBFV	Spectral approach	RMSE	5.1
	5 adult subjects				4.5
Proposed	13 subjects, CHARISD, PhysioNet	ECG	MAUDCAE	RMSE MAE R	4.582 ± 0.044 2.404 ± 0.043 0.914 ± 0.003

The association between ECG and hemodynamics is related to the fact that ECG and ABP have a very close relationship in time-related terms, especially in the evaluation of the heart-related system. This may have some similarity in CVP, PAP, and ICP due to the blood circulation. Furthermore, the systolic peak and the R peak intervals are shifted in some transmitting time. However, more investigation is required to be performed morphologically for the full cycle between the ECG and cardiovascular and cerebral hemodynamics signals, especially for CVP and ICP.

In order to evaluate the network performance, an ablation study was performed. The autoencoder structure was utilized based on [10]. From this previous study, specifically for hemodynamics, the UNET-based model provided a better result compared to the classical autoencoder network. Hence, based on reference [10], the structure was modified by utilizing the multi-atrous system. For simplification, this study investigates the results only at 25 and 5 epochs, respectively, for the cardiovascular and cerebral hemodynamics systems.

The ablation network evaluation was also conducted in our study, given in Figure 10. For the cardiovascular system, the ablated network convergence is shown in Figure 10a. Several ablations were conducted. The decreased layers are sequentially the layer connections between DP_04 and CV_05, DP_03 and CV_06, DP_02 and CV_07, DP_01 and CV_08, and the input layer and CV_09. From this figure, it can be seen that the first ablated layer attaching between DP_03 and CV_06 tends to have a faster convergence rate compared to others. However, the results are relatively similar at the 25th epoch. For testing, the deleted connection systems tend to have some oscillation during this early period compared to the non-ablated networks.

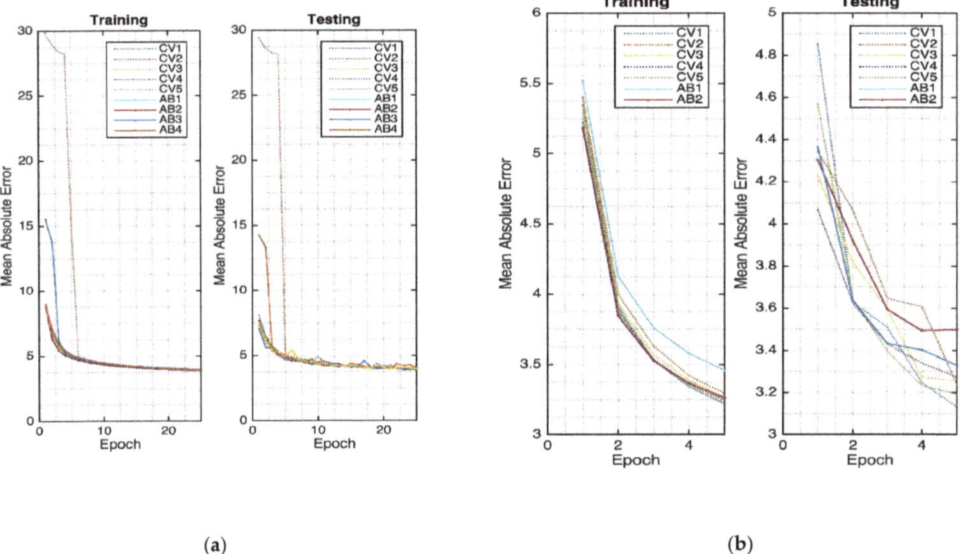

(a) (b)

Figure 10. Ablation network: (**a**) cardiovascular hemodynamics; (**b**) cerebral hemodynamics.

The cerebral hemodynamics ablation study was conducted by removing the concatenating layers. The first deleted layer is the connection between ME_02 and CV_04. The next one is between ME_01 and CV_05. The results are shown in Figure 10b, where in the training phase, most of the cross-validation models without ablation have relatively lower MAE values at the 5th epoch compared to the ablated models. In this system, the deleted connection models tend to saturate earlier compared to the system without any ablated connection. Furthermore, the ablated networks also have a slower convergence rate at the testing.

This study has several limitations due to the selection of systolic and diastolic techniques being sensitive to noise. In more detail, the noise has a positive effect on the systolic and a negative effect on the diastolic pressure. This phenomenon will generate higher error during the evaluation.

Due to the data limitations, one of the main limitations of this study is the testing data separated randomly from all patients instead of patient-based partition. Even though the data were not fully interpreted, this strategy will let the model learn and memorize some patterns of the output through the input. Another limitation is the high dependency on the quality of the ECG signal. The mitigations of the dataset unbalancing can be investigated further [49] and other deep learning structures can be applied in future [50–52].

There are also concerns regarding the evaluations of arrhythmia conditions. In this study, some of the arrhythmia ECG factors affecting the cardiovascular and cerebral hemodynamics systems have been investigated. In addition, the cardiovascular hemodynamics dataset was well annotated for some arrhythmia conditions. However, for the cerebral hemodynamics, the heartbeat in the dataset was not labeled. Nevertheless, it is still possible for the rapid and irregular beats to be investigated. The premature ventricular contraction (PVC) and supraventricular premature/ectopic beats were evaluated. Figure 11 shows how arrhythmia affects the cardiovascular hemodynamics. It can be seen that most of the abnormal ECG signals are relatively good in generating the ABP. However, there are some shifts in the SBP and DBP, as shown in Figure 11a,f. Bigger differences are shown for CVP and PAP signals. The generated ICP from the abnormal ECG is shown in Figure 12. From Figure 12a,b, the R-R interval irregularity has a worse effect on the ICP prediction compared to the arrhythmia generated from the rapid R interval. However, deeper evaluation with many more additional arrhythmia cases to investigate the effect of arrhythmia on the hemodynamics should be performed in the future.

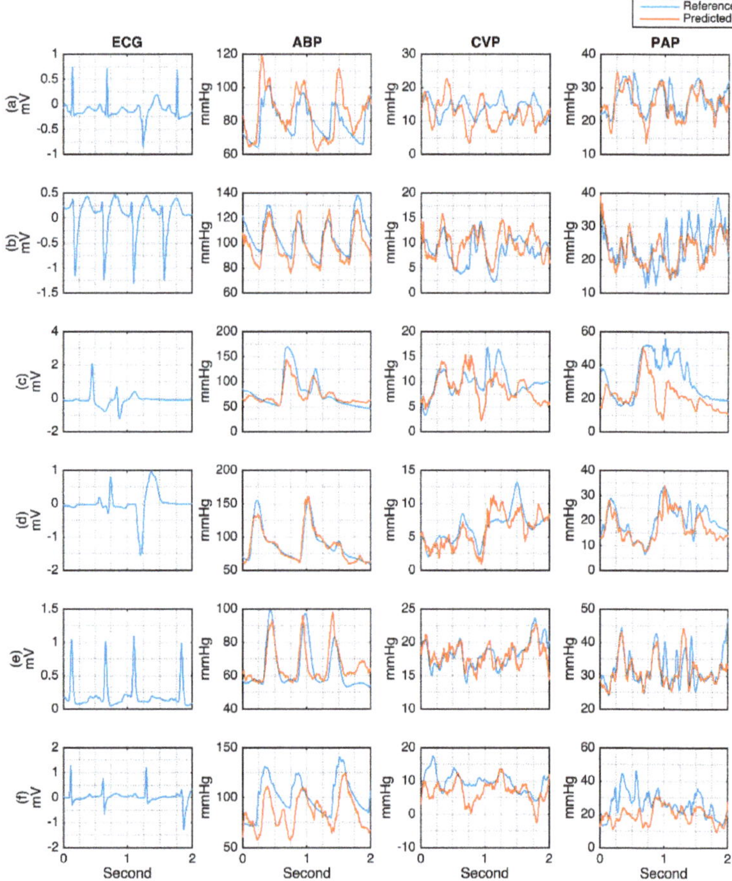

Figure 11. Arrhythmic ECG predictions for cardiovascular hemodynamics. (**a**) rapid R waves with premature beat; (**b**) rapid heart rate with downward R waves; (**c**) Slow heart rate with relatively small downward R wave; (**d**) Slow heart rate with bigger downward R wave; (**e**) Rapid upward R wave with irregular interval; (**f**) Rapid heart rate with multiple downward R wave.

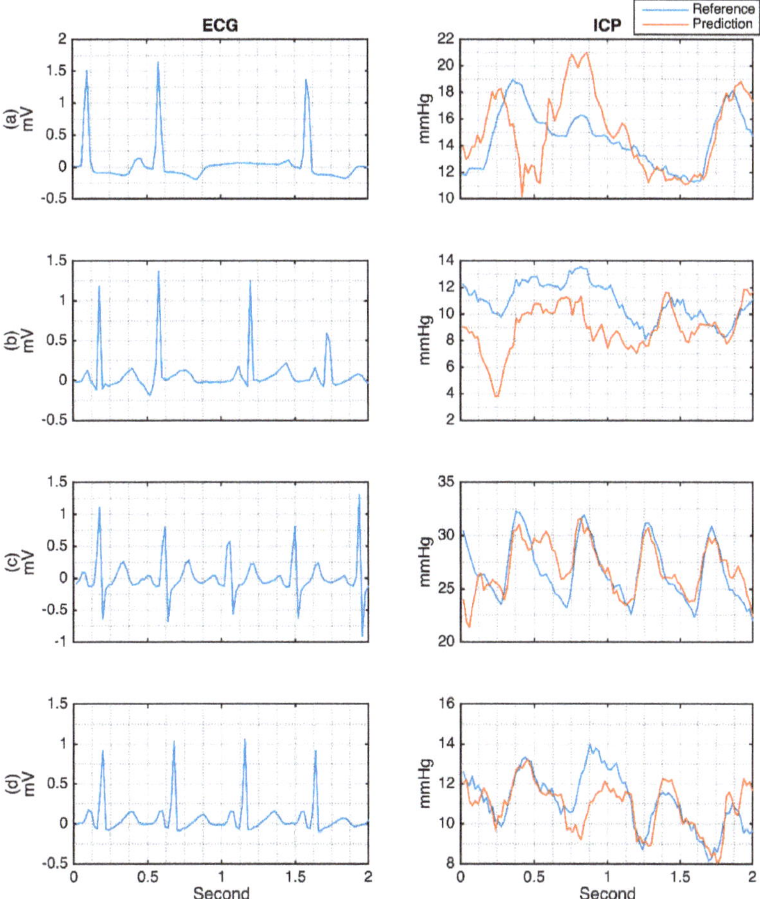

Figure 12. Arrhythmic ECG evaluations for the intracranial pressure. (**a**) Irregular interval heart rate; (**b**) Irregular and rapid heart rate; (**c**) Rapid heart rate with downward shifted R waves; (**d**) Rapid heart rate.

Since this study is preliminary and a proof of concept, it may not provide superior results compared to previous studies that used multi-input signals such as the combination of PPG and ECG, in which the shape of the PPG signal is much more identical to ABP signal compared to the ECG and ABP signal. However, as with preliminary and proof of concept studies, this study can be a finding of using the ECG in hemodynamics investigations. Finally, the cross-validation test using the same group of subjects in our current study was inner loop cross-validation, which only rotates the validation data and training data. However, to make the results more convincing, the testing data need to be rotated into training and validation, which is known as outer loop cross-validation, and is to be considered in future works. In addition, the pre-processing of the data was initially filtered manually by visualization. This manual filter may not be practical for the study. We still need further investigation about using automatic filtering to filter all these vital signs for future work.

5. Conclusions

In order to design the most precise health monitoring system to solve the hemodynamics system in ICU, in this study, as a proof of concept, a deep convolutional autoencoder system has been implemented for a non-invasive system by only using a single ECG signal and utilizing two databases for cardiovascular and cerebrovascular hemodynamics systems. For the preliminary result, it can be seen that ECG has great potential in generating the ABP, CVP, PAP, and ICP waveform, as well as their essential information for the extensive evaluations.

Author Contributions: Conceptualization, M.S., H.-T.H., S.-Z.F., M.F.A. and J.-S.S.; Formal analysis, M.S.; Investigation, M.S., H.-T.H., S.-Z.F., M.F.A. and J.-S.S.; Methodology, M.S., Y.-T.L., M.F.A. and J.-S.S.; Project administration, J.-S.S.; Software, M.S., C.-H.L. and B.M.; Supervision, Y.-T.L., C.-H.L., H.-T.H., S.-Z.F., M.F.A. and J.-S.S.; Validation, M.S.; Visualization, M.S.; Writing—original draft, M.S., M.F.A. and J.-S.S.; Writing—review and editing, M.S., M.F.A. and J.-S.S. All authors have read and agreed to the published version of the manuscript.

Funding: This research received no external funding.

Institutional Review Board Statement: Not applicable.

Informed Consent Statement: Not applicable.

Data Availability Statement: This study utilizes the publicly available dataset, from https://physionet.org, accessed on 14 January 2021.

Conflicts of Interest: The authors declare no conflict of interest.

References

1. Chesnut, R.M.; Temkin, N.; Carney, N.; Dikmen, S.; Rondina, C.; Videtta, W.; Petroni, G.; Lujan, S.; Pridgeon, J.; Barber, J.; et al. A trial of intracranial-pressure monitoring in traumatic brain injury. *N. Engl. J. Med.* **2012**, *367*, 2471–2481. [CrossRef] [PubMed]
2. Jaishankar, R.; Fanelli, A.; Filippidis, A.; Vu, T.; Holsapple, J.; Heldt, T. A spectral approach to model-based noninvasive intracranial pressure estimation. *IEEE J. Biomed. Health Inform.* **2019**, *24*, 2398–2406. [CrossRef] [PubMed]
3. Imaduddin, S.M.; Fanelli, A.; Vonberg, F.W.; Tasker, R.C.; Heldt, T. Pseudo-Bayesian model-based noninvasive intracranial pressure estimation and tracking. *IEEE Trans. Biomed. Eng.* **2019**, *67*, 1604–1615. [CrossRef]
4. Merrer, J.; De Jonghe, B.; Golliot, F.; Lefrant, J.Y.; Raffy, B.; Barre, E.; Rigaud, J.P.; Casciani, D.; Misset, B.; Bosquet, C.; et al. Complications of femoral and subclavian venous catheterization in critically ill patients: A randomized controlled trial. *JAMA* **2001**, *286*, 700–707. [CrossRef] [PubMed]
5. Sadrawi, M.; Shieh, J.S.; Fan, S.Z.; Lin, C.H.; Haraikawa, K.; Chien, J.C.; Abbod, M.F. Intermittent blood pressure prediction via multiscale entropy and ensemble artificial neural networks. In Proceedings of the 2016 IEEE EMBS Conference on Biomedical Engineering and Sciences (IECBES), Kuala Lumpur, Malaysia, 4–7 December 2016; pp. 356–359.
6. Tanveer, M.S.; Hasan, M.K. Cuffless blood pressure estimation from electrocardiogram and photoplethysmogram using waveform based ANN-LSTM network. *Biomed. Signal Process. Control* **2019**, *51*, 382–392. [CrossRef]
7. Wu, H.; Ji, Z.; Li, M. Non-Invasive Continuous Blood-Pressure Monitoring Models Based on Photoplethysmography and Electrocardiography. *Sensors* **2019**, *19*, 5543. [CrossRef]
8. Eom, H.; Lee, D.; Han, S.; Hariyani, Y.S.; Lim, Y.; Sohn, I.; Park, K.; Park, C. End-To-End Deep Learning Architecture for Continuous Blood Pressure Estimation Using Attention Mechanism. *Sensors* **2020**, *20*, 2338. [CrossRef]
9. Sideris, C.; Kalantarian, H.; Nemati, E.; Sarrafzadeh, M. Building continuous arterial blood pressure prediction models using recurrent networks. In Proceedings of the 2016 IEEE International Conference on Smart Computing (SMARTCOMP), St. Louis, MO, USA, 18–20 May 2016; pp. 1–5.
10. Sadrawi, M.; Lin, Y.T.; Lin, C.H.; Mathunjwa, B.; Fan, S.Z.; Abbod, M.F.; Shieh, J.S. Genetic Deep Convolutional Autoencoder Applied for Generative Continuous Arterial Blood Pressure via Photoplethysmography. *Sensors* **2020**, *20*, 3829. [CrossRef]
11. Sadrawi, M.; Shieh, J.S.; Haraikawa, K.; Chien, J.C.; Lin, C.H.; Abbod, M.F. Ensemble empirical mode decomposition applied for PPG motion artifact. In Proceedings of the 2016 IEEE EMBS Conference on Biomedical Engineering and Sciences (IECBES), Kuala Lumpur, Malaysia, 4–7 December 2016; pp. 266–269.
12. Salehizadeh, S.; Dao, D.; Bolkhovsky, J.; Cho, C.; Mendelson, Y.; Chon, K.H. A novel time-varying spectral filtering algorithm for reconstruction of motion artifact corrupted heart rate signals during intense physical activities using a wearable photoplethysmogram sensor. *Sensors* **2016**, *16*, 10. [CrossRef]
13. Thalhammer, C.; Aschwanden, M.; Odermatt, A.; Baumann, U.A.; Imfeld, S.; Bilecen, D.; Marsch, S.C.; Jaeger, K.A. Noninvasive central venous pressure measurement by controlled compression sonography at the forearm. *J. Am. Coll. Cardiol.* **2007**, *50*, 1584–1589. [CrossRef]
14. Szymczyk, T.; Sauzet, O.; Paluszkiewicz, L.J.; Costard-Jäckle, A.; Potratz, M.; Rudolph, V.; Gummert, J.F.; Fox, H. Non-invasive assessment of central venous pressure in heart failure: A systematic prospective comparison of echocardiography and Swan-Ganz catheter. *Int. J. Cardiovasc. Imaging* **2020**, *36*, 1821–1829. [CrossRef]

15. Proença, M.; Braun, F.; Lemay, M.; Solà, J.; Adler, A.; Riedel, T.; Messerli, F.H.; Thiran, J.P.; Rimoldi, S.F.; Rexhaj, E. Non-invasive pulmonary artery pressure estimation by electrical impedance tomography in a controlled hypoxemia study in healthy subjects. *Sci. Rep.* **2020**, *10*, 1–8. [CrossRef]
16. Sadrawi, M.; Lin, C.H.; Lin, Y.T.; Hsieh, Y.; Kuo, C.C.; Chien, J.C.; Haraikawa, K.; Abbod, M.F.; Shieh, J.S. Arrhythmia evaluation in wearable ECG devices. *Sensors* **2017**, *17*, 2445. [CrossRef] [PubMed]
17. Liu, Q.; Ma, L.; Chiu, R.C.; Fan, S.Z.; Abbod, M.F.; Shieh, J.S. HRV-derived data similarity and distribution index based on ensemble neural network for measuring depth of anaesthesia. *PeerJ* **2017**, *5*, e4067. [PubMed]
18. Sadrawi, M.; Fan, S.Z.; Abbod, M.F.; Jen, K.K.; Shieh, J.S. Computational depth of anesthesia via multiple vital signs based on artificial neural networks. *BioMed Res. Int.* **2015**, *2015*, 536863. [CrossRef] [PubMed]
19. Varon, C.; Caicedo, A.; Testelmans, D.; Buyse, B.; Van Huffel, S. A novel algorithm for the automatic detection of sleep apnea from single-lead ECG. *IEEE Trans. Biomed. Eng.* **2015**, *62*, 2269–2278. [CrossRef]
20. Penzel, T.; Kantelhardt, J.W.; Lo, C.C.; Voigt, K.; Vogelmeier, C. Dynamics of heart rate and sleep stages in normals and patients with sleep apnea. *Neuropsychopharmacology* **2003**, *28*, S48–S53. [CrossRef] [PubMed]
21. Penzel, T.; Kantelhardt, J.W.; Grote, L.; Peter, J.H.; Bunde, A. Comparison of detrended fluctuation analysis and spectral analysis for heart rate variability in sleep and sleep apnea. *IEEE Trans. Biomed. Eng.* **2003**, *50*, 1143–1151. [CrossRef] [PubMed]
22. Ross, G. Effect of hypertension on the P wave of the electrocardiogram. *Br. Heart J.* **1963**, *25*, 460. [CrossRef]
23. Tarazi, R.C.; Miller, A.; Froklich, E.D.; Dustan, H.P. Electrocardiographic changes reflecting left atrial abnormality in hypertension. *Circulation* **1966**, *34*, 818–822. [CrossRef]
24. Abou Farha, K.; van Vliet, A.; van Marle, S.; Vrijlandt, P.; Westenbrink, D. Hypertensive crisis-induced electrocardiographic changes: A case series. *J. Med. Case Rep.* **2009**, *3*, 1–6.
25. Hedblad, B.; Janzon, L. Hypertension and ST segment depression during ambulatory electrocardiographic recording. Results from the prospective population study'men born in 1914'from Malmö, Sweden. *Hypertension* **1992**, *20*, 32–37. [CrossRef] [PubMed]
26. Van Der Ende, M.Y.; Hendriks, T.; Van Veldhuisen, D.J.; Snieder, H.; Verweij, N.; Van Der Harst, P. Causal Pathways from Blood Pressure to Larger QRS Amplitudes: A Mendelian Randomization Study. *Sci. Rep.* **2018**, *8*, 1–7.
27. Kovacs, G.; Avian, A.; Foris, V.; Tscherner, M.; Kqiku, X.; Douschan, P.; Bachmaier, G.; Olschewski, A.; Matucci-Cerinic, M.; Olschewski, H. Use of ECG and other simple non-invasive tools to assess pulmonary hypertension. *PLoS ONE* **2016**, *11*, e0168706. [CrossRef]
28. Waligóra, M.; Tyrka, A.; Podolec, P.; Kopeć, G. ECG markers of hemodynamic improvement in patients with pulmonary hypertension. *BioMed Res. Int.* **2018**, *2018*, 4606053. [CrossRef]
29. Rajput, J.S.; Sharma, M.; San Tan, R.; Acharya, U.R. Automated detection of severity of hypertension ECG signals using an optimal bi-orthogonal wavelet filter bank. *Comput. Biol. Med.* **2020**, *123*, 103924. [CrossRef]
30. Fan, X.; Wang, H.; Zhao, Y.; Li, Y.; Tsui, K.L. An Adaptive Weight Learning-Based Multitask Deep Network for Continuous Blood Pressure Estimation Using Electrocardiogram Signals. *Sensors* **2021**, *21*, 1595. [CrossRef] [PubMed]
31. Hersch, C. Electrocardiographic changes in head injuries. *Circulation* **1961**, *23*, 853–860. [CrossRef] [PubMed]
32. Wittebole, X.; Hantson, P.; Laterre, P.F.; Galvez, R.; Duprez, T.; Dejonghe, D.; Renkin, J.; Gerber, B.L.; Brohet, C.R. Electrocardiographic changes after head trauma. *J. Electrocardiol.* **2005**, *38*, 77–81. [CrossRef] [PubMed]
33. Chatterjee, S. ECG changes in subarachnoid haemorrhage: A synopsis. *Neth. Heart J.* **2011**, *19*, 31–34. [CrossRef] [PubMed]
34. Collier, B.R.; Miller, S.L.; Kramer, G.S.; Balon, J.A.; Gonzalez, L.S., III. Traumatic subarachnoid hemorrhage and QTc prolongation. *J. Neurosurg. Anesthesiol.* **2004**, *16*, 196–200. [CrossRef]
35. Jachuck, S.J.; Ramani, P.S.; Clark, F.; Kalbag, R.M. Electrocardiographic abnormalities associated with raised intracranial pressure. *Br. Med. J.* **1975**, *1*, 242–244. [CrossRef]
36. Milewska, A.; Guzik, P.; Rudzka, M.; Baranowski, R.; Jankowski, R.; Nowak, S.; Wysocki, H. J-wave formation in patients with acute intracranial hypertension. *J. Electrocardiol.* **2009**, *42*, 420–423. [CrossRef] [PubMed]
37. Lenstra, J.J.; Kuznecova-Keppel Hesselink, L.; la Bastide-van Gemert, S.; Jacobs, B.; Nijsten, M.W.N.; van der Horst, I.C.C.; van der Naalt, J. The association of early electrocardiographic abnormalities with brain injury severity and outcome in severe traumatic brain injury. *Front. Neurol.* **2021**, *11*, 1840. [CrossRef]
38. Goldberger, A.L.; Amaral, L.A.; Glass, L.; Hausdorff, J.M.; Ivanov, P.C.; Mark, R.G.; Mietus, J.E.; Moody, G.B.; Peng, C.K.; Stanley, H.E. PhysioBank, PhysioToolkit, and PhysioNet: Components of a new research resource for complex physiologic signals. *Circulation* **2000**, *101*, e215–e220. [CrossRef] [PubMed]
39. Welch, J.; Ford, P.; Teplick, R.; Rubsamen, R. The Massachusetts General Hospital-Marquette Foundation hemodynamic and electrocardiographic database–comprehensive collection of critical care waveforms. *J. Clin. Monit.* **1991**, *7*, 96–97.
40. Kim, N.; Krasner, A.; Kosinski, C.; Wininger, M.; Qadri, M.; Kappus, Z.; Danish, S.; Craelius, W. Trending autoregulatory indices during treatment for traumatic brain injury. *J. Clin. Monit. Comput.* **2016**, *30*, 821–831. [CrossRef]
41. Ronneberger, O.; Fischer, P.; Brox, T. U-net: Convolutional networks for biomedical image segmentation. In Proceedings of the 18th International Conference, Munich, Germany, 5–9 October 2015.
42. Meek, S.; Morris, F. Introduction. I—Leads, rate, rhythm, and cardiac axis. *BMJ* **2002**, *324*, 415–418. [CrossRef]
43. Abadi, M.; Barham, P.; Chen, J.; Chen, Z.; Davis, A.; Dean, J.; Devin, M.; Ghemawat, S.; Irving, G.; Isard, M.; et al. Tensorflow: A system for large-scale machine learning. In Proceedings of the 12th USENIX Symposium on Operating Systems Design and Implementation (OSDI 16), Savannah, GA, USA, 2–4 November 2016; pp. 265–283.

44. Avolio, A.P.; Butlin, M.; Walsh, A. Arterial blood pressure measurement and pulse wave analysis—their role in enhancing cardiovascular assessment. *Physiol. Meas.* **2009**, *31*, R1. [CrossRef] [PubMed]
45. Slapničar, G.; Mlakar, N.; Luštrek, M. Blood pressure estimation from photoplethysmogram using a spectro-temporal deep neural network. *Sensors* **2019**, *19*, 3420. [CrossRef]
46. Chowdhury, M.H.; Shuzan, M.N.I.; Chowdhury, M.E.; Mahbub, Z.B.; Uddin, M.M.; Khandakar, A.; Reaz, M.B.I. Estimating Blood Pressure from the Photoplethysmogram Signal and Demographic Features Using Machine Learning Techniques. *Sensors* **2020**, *20*, 3127. [CrossRef]
47. Zadi, A.S.; Alex, R.; Zhang, R.; Watenpaugh, D.E.; Behbehani, K. Arterial blood pressure feature estimation using photoplethysmography. *Comput. Biol. Med.* **2018**, *102*, 104–111. [CrossRef] [PubMed]
48. Aguirre, N.; Grall-Maës, E.; Cymberknop, L.J.; Armentano, R.L. Blood pressure morphology assessment from photoplethysmogram and demographic information using deep learning with attention mechanism. *Sensors* **2021**, *21*, 2167. [CrossRef] [PubMed]
49. Sadrawi, M.; Sun, W.Z.; Ma, M.H.M.; Yeh, Y.T.; Abbod, M.F.; Shieh, J.S. Ensemble genetic fuzzy neuro model applied for the emergency medical service via unbalanced data evaluation. *Symmetry* **2018**, *10*, 71. [CrossRef]
50. Litjens, G.; Ciompi, F.; Wolterink, J.M.; de Vos, B.D.; Leiner, T.; Teuwen, J.; Išgum, I. State-of-the-art deep learning in cardiovascular image analysis. *JACC Cardiovasc. Imaging* **2019**, *12 Pt 1*, 1549–1565. [CrossRef]
51. Wang, J.; Ding, H.; Bidgoli, F.A.; Zhou, B.; Iribarren, C.; Molloi, S.; Baldi, P. Detecting cardiovascular disease from mammograms with deep learning. *IEEE Trans. Med. Imaging* **2017**, *36*, 1172–1181. [CrossRef] [PubMed]
52. Munir, K.; Elahi, H.; Ayub, A.; Frezza, F.; Rizzi, A. Cancer diagnosis using deep learning: A bibliographic review. *Cancers* **2019**, *11*, 1235. [CrossRef]

Article

Automatic Classification of Myocardial Infarction Using Spline Representation of Single-Lead Derived Vectorcardiography

Yu-Hung Chuang , Chia-Ling Huang, Wen-Whei Chang * and Jen-Tzung Chien

Institute of Electrical and Computer Engineering, National Chiao-Tung University, Hsinchu 30010, Taiwan; yuhung1206.eed09g@nctu.edu.tw (Y.-H.C.); alin0624.eed05@nctu.edu.tw (C.-L.H.); jtchien@nctu.edu.tw (J.-T.C.)
* Correspondence: wwchang@cc.nctu.edu.tw

Received: 18 November 2020; Accepted: 14 December 2020; Published: 17 December 2020

Abstract: Myocardial infarction (MI) is one of the most prevalent cardiovascular diseases worldwide and most patients suffer from MI without awareness. Therefore, early diagnosis and timely treatment are crucial to guarantee the life safety of MI patients. Most wearable monitoring devices only provide single-lead electrocardiography (ECG), which represents a major limitation for their applicability in diagnosis of MI. Incorporating the derived vectorcardiography (VCG) techniques can help monitor the three-dimensional electrical activities of human hearts. This study presents a patient-specific reconstruction method based on long short-term memory (LSTM) network to exploit both intra- and inter-lead correlations of ECG signals. MI-induced changes in the morphological and temporal wave features are extracted from the derived VCG using spline approximation. After the feature extraction, a classifier based on multilayer perceptron network is used for MI classification. Experiments on PTB diagnostic database demonstrate that the proposed system achieved satisfactory performance to differentiating MI patients from healthy subjects and to localizing the infarcted area.

Keywords: electrocardiography; vectorcardiography; myocardial infarction; long short-term memory; spline; multilayer perceptron

---

1. Introduction

Myocardial infarction (MI) has long been recognized as the main cause of death worldwide. According to the data from the World Health Organization (WHO) [1], cardiovascular diseases, including MI, were estimated to account for 31% of deaths worldwide in 2017. In the United States, about 110,000 Americans died of MI in 2015 and the estimated annual incidence of MI is 605,000 new attacks [2]. MI results from an occlusion of the coronary artery and insufficient blood supply to the myocardium. It can be further classified into various subtypes depending on the localization of infarcted area. In clinical setting, MI is diagnosed using 12-lead electrocardiography (ECG) [3] as well as 3-lead vectorcardiography (VCG) [4]. ECG signals are recorded from different locations of the body to capture the three-dimensional view of the human heart. The standard ECG has 12 leads, including six limb leads (I, II, III, aVR, aVL, aVF) and six chest leads (V_1 to V_6). Figure 1 shows the three-dimensional view of 12 standard leads on the xyz-coordinate axis system. According to electrode positioning, the 12 ECG leads can be used to localize different types of MI, such as inferior leads (II, III, aVF), septal leads (V_1, V_2), anterior leads (V_3, V_4), and lateral leads (I, aVL, V_5, V_6). A typical waveform of the ECG beat consists of a P wave, a QRS-complex, and a T wave. These characteristic waves correspond to the sequence of depolarization and repolarization of the atria and ventricles. ECG signs suggestive of MI include ST-segment deviation or changes in the shapes of Q-wave and T-wave, using which physicians can localize damage to specific areas of the heart. However, it may be noted that 12-lead ECG requires ten electrodes for recording and some of the leads contain redundant information. Instead, VCG requires a minimum of four electrodes and it monitors cardiac electrical activity in three orthogonal planes of the body [5]. Generally, Frank leads (V_x, V_y, V_z) scanned in orthogonal xyz axes are used for VCG measurements. The main advantage of VCG is that it uses fewer leads than 12-lead ECG for medical diagnostic applications. Moreover, different studies [6–8] have demonstrated that VCG provides a higher sensitivity for the diagnosis of MI as well as ischemic heart diseases. In this study, VCG signal is processed to extract clinically significant features that will allow for MI classification.

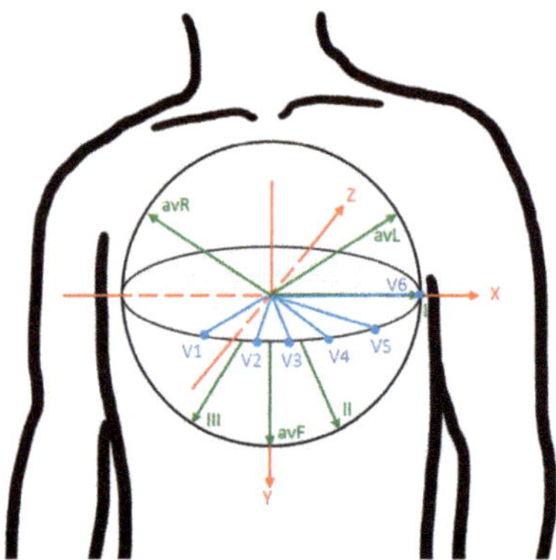

Figure 1. The three-dimensional view of 12 ECG leads on the xyz-coordinate axis system.

MI is also known as a silent heart attack that usually occurs without clear symptoms. Hence, early diagnosis and timely treatment are crucial to improve the recovery rate of MI patients. In recent years, several computer-aided diagnostic methods have been proposed for automatic MI detection and localization [9–18]. Most of these approaches extract the clinically significant features from the ECG signal and then apply an appropriate classifier in the classification stage. Various informative features have been extracted to represent the ECG beats, such as morphological features [10] as well as frequency and wavelet-based features [11,12]. Moreover, some studies have attempted to use directly measured or derived VCG to identify changes in the VCG morphology such as the QRS and T-wave loops [16–18]. For classification, different machine learning algorithms have been investigated, including k-nearest neighbors (KNN) [10,12], artificial neural network (ANN) [11], recurrent neural network (RNN) [13] and convolutional neural network (CNN) [14,15]. Furthermore, several researchers [13–15] have proposed end-to-end approaches for MI detection and localization. These methods obviate the need to extract features at the cost of higher computational complexity. ECG abnormalities due to MI may be observed in the ST-segment deviation or changes in the shapes of T-wave and Q-wave. Generally, it is a prerequisite to identify characteristic waves of ECG beats before performing the feature extraction. Although various methods have been proposed for ECG wave delineation [19–22], they still have some limitations for characterization of MI beats. To address this constraint, we apply spline curve fitting [23,24] to the entire heartbeat to model all of the characteristic waves and use fitted coefficients as features. The advantage of using the entire heartbeat is that the QRS complexes and P and T waves can be included in the curve fitting so that poor quality features resulting from delineation errors can be avoided. Moreover, the VCG signal is semiperiodic in nature and has numerous clinically relevant turning points in each heartbeat. Such signals require a higher-order polynomial to fit, leading to severe oscillations of the fitted curve which cause the overfitting problem [25]. By contrast, the spline's flexibility in approximating curves with different degrees of smoothness at different locations is ideal for representing the semiperiodic VCG signal.

Another problem which requires further investigation is to test the feasibility of single-lead ECG in classifying different types of MI. Several wearable devices which use single-lead ECG to facilitate continuous ambulatory monitoring have recently appeared on the market [26]. While these devices make regular ECG recording possible, their practical applicability for cardiac diagnostics remain limited. This is because physicians need checking ECG patterns to diagnose by correlating information from two or more ECG leads. For example, abnormalities in chest leads (V_1 to V_4) are suggestive of a problem in the posterior wall of the heart and no abnormalities will be detected by a single lead [27]. The ability to transform from single-lead ECG to 12-lead ECG enables the wider use of wearable devices for clinical diagnostic applications. However, prior attempts to synthesize 12-lead ECG or 3-lead VCG from a single lead have not been successful. Most existing lead transformation approaches require at least two synchronously acquired leads [28–40], hampering their applicability to the present context. This has motivated our investigation into trying to synthesize the 3-lead VCG from single-lead ECG signal. Since lead I is provided by most wearable devices, we propose a derived VCG system by considering the lead I ECG signal as input and three Frank leads as output of the system.

A lot of emphases have been recently put on derived ECG systems due to the increasing demand of personalized healthcare applications. The methods of lead synthesis can be categorized in terms of reconstruction algorithms and lead configuration. The lead configuration for ECG synthesis can be divided into two groups: use of subsets of 12-lead ECG [28–31] and use of Frank VCG leads [32–40]. A common assumption in previous works was that the heart-torso electrical system is linear and quasi-static, which allows for the use of linear transformation to derive the 3-lead VCG from reduced-lead set of the 12-lead ECG. These can either be patient-specific or generic transformation of which the former is learned using data from a single patient, while the latter requires data from a group of patients. Previous studies have shown the possibility to derive the 12-lead ECG from the three Frank XYZ leads through Dower transformation [34] and vice versa through the inverse Dower transformation [35]. Similarly, Kors et al. [36] derived the transformation matrix using the regression analysis method. In [37], Dawson et al. derived the linear affine transformation between 3-lead VCG and 12-lead ECG, which achieved higher accuracy than Kors and inverse Dower transformation. Another strategy can be seen in [31,32], where nonlinear methods such as ANN were used to synthesize the 12-lead ECG and 3-lead VCG from leads I, II, and V_2. It was found that nonlinear transformation are appropriate for ECG data with diversity resulting from variation in individuals and measurement positions. A weakness for majority of the reviewed methods is that they only exploited the inter-lead correlation between spatially aligned samples of the lead signals. It is important to note that, in addition to spatially correlated information in different leads, temporally correlated information can also be found between different waves within a single lead. System design approaches that consider both intra- and inter-lead correlation are expected to provide better solutions to the VCG synthesis problem. This task can be accomplished by using RNN [41] as it can use the learning capabilities of ANN and could further improve it by representing the spatio-temporal correlations between the lead signals. In this work, we proposed a patient-specific transformation for VCG synthesis by applying a long short-term memory (LSTM) network [42] with sliding window approach.

This study focuses on two issues: synthesis of 3-lead VCG and extraction of VCG features, to develop an MI classifier that is suitable for wearable devices with only a single lead recording. The first part of this study focuses on developing a method of VCG reconstruction from lead I ECG using a LSTM network to exploit both intra- and inter-lead correlations of ECG signals. The second part of this study develops a novel spline framework for parametrically representing the derived Frank lead signals. After extracting features by the spline approximation, a classifier is used for the classification of healthy and 11 types of MI.

2. Methods

This study proposes a new method for automatic MI classification using the single-lead derived VCG. As shown in Figure 2, the proposed method consists of four stages, i.e., preprocessing,

VCG synthesis, feature extraction, and classification. The raw ECG signals are preprocessed to remove various kinds of noise associated with them. Next, a patient-specific reconstruction method is used to synthesize the 3-lead VCG from lead I ECG. In the feature extraction stage, the clinically significant features are extracted from three derived Frank leads that quantify the VCG abnormalities due to MI. Later in the classification stage, the most likely ECG class has to be predicted from the analysis of the feature data.

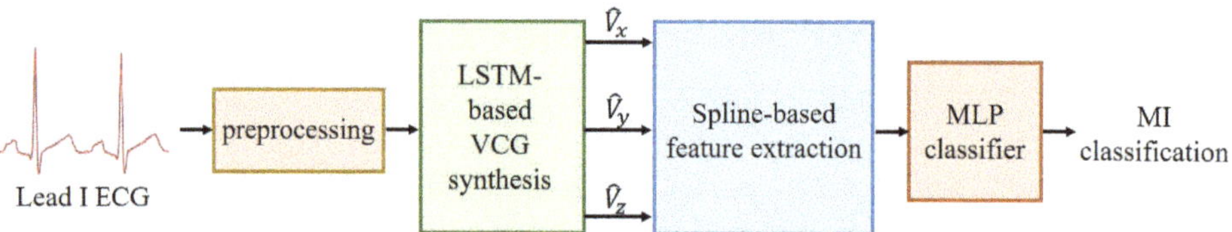

Figure 2. Block diagram of the proposed MI classification system.

2.1. Preprocessing

The raw ECG signal is typically contaminated by high-frequency noises caused by power-line interference, electromyographic noises due to muscle activity, motion artifacts caused by patient's movements, and radio frequency noises from other equipments. Moreover, baseline wander is low-frequency (0–0.5 Hz) interference in the ECG signal caused by respiration, body movement and changes in electrode impedance. These noises degrade the quality of ECG signals and introduce ambiguity in the MI classification. Hence, the preprocessing is generally performed to to remove various types of noises associated with the input signal. The guidelines for the standardization and interpretation of ECG, published by the American Heart Association [43], advise using a cutoff frequency of 0.05 Hz for the high-pass filter and 150 Hz for the low-pass filter in adults. Thus, in this study, the raw ECG signal is down-sampled to 500 Hz and then filtered using a band-pass filter with a bandwidth between 0.5 and 150 Hz to remove noise and baseline wander. A similar approach has been used in several other studies [16,44].

2.2. VCG Synthesis

Synthesis of 3-lead VCG from reduced-lead set of 12-lead ECG [32,35–40] has been investigated in the past to satisfy the need for more wearing comfort and ambulatory situations. Most methods [35–40] are based on linear transformation and the differences between them are in coefficients of transformation matrices. In [32], Vozda et al. used nonlinear methods such as ANN to synthesize the 3-lead VCG from quasi-orthogonal leads I, II, and V_2. Most current approaches to VCG synthesis focus on the inter-lead correlation, with less emphasis placed on the intra-lead correlation. The ECG signals from leads I, V_x, V_y, and V_z are shown in Figure 3. It can be observed that, in addition to spatially correlated information in different leads, temporally correlated information can also be found between different waves within a single lead. The lead signals are narrow angle projections of the same electric heart vector and hence correlations can be found among the signals of various leads. Moreover, the cardiac cycle is quasi-periodic in nature and hence intra-correlations are evident between different characteristic waves. A model which can simultaneously learn the intra- and inter-lead correlations of ECG signals is expected to further improve the reconstruction accuracy. This is because synthesizing a VCG lead essentially involves estimating morphology of the waveform and timings of the characteristic waves. The morphology information holds significant similarity within a lead and hence it can be obtained by exploiting the intra-lead correlation. Similarly, inter-lead correlation can be used to derive the temporal information because timings of the characteristic waves are highly correlated between synchronously recorded leads. This can be achieved by using RNN [41] based models as they can combine information from the present and previous inputs to decide the present output. Recognizing this, we propose a patient-specific VCG synthesis method based on a sliding-window approach together with LSTM network [42]. At the model estimation stage, the LSTM parameters were estimated for each individual by considering the lead I ECG as input and Frank XYZ leads as output of the model.

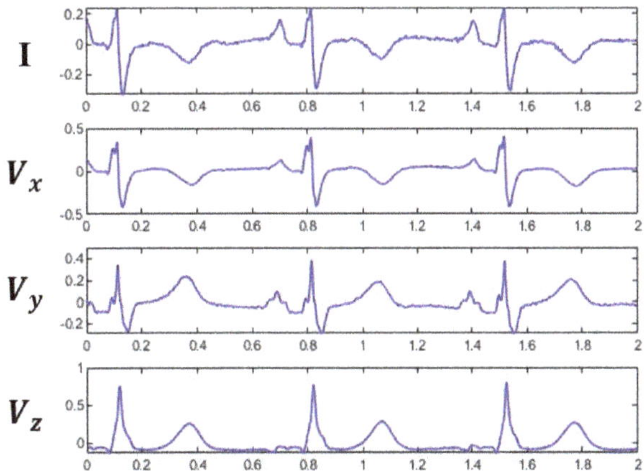

Figure 3. ECG waveforms of measured lead I and Frank XYZ leads.

The LSTM network is commonly used for time series modeling because it solves the gradient vanishing problem by incorporating gate units and memory cells. In an LSTM, the error information is preserved and is back-propagated through the layers which essentially helps the model to learn over a large number of time-steps. The system architecture of the proposed VCG synthesizer is shown in Figure 4. The system starts by applying a sliding window which spreads a segment of currently available lead I data across the input neurons of LSTM. Then, we use an LSTM network to reconstruct three Frank leads by applying a transformation based on the data series in each window. Let x_t denote the lead I ECG data at time t and let $y_t^{(1)}, y_t^{(2)}, y_t^{(3)}$ denote the Frank X, Y, Z lead data, respectively. For a sliding window of size L, suppose that the pair (s_t, y_t) at time t contains the data series $s_t = \{x_{t-L+1}, x_{t-L+2}, \ldots, x_t\}$ and its corresponding target output $y_t = \{y_t^{(1)}, y_t^{(2)}, y_t^{(3)}\}$. Given a set of T training data pairs $\{(s_t, y_t), t = 1, 2, \ldots, T\}$, learning the derived VCG model consists of finding a function F which minimizes the mean square error between the original signal y_t and its reconstructed signal $\hat{y}_t = F(s_t)$. Proceeding in this way, we transform the VCG synthesis problem into a supervised learning problem.

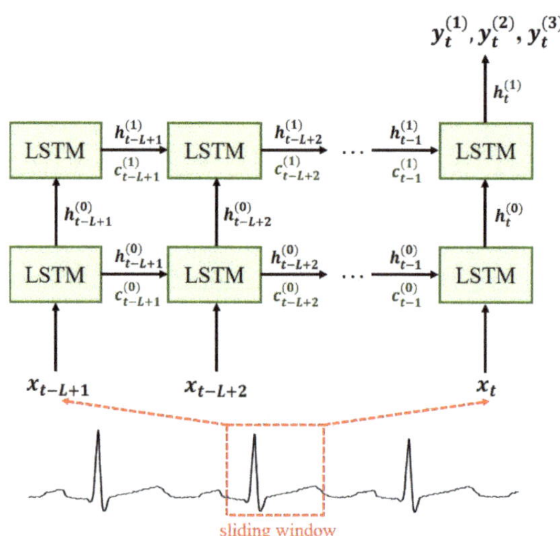

Figure 4. System architecture of the proposed VCG synthesizer.

An LSTM model has the units composed of a memory cell, an input gate, an output gate and a forget gate. The structure of the LSTM unit is shown in Figure 5. An LSTM unit computes a mapping from the input x_t to output y_t by calculating the network unit activations using Equations (1) to (5) iteratively from $t = 1$ to T.

$$f_t = \sigma(W_f x_t + U_f h_{t-1} + b_f) \tag{1}$$

$$i_t = \sigma(W_i x_t + U_i h_{t-1} + b_i) \tag{2}$$

$$c_t = f_t \circ c_{t-1} + i_t \circ \tanh(W_c x_t + U_c h_{t-1} + b_c) \tag{3}$$

$$o_t = \sigma(W_o x_t + U_o h_{t-1} + b_o) \tag{4}$$

$$y_t = h_t = o_t \circ \tanh(c_t) \tag{5}$$

where W, U, and b denote the weight matrices and bias vectors which need to be learned during training. The operator $\circ$ denotes the element-wise product and σ is the sigmoid function. c_t is the cell state, h_t is the hidden state, and f_t, i_t, o_t represent the forget gate, input gate and output gate, respectively. A series of experiments were performed to optimize the LSTM topology used for the VCG synthesizer. The networks with 1, 2, and 3 hidden layers and different number of neurons in hidden layers were tested. It was found that a network with two hidden layers and 30 neurons in each hidden layer achieved the best accuracy of transformation. The LSTM was trained using backpropagation through time (BPTT) algorithm [45], combined with the stochastic gradient descent algorithm. Adam optimizer was used in the model fine-tuning phase to further determine the LSTM parameters. Selected through iterative experiments, a time-step of 1, a mini-batch size of 128, and an epoch number of 300 were used to minimize the mean square error of the VCG synthesizer.

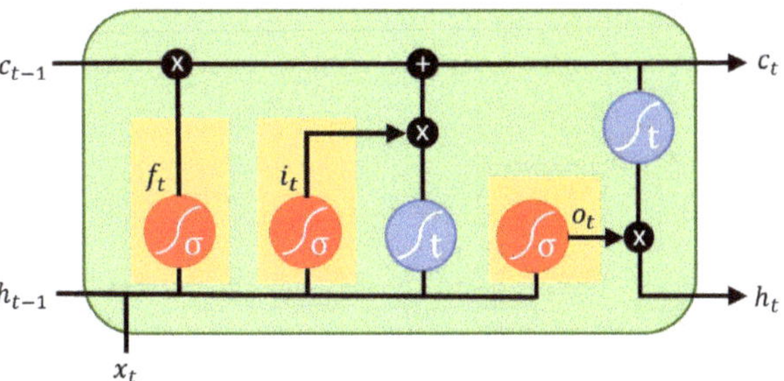

Figure 5. Structure of the LSTM unit.

2.3. Feature Extraction

In the feature extraction stage, Frank XYZ leads of the derived VCG were individually processed in the following steps. First, we detect the R peak in each QRS complex using the Pan-Tompkins algorithm [20] and split the signals into heartbeat segments between two neighboring R peaks. Since the heartbeats may have different lengths, each heartbeat is period normalized to a fixed length of 400 samples via cubic spline interpolation. This choice was based on the observation that the average heartbeat length is about 0.8 s, which corresponds to 400 samples for a sampling frequency of 500 Hz. To make different lead signals comparable to each other, the min-max normalization was applied to scale both the amplitude and time in the range of [0,1], as described in [46]. For the i-th heartbeat with length N_i, let $\alpha_i = N_i/400$ denote the time scaling factor and let $\beta_i^{(1)}, \beta_i^{(2)}, \beta_i^{(3)}$ denote the amplitude scaling factor of Frank X, Y, Z lead, respectively. Once the heartbeats have been segmented and normalized, spline curve fitting [23] is applied to the entire heartbeat to model all of the characteristic waves and fitted coefficients are used as VCG representing features. Two advantages are provided. First, by using the entire beat, the method not only obviates the need for ECG wave delineation but also provides better representation of all regions of ECG beats for MI classification. Second, splines provide an efficient and accurate representation of VCG signals with semiperiodic patterns. VCG signals are a special type of semiperiodic signal which exhibits different degrees of smoothness in different intervals. Such signals require a higher-order polynomial to fit, leading to severe oscillations of the fitted curve which cause the overfitting problem [25]. To address this problem, we develop a framework for an efficient representation of Frank lead signals using splines.

Splines are piecewise polynomial approximations of a signal defined by constraint points on each piecewise segment known as knots. Since VCG signal has numerous clinically relevant turning points, the spline represented as a linear combination of p-degree B-spline basis function has been chosen as the approximation function. The knot vector $\{\zeta_j\}_0^m = \{\zeta_j, 0 \leq j \leq m\}$ is a non-decreasing sequence, where the first $(p+1)$ knots are all equal to 0.0025 and the last $(p+1)$ knots are all equal to 1. The knots from ζ_{p+1} to ζ_{m-p-1} correspond to interior knots which are generated via the knot averages [25] according to Equation (6).

$$\zeta_k = \frac{(\tau_{k+1} + \tau_{k+2} + \cdots + \tau_{k+p})}{p}, \quad p+1 \leq k \leq m-p-1 \tag{6}$$

where $\{\tau_{p+1}, \tau_{p+2}, \ldots, \tau_m\}$ is an arithmetic sequence with the first term $\tau_{p+1} = 0.0025$ and the last term $\tau_m = 1$. The spline curve approximation can be expressed in the form of Equation (7).

$$u(t) = \sum_{i=0}^{n} a_i B_{i,p}(t), \tag{7}$$

where $n = m - p - 1$ and a_i represents the i-th B-spline coefficient. $B_{i,p}(t)$ denotes the i-th p-degree B-spline basis function which is computed recursively [25] using Equations (8) and (9).

$$B_{i,0}(t) = \begin{cases} 1, & \zeta_i \leq t \leq \zeta_{i+1} \\ 0, & \text{otherwise} \end{cases} \tag{8}$$

$$B_{i,j}(t) = \frac{t - \zeta_i}{\zeta_{i+j} - \zeta_i} B_{i,j-1}(t) + \frac{\zeta_{i+j+1} - t}{\zeta_{i+j+1} - \zeta_i} B_{i+1,j-1}(t) \tag{9}$$

The vector of coefficients $\{a_i, 0 \leq i \leq n\}$ is calculated by using the least square spline approximation. Generally, the B-spline approximation of VCG signal yielded better performance with an increase in the value of n. Figure 6 shows the original heartbeat and the spline fitting curve with $n = 23$ using one MI sample and one healthy sample. Experimentally, it was found that the use of $n = 15$ gives a good trade-off between computational efficiency and the quality of fit. Each normalized heartbeat is transformed into 16 features $\{a_0, a_1, \ldots, a_{15}\}$, and three VCG leads during the time of a given heartbeat have 48 features. Together with the time scaling factor α_i and amplitude scaling factors $\{\beta_i^{(1)}, \beta_i^{(2)}, \beta_i^{(3)}\}$, the complete heartbeat of 3-lead VCG is transformed as a 52-dimensional feature vector.

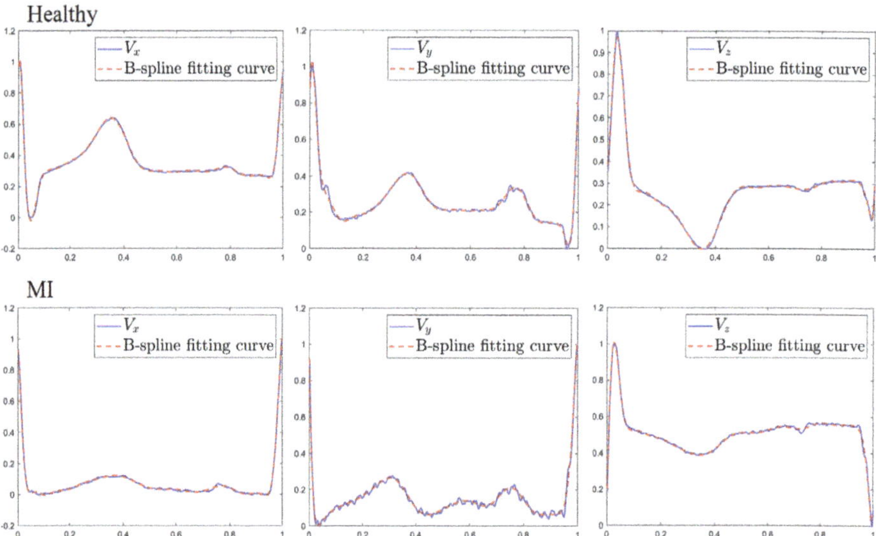

Figure 6. Comparison between original heartbeats (**blue**) and fitting curves (**red**) for healthy and MI subjects.

2.4. Classification

The system performance of MI classification depends critically on the underlying classifier, which builds a model of how to best predict which class a test ECG beat belongs. In this study, a classifier based on multilayer perceptron network (MLP) is used for classification into 12 classes of ECG beats. The MLP is a class of feedforward ANN model and widely used in many fields, such as object recognition, pattern classification, and biological data analysis. Among the reasons for this popularity are its nonlinearity, parallelism, learning and generalization capabilities [47]. A MLP is a network composed of parallel layers of neurons. In building MI classifiers, the input layer receives spline-fitted features from the derived VCG, and the output layer provides the predicted

ECG classes. The relations between the input and output layers are expressed through the weights and biases of the hidden layer. All of the weights were initialized to small random numbers and then subjected to incremental changes by the error backpropagation algorithm based on the cross-entropy loss function [48]. To optimize the classifier design, we tested the MLP with 1, 2, and 3 hidden layers and the number of neurons in each hidden layer was tuned by a grid search from 50 to 500 in steps of 25. Based on the results, we chose the MLP network with 52 input nodes (one for each spline-fitted feature), 12 output nodes (one for each ECG class) and two hidden layers which had 300 and 275 nodes, respectively. To describe the intensity of neural firing, a neuron output was generally obtained by applying an activation function to the weighted sum of its inputs. Due to its ability to enable fast training, the rectified linear unit (ReLU) activation function [47] was used for the hidden layer. However, the ReLU nonlinearity is not applicable for the activation of the present output-layer neurons because their respective output values represent a categorical probability distribution. With this consideration, we applied the softmax function for the output layer to generate values which are in the unit interval and summed to one. Since MI diagnosis involves the simultaneous discrimination of several ECG classes, we considered the one-hot encoding [49] scheme for solving the categorical data classification problem. Specifically, the MLP outputs are represented as binary vectors, each vector consists of 0 s in all cells with the exception of a single 1 in an entry corresponding to the most likely class.

3. Evaluation Parameters

In this study, ECG records were taken from the Physikalisch-TechnischeBundesanstalt (PTB) [50] diagnostic database. The PTB database consists of 549 ECG records from 290 subjects and each record contains 12 ECG leads and 3 Frank VCG leads. From the database, a total of 26,080 heartbeats from 52 healthy subjects and 143 MI patients were included in the analysis. Table 1 shows the number of heartbeats for each type of MI and healthy subjects in this study. These data were further divided into 12 classes of ECG beats: anterior (AMI), anterior-lateral (ALMI), anterior-septal (ASMI), anterior-septal-lateral (ASLMI), inferior (IMI), inferior-lateral (ILMI), inferior-posterior (IPMI), inferior-posterior-lateral (IPLMI), lateral (LMI), posterior (PMI), posterior-lateral (PLMI), and healthy control (HC).

Table 1. Number of beats for different types of MI and healthy subjects in this study.

Class	Number of Beats
Anterior (AMI)	2800
Anterior-Lateral (ALMI)	2534
Anterior-Septal (ASMI)	4114
Anterior-Septal-Lateral (ASLMI)	134
Inferior (IMI)	4569
Inferior-Lateral (ILMI)	3143
Inferior-Posterior (IPMI)	336
Inferior-Posterior-Lateral (IPLMI)	1063
Lateral (LMI)	159
Posterior (PMI)	137
Posterior-Lateral (PLMI)	288
Healthy Control (HC)	6803

Root-mean-square-error (RMSE) and correlation coefficient (CC) were chosen to test the accuracy of derived VCG by the individual methods in relation to the measured VCG. RMSE measures the similarity of two recordings and it is defined as Equation (10), where V is the original value of the measured VCG, $\hat{V}$ is the value of the derived VCG, and N is the number of samples. Instead, CC is a statistic that measures the correlation between two recordings, which is defined in Equation (11).

$$RMSE = \sqrt{\frac{1}{N}\sum_{i=1}^{N}(V_i - \hat{V}_i)^2} \tag{10}$$

$$CC = \frac{\sum_{i=1}^{N}V_i \cdot \hat{V}_i}{\sqrt{\sum_{i=1}^{N}V_i^2 \sum_{i=1}^{N}\hat{V}_i^2}} \tag{11}$$

To test the feasibility of the proposed MI classifiers, the performance analysis is based on the accuracy (ACC), sensitivity (SEN), and specificity (SPE) represented in the form of confusion matrix [51]. These performance metrics are related to the number of true positives (TP), true negatives (TN), false positives (FP), and false negatives (FN). The accuracy is the proportion of correctly classified samples to the total number of samples, and it is defined as Equation (12). Sensitivity, defined in Equation (13), measures the proportion of positives that are correctly identified. Instead, specificity measures the proportion of negatives that are correctly identified and defined as Equation (14).

$$ACC = \frac{TP + TN}{TP + FP + TN + FN} \tag{12}$$

$$SEN = \frac{TP}{TP + FN} \tag{13}$$

$$SPE = \frac{TN}{FP + TN} \tag{14}$$

4. Results

Computer simulations were conducted to evaluate the validity of the proposed method in differentiating 11 types of MI and healthy subjects. A preliminary experiment was first conducted to examine the performance dependence of VCG synthesis on the sliding window size L employed in constructing the LSTM models. In this experiment, ECG recordings from 20 HC subjects and 20 MI patients were used. Table 2 presents the RMSE and CC between measured and derived Frank XYZ leads. Based on the results, we empirically chose $L = 150$ in the sequel.

Table 2. RMSE and CC between measured and derived Frank XYZ leads.

Window Size	Performance	$\hat{V}_x$	$\hat{V}_y$	$\hat{V}_z$
50	CC	0.9939	0.9789	0.9890
	RMSE	13.2215	16.9526	14.2964
100	CC	0.9956	0.9843	0.9933
	RMSE	11.6335	14.6710	10.8368
150	CC	0.9963	0.9862	0.9940
	RMSE	11.0374	14.0393	10.0226
200	CC	0.9962	0.9855	0.9939
	RMSE	11.2348	14.3675	10.2153

We next compare the reconstruction performance of using MLP [32] and LSTM for learning the derived VCG models. All the experiments were based on the evaluation of RMSE and CC and experimental results were obtained by five-fold cross-validation. ECG recordings from 52 HC subjects are denoted as dataset DS1, and ECG recordings from 143 MI patients are denoted as dataset DS2. For comparison purposes, the MLP consists of one input layer with 150 neurons, one output layer with three neurons, two hidden layers and 150 neurons per hidden layer. The results of VCG synthesis by five-fold cross-validation are presented in Table 3. The results clearly demonstrate that the LSTM is preferred to MLP for use in constructing the VCG synthesizer because the LSTM can exploit both intra-

and inter-lead correlations of ECG signals. Further analysis indicates that the average CC of three Frank leads using LSTM were 0.9943 and 0.9807 for dataset DS1 and DS2, respectively, suggesting that the MI patient data was less accurately reconstructed than HC subjects. Visual inspection of the reconstructed signals showed that the derived VCG signals were not significantly different from the measured signals. A typical example for measured and derived Frank XYZ leads is depicted in Figure 7.

Table 3. Five-fold cross validation and average CC and RMSE between measured and derived Frank leads for the MLP and LSTM models.

Folds	Performance	Model	DS1			DS2		
			$\hat{V}_x$	$\hat{V}_y$	$\hat{V}_z$	$\hat{V}_x$	$\hat{V}_y$	$\hat{V}_z$
Fold 1	CC	MLP	0.9947	0.9732	0.9830	0.9815	0.9396	0.9716
		LSTM	0.9977	0.9881	0.9941	0.9909	0.9680	0.9885
	RMSE	MLP	17.2434	21.4437	21.4656	24.4987	28.9382	28.5079
		LSTM	11.4825	14.3947	10.2566	16.9540	20.7532	16.9667
Fold 2	CC	MLP	0.9949	0.9745	0.9835	0.9784	0.9388	0.9718
		LSTM	0.9981	0.9898	0.9963	0.9895	0.9686	0.9879
	RMSE	MLP	17.3081	20.4252	21.2486	24.3083	29.0842	28.3771
		LSTM	11.1590	12.6073	9.7404	15.9387	20.9914	17.4463
Fold 3	CC	MLP	0.9946	0.9719	0.9835	0.9773	0.9398	0.9720
		LSTM	0.9979	0.9878	0.9959	0.9864	0.9659	0.9875
	RMSE	MLP	17.6641	20.9450	21.6205	24.2560	29.7097	28.2240
		LSTM	11.1632	13.1473	10.0268	16.9523	21.6060	16.9857
Fold 4	CC	MLP	0.9952	0.9746	0.9843	0.9798	0.9390	0.9716
		LSTM	0.9986	0.9901	0.9966	0.9880	0.9695	0.9883
	RMSE	MLP	16.6822	20.1005	20.9287	24.3547	29.1888	28.4182
		LSTM	10.0381	12.0188	9.3277	16.0953	20.4111	16.7269
Fold 5	CC	MLP	0.9950	0.9758	0.9800	0.9784	0.9349	0.9692
		LSTM	0.9985	0.9922	0.9929	0.9847	0.9618	0.9844
	RMSE	MLP	16.9020	20.3295	21.7694	25.3288	30.9232	29.7329
		LSTM	10.0791	11.5885	9.6506	17.9217	22.7715	18.0954
Mean	CC	MLP	0.9949	0.9740	0.9829	0.9791	0.9384	0.9712
		LSTM	0.9982	0.9896	0.9952	0.9879	0.9668	0.9873
	RMSE	MLP	17.1600	20.6488	21.4066	24.5493	29.5688	28.6520
		LSTM	10.7844	12.7513	9.8004	16.7724	21.3066	17.2442

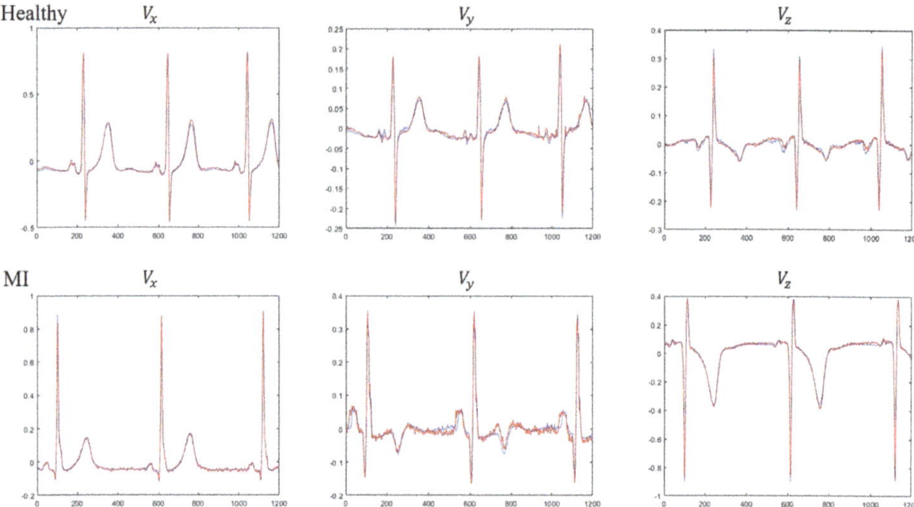

Figure 7. Comparison between measured (**blue**) and derived (**red**) Franks leads for healthy and MI subjects.

Next, we assess the performance of MLP classifiers for the classification of normal and 11 MI classes. One problem with the PTB database is the high imbalance between the number of heartbeats belonging to each ECG class. Training an MLP classifier with unbalanced data usually leads to a certain bias towards the majority class. Recognizing this, we applied the Synthetic Minority Over-sampling Technique (SMOTE) [52] before starting the training process. Moreover, we used 5-fold cross-validation technique to train and test the MLP classifiers. We began testing the MLP classifiers for the situation where MI classes were identified solely by means of single-lead feature data. The classification performance for each ECG class is summarized in Table 4. Simulation results indicated that using lead I yielded an overall accuracy of 50.72%, suggesting that it cannot provide sufficient cues for reliable classification. To elaborate further, we show in Table 5 the confusion matrix of all the 12 classes for lead I ECG beats. It was found that the notably low classification accuracy can be attributed to the high confusions made across anterior MI group (AMI, ALMI, ASMI, ASLMI) and inferior MI group (IMI, ILMI, IPMI, IPLMI). For instance, 14.71% of the ECG beats notated in ASMI were classified as representing IMI and 8.84% of the IMI beats were classified as being ASMI. Furthermore, results indicate that derived Frank leads $\hat{V}_y$ and $\hat{V}_z$ are preferred to lead I ECG for use in constructing the MI classifier. Notably, the use of lead $\hat{V}_z$ yielded an overall accuracy of 82.09%, compared with 50.72% for lead I and 81.45% for $\hat{V}_y$. The confusion matrices obtained using derived Frank lead $\hat{V}_y$ and $\hat{V}_z$ are shown in Tables 6 and 7, respectively. Further analysis indicates that anterior and inferior MI groups are dominant in the MI groups that benefited the most from exploitation of derived VCG leads. In case of inferior MI group, the average sensitivity has increased from 56.9% in lead I to 87.1% in lead $\hat{V}_y$. Similarly in case of anterior MI group, average sensitivity obtained for lead I and $\hat{V}_z$ is 60.25% and 88.9%, respectively. We speculate that this might be attributed to the difference in closeness between Frank leads and 12 ECG leads. Support for such a speculation can be found in [39], where the authors showed that Frank lead V_y is most likely associated with inferior leads (II, III, aVF), and Frank lead V_z is closest to subset of anteroseptal leads (V_1, V_2, V_3). We can also see from Figure 1 that leads V_1, V_2 and V_3 are located near the negative Z-axis in the sagittal plane. Similarly, it can be found that leads II, III and aVF are oriented along the Y-axis.

Table 4. Classification results of MLP classifier with single-lead signal.

Classes	I			$\hat{V}_x$			$\hat{V}_y$			$\hat{V}_z$		
	ACC(%)	SEN(%)	SPE(%)	ACC(%)	SEN(%)	SPE(%)	ACC(%)	SEN(%)	SPE(%)	ACC(%)	SEN(%)	SPE(%)
AMI	90.55	51.89	95.20	91.74	58.39	95.75	95.73	79.71	97.66	95.76	79.57	97.70
ALMI	91.33	51.10	95.66	93.01	57.54	96.83	96.68	87.10	97.72	97.58	88.60	98.54
ASMI	85.28	37.99	94.13	88.35	56.73	94.27	93.60	75.62	96.96	95.41	87.29	96.93
ASLMI	99.40	100.00	99.40	99.75	100.00	99.75	99.89	98.51	99.90	99.99	100.00	99.99
IMI	80.87	38.26	89.93	81.94	33.44	92.24	93.49	81.02	96.14	92.08	73.95	95.93
ILMI	86.19	54.82	90.49	87.81	61.47	91.42	95.82	84.16	97.41	94.69	83.33	96.25
IPMI	99.09	92.26	99.18	99.36	89.88	99.48	99.78	94.94	99.84	99.70	93.15	99.78
IPLMI	93.95	42.24	96.15	93.17	35.18	95.63	98.76	88.33	99.20	97.67	81.84	98.34
LMI	98.12	98.11	98.12	99.00	99.37	99.00	99.69	100.00	99.68	99.83	98.11	99.84
PMI	99.24	100.00	99.23	98.49	100.00	98.48	99.51	100.00	99.51	99.89	100.00	99.89
PLMI	96.92	90.63	97.00	97.25	87.85	97.36	99.65	98.96	99.65	99.76	96.88	99.79
HC	80.49	58.78	88.15	85.20	69.98	90.57	90.30	78.99	94.29	91.83	80.21	95.93

Table 5. Confusion matrix for MI classification using measured lead I ECG.

Notated	Predicted												Total	ACC(%)	SEN(%)	SPE(%)
	AMI	ALMI	ASMI	ASLMI	IMI	ILMI	IPMI	IPLMI	LMI	PMI	PLMI	Norm				
AMI	1453	255	150	0	335	228	21	97	44	2	39	176	2800	90.55	51.89	95.2
ALMI	154	1295	219	54	228	141	37	57	2	128	129	90	2534	91.33	51.1	95.66
ASMI	273	217	1563	25	605	555	41	101	129	10	125	470	4114	85.28	37.99	94.13
ASLMI	0	0	0	134	0	0	0	0	0	0	0	0	134	99.4	100	99.4
IMI	219	279	404	65	1748	372	72	144	51	35	177	1003	4569	80.87	38.26	89.93
ILMI	173	82	113	0	281	1723	26	204	122	3	5	411	3143	86.19	54.82	90.49
IPMI	5	8	3	0	9	0	310	0	0	0	0	1	336	99.09	92.26	99.18
IPLMI	31	68	77	1	84	155	2	449	53	1	19	123	1063	93.95	42.24	96.15
LMI	2	0	0	0	0	1	0	0	156	0	0	0	159	98.12	98.11	98.12
PMI	0	0	0	0	0	0	0	0	0	137	0	0	137	99.24	100	99.23
PLMI	0	1	1	0	7	3	0	5	0	0	261	10	288	96.92	90.63	97
HC	260	111	322	11	618	726	12	356	87	20	281	3999	6803	80.49	58.78	88.15

Table 6. Confusion matrix for MI classification using derived Frank Y lead.

Notated	Predicted												Total	ACC(%)	SEN(%)	SPE(%)
	AMI	ALMI	ASMI	ASLMI	IMI	ILMI	IPMI	IPLMI	LMI	PMI	PLMI	Norm				
AMI	2232	44	113	0	137	55	4	32	5	32	4	142	2800	95.73	79.71	97.66
ALMI	35	2207	49	15	87	25	3	9	0	34	1	69	2534	96.68	87.1	97.72
ASMI	161	78	3111	9	176	103	12	25	0	0	36	403	4114	93.6	75.62	96.96
ASLMI	0	1	0	132	0	0	0	0	0	0	1	0	134	99.89	98.51	99.9
IMI	95	116	157	2	3702	136	6	72	12	2	38	231	4569	93.49	81.02	96.14
ILMI	8	15	50	1	130	2645	4	21	26	0	3	240	3143	95.82	84.16	97.41
IPMI	3	0	4	0	5	1	319	0	0	0	0	4	336	99.78	94.94	99.84
IPLMI	26	7	5	0	43	27	2	939	2	1	0	11	1063	98.76	88.33	99.2
LMI	0	0	0	0	0	0	0	0	159	0	0	0	159	99.69	100	99.68
PMI	0	0	0	0	0	0	0	0	0	137	0	0	137	99.51	100	99.51
PLMI	0	1	0	0	0	2	0	0	0	0	285	0	288	99.65	98.96	99.65
HC	217	276	289	0	252	244	10	40	37	58	6	5374	6803	90.3	78.99	94.29

Next, we examine whether combining multiple derived Frank leads would improve the classification performance. Table 8 shows the MLP classifier results for healthy and 11 types of MI ECG beats obtained using various lead configurations. Our proposed method yielded the best performance with an overall accuracy of 99.15%, sensitivity of 99.16% and specificity of 99.92% in MI classification, by using 52 features obtained from the derived Frank XYZ leads. The results also indicate that the ability of derived VCG to correctly identify the MI classes is almost identical to that of measured VCG. Table 9 shows the confusion matrix of all classes obtained using MLP classifier on the derived Frank XYZ leads. A comparison between Tables 5 and 9 indicates that the improvement can be seen in the following areas. First, the derived VCG can reduce a significant portion of confusions across anterior and inferior MI groups. For instance, only 0.15% of ECG beats notated in ASMI were misclassified as representing IMI, the corresponding value for lead I being 14.71%. Second, the derived VCG significantly increased the sensitivity of healthy subjects to 99.68%, compared with 58.78% for lead I, 78.99% for $\hat{V}_y$, and 80.21% for $\hat{V}_z$. The results clearly demonstrate that MI classification by computational means is significantly improved when clinically significant features relating to the derived VCG are taken into account.

Table 7. Confusion matrix for MI classification using derived Frank Z lead.

Notated	Predicted												Total	ACC(%)	SEN(%)	SPE(%)
	AMI	ALMI	ASMI	ASLMI	IMI	ILMI	IPMI	IPLMI	LMI	PMI	PLMI	Norm				
AMI	2228	74	272	1	76	24	8	14	5	6	9	83	2800	95.76	79.57	97.7
ALMI	45	2245	96	0	38	11	18	7	1	2	1	70	2534	97.58	88.6	98.54
ASMI	103	76	3591	2	105	87	4	13	0	0	0	133	4114	95.41	87.29	96.93
ASLMI	0	0	0	134	0	0	0	0	0	0	0	0	134	99.99	100	99.99
IMI	147	94	186	0	3379	254	18	80	14	3	28	366	4569	92.08	73.95	95.93
ILMI	18	17	35	0	202	2619	2	109	11	2	10	118	3143	94.69	83.33	96.25
IPMI	7	3	1	0	5	5	313	1	0	0	0	1	336	99.7	93.15	99.78
IPLMI	3	2	3	0	137	31	3	870	0	0	0	14	1063	97.67	81.84	98.34
LMI	0	0	0	0	2	0	0	1	156	0	0	0	159	99.83	98.11	99.84
PMI	0	0	0	0	0	0	0	0	0	137	0	0	137	99.89	100	99.89
PLMI	0	0	0	0	0	0	0	6	3	0	279	0	288	99.76	96.88	99.79
HC	212	77	82	0	311	448	3	184	8	15	6	5457	6803	91.83	80.21	95.93

Table 8. Classification results of MLP classifier with various lead configurations.

Leads	ACC(%)	SEN(%)	SPE(%)
I	50.72	68.01	95.22
$\hat{V}_x$	57.54	70.82	95.90
$\hat{V}_y$	81.45	88.95	98.17
$\hat{V}_z$	82.09	88.58	98.24
$\hat{V}_x + \hat{V}_y$	93.36	95.52	99.34
$\hat{V}_y + \hat{V}_z$	96.99	97.74	99.70
$\hat{V}_x + \hat{V}_z$	83.68	89.80	98.42
$\hat{V}_x + \hat{V}_y + \hat{V}_z$	99.15	99.16	99.92
$V_x + V_y + V_z$	99.14	99.39	99.92

Table 9. Confusion matrix for MI classification using derived Frank XYZ leads.

Notated	Predicted												Total	ACC(%)	SEN(%)	SPE(%)
	AMI	ALMI	ASMI	ASLMI	IMI	ILMI	IPMI	IPLMI	LMI	PMI	PLMI	Norm				
AMI	2762	5	9	0	7	4	0	1	0	0	3	9	2800	99.72	98.64	99.85
ALMI	10	2504	6	0	2	2	2	0	1	0	0	7	2534	99.78	98.82	99.88
ASMI	9	12	4078	1	6	2	3	3	0	0	0	0	4114	99.73	99.12	99.85
ASLMI	0	0	0	134	0	0	0	0	0	0	0	0	134	100	100	100
IMI	5	3	4	0	4528	11	1	7	0	0	4	6	4569	99.7	99.1	99.83
ILMI	3	3	8	0	3	3119	1	1	1	0	1	3	3143	99.8	99.24	99.88
IPMI	3	0	0	0	3	0	329	0	0	0	0	1	336	99.94	97.92	99.97
IPLMI	2	3	3	0	6	8	0	1039	0	1	0	1	1063	99.85	97.74	99.94
LMI	0	0	0	0	0	0	0	0	159	0	0	0	159	99.99	100	99.99
PMI	0	0	0	0	0	0	0	0	0	137	0	0	137	99.99	100	99.99
PLMI	0	0	0	0	0	0	0	1	0	0	287	0	288	99.97	99.65	99.97
HC	2	2	4	0	10	1	1	1	0	1	0	6781	6803	99.81	99.68	99.86

5. Discussion

In recent years, numerous approaches were proposed to identify various types of MI from ECG records. The numbers of ECG leads and MI classes are important factors correlated with diagnosis efficiency, and should be noted when comparing their relative performances. Table 10 summarizes the studies employing different techniques in MI classification with the same PTB database. Arif et al. [10] used 12 lead ECG signal and time domain features such as T-wave amplitude, Q-wave and ST-level elevation, reporting overall accuracy of 98.8% on ten different MI classes with a KNN classifier. Alternatively, Noorian et al. [11] used ANN classifier and wavelet coefficients as features extracted from the derived VCG. Acharya et al. [12] have evaluated ten MI classes with 12 types of nonlinear features based on wavelet transform. They obtained an accuracy of 98.74%, sensitivity of 99.55%, and specificity of 99.16% by only using lead V3 ECG signal. Lui et al. [13] combined the power of CNN and RNN, and achieved 92.4% sensitivity and 97.7% specificity for classification of MI as well as other cardiovascular diseases. Baloglu et al. [14] proposed an end-to-end approach based on deep CNN and reported an overall accuracy of 99.78% by using 12 lead ECG signal for classification into 11 types of ECG beats. In [15], a multi-lead attention mechanism integrated with CNN and bidirectional gated recurrent unit was applied for MI classification based on six classes of 12-lead ECG records, namely HC, AMI, ALMI, ASMI, IMI, and ILMI. Towards addressing the challenges in identifying MIs using wearable devices, our work, as well as some earlier studies [12,13], was focused on single-lead rather than 12-lead exploration. Results reported in this paper are generally better than those of MI classifiers in the literature, with its performance only slightly lower than that of [14]. However, our proposed method applies single-lead derived VCG for classification into 12 types of ECG beats, in which ASLMI with larger necrotic area is ignored in [14]. Overall, the proposed method obtained an accuracy of 99.15%, sensitivity of 99.16% and specificity of 99.92%. With this performance, our proposed model has the potential to provide an early and accurate diagnosis of MI in wearable ECG monitoring devices.

Table 10. Comparison of this study with other studies using the PTB diagnostic database.

Ref	Leads	No. of Classes	ACC(%)	SEN(%)	SPE(%)
Arif et al. (2012) [10]	12 leads	11	98.80%	98.67%	98.71%
Noorian et al. (2014) [11]	12 leads	10	95.35%	99.09%	94.23%
Acharya et al. (2016) [12]	V_3	11	98.74%	99.55%	99.16%
Lui nad Chow (2018) [13]	I	4	95.25%	92.40%	97.70%
Baloglu et al. (2019) [14]	12 leads	11	99.78%	99.84%	99.98%
Fu et al. (2020) [15]	12 leads	6	99.11%	99.02%	99.10%
Proposed method	I	12	99.15%	99.16%	99.92%

6. Conclusions

This paper proposed a new method for automatic MI classification using single-lead derived VCG. We first emphasized the importance of exploiting both intra-lead and inter-lead correlation for learning the derived VCG models. This task was accomplished by using a patient-specific transformation based on LSTM network with sliding window approach. Performance is further enhanced by using B-spline curve fitting to extract clinically significant features from the three derived Frank leads. After feature extraction, a classifier based on MLP network is used for classification into 12 types of ECG beats. Combined performance from 52 healthy subjects and 143 MI patients demonstrate the validity of the proposed MI classification system with an accuracy of 99.15%, sensitivity of 99.16% and specificity of 99.92%.

Author Contributions: Conceptualization, W.-W.C. and J.-T.C.; methodology, Y.-H.C. and C.-L.H.; software, Y.-H.C. and C.-L.H.; writing-original draft preparation, Y.-H.C. and W.-W.C.; writing-review and editing, W.-W.C. and J.-T.C.; visualization, Y.-H.C. and C.-L.H.; supervision, W.-W.C. and J.-T.C. All authors have read and agreed to the published version of the manuscript.

Funding: This research was supported by the Ministry of Science and Technology, Taiwan, under Grant MOST 109-2634-F-009-024.

Conflicts of Interest: The authors declare no conflicts of interest.

References

1. Kaptoge, S.; Pennells, L.; De Bacquer, D.; Cooney, M.T.; Kavousi, M.; Stevens, G.; Riley, L.M.; Savin, S.; Khan, T.; Altay, S. World Health Organization cardiovascular disease risk charts: Revised models to estimate risk in 21 global regions. *Lancet Glob. Health* **2019**, *7*, e1332–e1345. [CrossRef]
2. Benjamin, E.J.; Muntner, P.; Alonso, A.; Bittencourt, M.S.; Callaway, C.W.; Carson, A.P.; Chamberlain, A.M.; Chang, A.R.; Cheng, S.; Das, S.R.; et al. Heart disease and stroke statistics-2019 update: A report from the American Heart Association. *Circulation* **2019**, *139*, e56–e528. [CrossRef] [PubMed]
3. Plonsey, J.M. 12-lead ECG system. *Bioelectromagnetism* **1995**, *15*, 23–34.
4. Frank, E. An accurate clinically practical system for spatial vectorcardiography. *Circulation* **1956**, *13*, 737–749. [CrossRef] [PubMed]
5. Ghista, D.N.; Acharya, R.; Nagenthiran, T. Frontal plane vectorcardiograms: Theory and graphics visualization of cardiac health status. *J. Med. Syst.* **2010**, *34*, 445–458. [CrossRef] [PubMed]
6. Bortolan, G.; Christov, I. Myocardial infarction and ischemia characterization from T-loop morphology in VCG. *Comput. Cardiol.* **2001**, 633–636. [CrossRef]
7. Ge, D. Detecting myocardial infarction using vcg leads. In Proceedings of the 2008 International Conference on Bioinformatics and Biomedical Engineering, Shanghai, China, 16–18 May 2008; pp. 2217–2220. [CrossRef]
8. Panagiotou, C.; Dima, S.-M.; Mazomenos, E.B.; Rosengarten, J.; Maharatna, K.; Gialelis, J.; Morgan, J. Detection of myocardial scar from the vcg using a supervised learning approach. *Annu. Int. Conf. IEEE Eng. Med. Biol. Soc.* **2013**, 7326–7329. [CrossRef]
9. Ansari, S.; Farzaneh, N.; Duda, M.; Horan, K.; Anderson, H.B.; Goldberger, Z.D.; Nallamothu, B.K.; Najarian, K. A review of automated methods for detection of myocardial ischemia and infarction using electrocardiogram and electronic health records. *IEEE Rev. Biomed. Eng.* **2017**, *10*, 264–298. [CrossRef]
10. Arif, M.; Malagore, I.A.; Afsar, F.A. Detection and localization of myocardial infarction using k-nearest neighbor classifier. *J. Med. Syst.* **2012**, *36*, 279–289. [CrossRef]
11. Noorian, A.; Dabanloo, N.J.; Parvanch, S. Wavelet based method for localization of myocardial infarction using the electrocardiogram. *Proc. Comput. Cardiol.* **2014**, *2014*, 645–648.
12. Acharya, U.R.; Fujita, H.; Sudarshan, V.K.; Oh, S.L.; Adam, M.; Koh, J.E.W.; Tan, J.H.; Ghista, D.N.; Martis, R.J.; Chua, C.K.; et al. Automated detection and localization of myocardial infarction using electrocardiogram: A comparative study of different leads. *Knowl. Based Syst.* **2016**, *99*, 146–156. [CrossRef]
13. Lui, H.W.; Chow, K.L. Multiclass classification of myocardial infarction with convolutional and recurrent neural networks for portable ECG devices. *Informat. Med. Unlocked* **2018**, *13*, 26–33. [CrossRef]
14. Baloglu, U.B.; Talo, M.; Yildirim, O.; Tan, R.S.; Acharya, U.R. Classification of myocardial infarction with multi-lead ECG signals and deep CNN. *Pattern Recognit. Lett.* **2019**, *122*, 23–30. [CrossRef]
15. Fu, L.; Lu, B.; Nie, B.; Peng, Z.; Liu, H.; Pi, X. Hybrid network with attention mechanism for detection and location of myocardial infarction based on 12-lead electrocardiogram signals. *Sensors* **2020**, *20*, 1020. [CrossRef] [PubMed]
16. Correa, R.; Arini, P.; Valentinuzzi, M.E.; Laciar, E. Novel set of vectorcardiographic parameters for the identification of ischemic patients. *Med. Eng. Phys.* **2013**, *35*, 16–22. [CrossRef] [PubMed]
17. Yang, H.; Bukkapatnam, S.T.S.; Le, T.; Komanduri, R. Identification of myocardial infarction (MI) using spatio-temporal heart dynamics. *Med. Eng. Phys.* **2012**, *34*, 485–497. [CrossRef]
18. Aranda, A.; Bonizzi, P.; Karel, J.; Peeters, R. Performance of Dower's inverse transform and Frank lead system for identification of myocardial infarction. *Annu. Int. Conf. IEEE Eng. Med. Biol. Soc.* **2015**, 4495–4498. [CrossRef]
19. Banerjee, S.; Gupta, R.; Mitra, M. Delineation of ECG characteristic features using multiresolution wavelet analysis method. *Measurement* **2012**, *45*, 474–487. [CrossRef]
20. Pan, J.; Tompkins, W.J. A real-time QRS detection algorithm. *IEEE Trans. Biomed. Eng.* **1985**, *32*, 230–236. [CrossRef]
21. Chen, C.L.; Chuang, C.T. A QRS detection and R point recognition method for wearable single-lead ECG devices. *Sensors* **2017**, *17*, 1969. [CrossRef]
22. Zhang, Q.; Manriquez, A.I.; Medigue, C.; Papelier, Y.; Sorine, M. An algorithm for robust and efficient location of T-wave ends in electrocardiograms. *IEEE Trans. Biomed. Eng.* **2006**, *53 Pt 1*, 2544–2552. [CrossRef]
23. DeBoor, C. *A Practical Guide to Splines*; Springer: New York, NY, USA, 2001.

24. Guilak, F.G.; McNames, J. A Bayesian-optimized spline representation of the electrocardiogram. *Physiol. Meas.* **2013**, *34*, 1467–1482. [CrossRef] [PubMed]

25. Dung, V.T.; Tjahjowidodo, T. A direct method to solve optimal knots of B-spline curves: An application for non-uniform B-spline curves fitting. *PLoS ONE* **2017**, *12*. [CrossRef] [PubMed]

26. Baig, M.M.; Gholamhosseini, H.; Connolly, M.J. A comprehensive survey of wearable and wireless ECG monitoring systems for older adults. *Med. Biol. Eng. Comput.* **2013**, *51*, 485–495. [CrossRef] [PubMed]

27. Hong, S.; Zhou, Y.; Shang, J.; Xiao, C.; Sun, J. Opportunities and challenges of deep learning methods for electrocardiogram data: A systematic review. *Comput. Biol. Med.* **2020**, *122*, 103801. [CrossRef] [PubMed]

28. Sohn, J.; Yang, S.; Lee, J.; Ku, Y.; Kim, H.C. Reconstruction of 12-lead electrocardiogram from a 3-lead patch-type device using a LSTM network. *Sensors* **2020**, *20*, 3278. [CrossRef] [PubMed]

29. Tomasic, I.; Trobec, R. Electrocardiographic systems with reduced numbers of leads-synthesis of the 12-lead ECG. *IEEE Rev. Biomed. Eng.* **2014**, *7*, 126–142. [CrossRef]

30. Nelwan, S.P.; Kors, J.A.; Meij, S.H.; Van Bemmel, J.H.; Simoons, M.L. Reconstruction of the 12-lead electrocardiogram from reduced lead sets. *J. Electrocardiol.* **2004**, *37*, 11–18. [CrossRef]

31. Atoui, H.; Fayn, J.; Rubel, P. A novel neural-network model for deriving standard 12-lead ECGs from serial three-lead ECGs: Application to self-care. *IEEE Trans. Inf. Technol. Biomed.* **2010**, *14*, 883–890. [CrossRef]

32. Vozda, M.; Peterek, T.; Cerny, M. Novel Method for Deriving Vectorcardiographic Leads Based on Artificial Neural Networks. In Proceedings of the 41st International Congress on Electrocardiol, Bratislava, Slovakia, 4–7 June 2014.

33. Jaros, R.; Martinek, R.; Danys, L. Comparison of different electrocardiography with vectorcardiography transformations. *Sensors* **2019**, *19*, 3027. [CrossRef]

34. Dower, G.E.; Machado, H.B.; Osborne, J.A. On deriving the electrocardiogram from vectorcardiographic leads. *Clin. Cardiol.* **1980**, *3*, 87–95. [CrossRef] [PubMed]

35. Edenbrandt, L.; Pahlm, O. Vectorcardiogram synthesized from 12-lead ECG: Superiority of the inverse Dower matrix. *J. Electrocardiol.* **1988**, *21*, 361–367. [CrossRef]

36. Kors, J.; van Herpen, G.; Sittig, A.; Bemmel, J. Reconstruction of the Frank vectorcardiogram from standard electrocardiographic leads: Diagnostic comparison of different methods. *Eur. Heart J.* **1990**, *11*, 1083–1092. [CrossRef] [PubMed]

37. Dawson, D.; Yang, H.; Malshe, M.; Bukkapatnam, S.T.S.; Benjamin, B.; Komanduri, R. Linear affine transformations between 3-lead (Frank XYZ leads) vectorcardiogram and 12-lead electrocardiogram signals. *J. Electrocardiol.* **2009**, *42*, 622–630. [CrossRef]

38. Vozda, M.; Cerny, M. Methods for derivation of orthogonal leads from 12-lead electrocardiogram: A review. *Biomed. Signal Process. Control* **2015**, *19*, 23–24. [CrossRef]

39. Maheshwari, S.; Acharyya, A.; Schiariti, M.; Puddu, P.E. Frank vectorcardiographic system from standard 12 lead ECG: An effort to enhance cardiovascular diagnosis. *J. Electrocardiol.* **2016**, *49*, 231–242. [CrossRef]

40. Schreck, D.M.; Fishberg, R.D. Derivation of the 12-lead electrocardiogram and 3-lead vectorcardiogram. *Am. J. Emerg. Med.* **2013**, *31*, 1183–1190. [CrossRef]

41. Goodfellow, I.; Bengio, Y.; Courville, A. *Deep Learning*; MIT Press: Cambridge, MA, USA, 2016.

42. Hochreiter S.; Schmidhuber J. Long Short-Term Memory. *Neural Comput.* **1997**, *9*, 1735–1780. [CrossRef]

43. Kligfield, P.; Gettes, L.S.; Bailey, J.J.; Childers, R.; Deal, B.J.; Hancock, E.W.; van Herpen, G.; Kors, J.A.; Macfarlane, P.; Mirvis, D.M.; et al. Recommendations for the Standardization and Interpretation of the Electrocardiogram: Part I: The Electrocardiogram and Its Technology: A Scientific Statement From the American Heart Association Electrocardiography and Arrhythmias Committee, Council on Clinical Cardiology; the American College of Cardiology Foundation; and the Heart Rhythm Society Endorsed by the International Society for Computerized Electrocardiology. *Circulation* **2007**, *115*, 1306–1324.

44. Parola, F.; Garcua-Niebla, J. Use of high-pass and low-pass electrocardiographic filters in an international cardiological community and possible clinical effects. *Adv. J. Vasc. Med.* **2017**, *2*, 034–038.

45. Pineda, F.J. Generalization of back-propagation to recurrent neural networks. *Phys. Rev. Lett.* **1987**, *59*, 2229–2232. [CrossRef] [PubMed]

46. Yang, M.; Liu, B.; Zhao, M.; Li, F.; Wang, G.; Zhou, F. Normalizing electrocardiograms of both healthy persons and cardiovascular disease patients for biometric authentication. *PLoS ONE* **2013**, *8*, 1–7. [CrossRef] [PubMed]

47. Yegnanarayana, B. *Artificial Neural Networks*; PHI Learning Pvt. Ltd.: New Delhi, India, 1999.

48. Svozil, D.; Kvasnicka, V.; Pospichal, J. Introduction to multi-layer feed-forward neural networks. *Chemom. Intell. Lab. Syst.* **1997**, *39*, 43–62. [CrossRef]
49. Pai, C.; Potdar, K. A Comparative Study of Categorical Variable Encoding Techniques for Neural Network Classifiers. *Artic. Int. J. Comput. Appl.* **2017**, *175*, 7–9.
50. Moody, G.B.; Mark, R.G.; Goldberger, A.L. PhysioNet: A web-based resource for the study of physiologic signals. *IEEE Eng. Med. Biol. Mag.* **2001**, *20*, 70–75. [CrossRef]
51. Zhu, W.; Zeng, N.; Wang, N. Sensitivity, specificity, accuracy, associated confidence interval and roc analysis with practical sas implementations. In Proceedings of the NESUG: Health Care and Life Sciences, Baltimore, MD, USA, 14–17 November 2010; pp. 1–9.
52. Chawla, N.V.; Hall, L.O.; Bowyer, K.W.; Kegelmeyer, W.P. SMOTE: Synthetic Minority Oversampling Technique. *J. Artif. Intell. Res.* **2002**, *16*, 321–357.

Publisher's Note: MDPI stays neutral with regard to jurisdictional claims in published maps and institutional affiliations.

 *sensors*

Article

Genetic Deep Convolutional Autoencoder Applied for Generative Continuous Arterial Blood Pressure via Photoplethysmography

Muammar Sadrawi [1], Yin-Tsong Lin [2], Chien-Hung Lin [2], Bhekumuzi Mathunjwa [1], Shou-Zen Fan [3], Maysam F. Abbod [4] and Jiann-Shing Shieh [1,*]

[1] Department of Mechanical Engineering, Yuan Ze University, Taoyuan 32003, Taiwan; muammarsadrawi@yahoo.com (M.S); mathunjwabhekie@gmail.com (B.M.)
[2] AI R&D Department, New Era AI Robotic Inc., Taipei 105, Taiwan; lotusytlin@neweraai.com (Y.-T.L.); lance_lin@neweraai.com (C.-H.L.)
[3] Department of Anesthesiology, College of Medicine, National Taiwan University, Taipei 100, Taiwan; shouzen@gmail.com
[4] Department of Electronic and Computer Engineering, Brunel University London, Uxbridge UB8 3PH, UK; Maysam.Abbod@brunel.ac.uk
* Correspondence: jsshieh@saturn.yzu.edu.tw

Received: 9 June 2020; Accepted: 7 July 2020; Published: 9 July 2020

Abstract: Hypertension affects a huge number of people around the world. It also has a great contribution to cardiovascular- and renal-related diseases. This study investigates the ability of a deep convolutional autoencoder (DCAE) to generate continuous arterial blood pressure (ABP) by only utilizing photoplethysmography (PPG). A total of 18 patients are utilized. LeNet-5- and U-Net-based DCAEs, respectively abbreviated LDCAE and UDCAE, are compared to the MP60 IntelliVue Patient Monitor, as the gold standard. Moreover, in order to investigate the data generalization, the cross-validation (CV) method is conducted. The results show that the UDCAE provides superior results in producing the systolic blood pressure (SBP) estimation. Meanwhile, the LDCAE gives a slightly better result for the diastolic blood pressure (DBP) prediction. Finally, the genetic algorithm-based optimization deep convolutional autoencoder (GDCAE) is further administered to optimize the ensemble of the CV models. The results reveal that the GDCAE is superior to either the LDCAE or UDCAE. In conclusion, this study exhibits that systolic blood pressure (SBP) and diastolic blood pressure (DBP) can also be accurately achieved by only utilizing a single PPG signal.

Keywords: photoplethysmography; continuous arterial blood pressure; systolic blood pressure; diastolic blood pressure; deep convolutional autoencoder; genetic algorithm

---

1. Introduction

Blood pressure (BP) is the pressure driven by the blood circulation to the artery wall. Meanwhile, hypertension or high blood pressure (HBP) is an excessive amount of a given force against blood vessels. In addition, according to World Health Organization (WHO), HBP affects more than one billion people in the world [1].

With having an impact on many people, HBP can incite several diseases. It has a solid contribution to cardiovascular and renal diseases [2]. HBP also contributes to stroke and ischemic heart diseases [3]. Furthermore, HBP can generate vascular damage of the retina related to cardiovascular-based fatality [4]. These aforementioned studies make HBP-related inspection become significant.

Photoplethysmography (PPG), one of the vital signs, has been a solid indicator for some medical-related investigations. PPG has been deployed as the heart rate measurement in motion artifact-interfered conditions with an empirical mode decomposition-based filter and time-frequency evaluation [5]. It also has been utilized, alongside electrocardiography, for atrial fibrillation in acute stroke patients [6]. Another study involved the PPG morphological feature for hypertension early identification [7]. Moreover, Phillips et al. applied PPG sensors to non-invasively evaluate hemoglobin concentration [8]. Meanwhile, Perpetuini et al. supervised a general linear model-based PPG to evaluate the ankle–brachial index, which was initially measured using a commercial instrument as the gold standard [9]. In a recent study, entropy-based PPG evaluations have been successfully applied to distinguish between healthy and diabetic patients [10].

There are many previous studies that effectively demonstrate the substantial interconnection between PPG and BP. A study investigated the relationship of PPG with intermittent systolic and diastolic blood pressures using multi-scale entropy and ensemble neural network [11]. Sideris et al. evaluated continuous arterial blood pressure (ABP) using long short-term memory (LSTM) from patients in an intensive care unit (ICU) using only a PPG signal [12]. Furthermore, the hybrid of LSTM and artificial neural network (ANN) was performed via ECG and PPG to measure BP [13]. In addition, autoregressive moving average to investigate the blood pressure also by using features of the PPG signal related to the specific breathing conditions was performed and showed a quality evaluation [14]. Another study used multiple signals from ECG and PPG, and ballistocardiograms (BCG) were used to investigate systolic blood pressure (SBP) and diastolic blood pressure (DBP) by utilizing hybrid artificial intelligence (AI) methods [15]. Meanwhile, Slapničar et al. utilized 510 subjects of a single PPG signal with a ResNet deep learning model [16].

Generally, AI has been widely used in many fields. It has been used simultaneously with computational fluid dynamics in order to optimize the control scheme by adjusting the triangular membership function for the cooling system in a heat exchanger [17]. A hybrid AI, combining the extreme learning machine with the cuckoo search algorithm, was applied for biodiesel production [18]. Meanwhile, a study used neural network with the multi-armed bandit algorithm for solid oxide fuel cell problems [19]. Moreover, Zaidan et al. applied an AI-based model for gas turbine engine inspection [20].

Specific to medical-related studies, AI was utilized for detecting the depth of anesthesia, involving multi-vital signs [21]. Another study applied entropy-based calculation to extract the feature from one vital sign, which is the EEG. Meanwhile, the 5-s intermittent data from other vital signs were later combined with the extracted entropy value from the EEG. A wearable device-related study also utilized ANN in classifying arrhythmia [22]. Fast Fourier transform (FFT) was also administered to evaluate arrhythmia in the frequency domain. Moreover, the ANN model was also implemented to predict pneumonia [23,24].

Besides being widely utilized, AI has generalization problems [21]. The ensemble technique is likely used to help the model to deal with this difficulty and increase the accuracy of the models. However, selecting all the models for the ensemble system has not always been the best solution [25]. The combination of fuzzy clustering, ANN and the genetic algorithm (GA) was administered for the ensemble model for highly unbalanced data evaluation in emergency medical services [26]. The GA was called to investigate which models should be allocated to have a good ensemble system. Furthermore, this related study examined the quality of the model based on the area under the curve (AUC) from the receiver operating characteristic (ROC) as the fitness function. The result from this study [26] was convincingly supported by the study by Zhou et al. [25]. The ensemble model will definitely increase the accuracy of the result. Nevertheless, selecting several classifiers is likely to produce a better result than combining all of them [25].

Recently, ANN algorithms are moving towards a deeper structure, called deep neural network [27]. This system has been administered to substantial studies. Other methods, such as the convolutional neural network (CNN), have been used to predict arrhythmia with a very precise result with reference to a cardiologist [28]. Another powerful evidence by the CNN-based evaluation technique has also been performed to solve the seizure problem using encephalograms (EEG) [29]. Moreover, a study to evaluate the depth of anesthesia that utilized short-time Fourier transform (STFT) and CNN [30] was investigated to evaluate a four-class system classification in anesthesia from this related study in comparison with several CNN models.

As revealed in the aforementioned details, PPG, as one of the vital signs, is highly potentially able to estimate the blood pressure system. Further, the AI method, especially the deep neural network, has been very widely utilized in many areas particularly medical-related fields either in the classification or the regression system. Moreover, with the help of the GA, as the optimizer, the ensemble model of the deep learning algorithm is prospectively utilized. Hence, the aim of this paper is to investigate generative continuous ABP using deep neural network models via a deep convolutional autoencoder (DCAE) by utilizing only a single PPG sensor. Finally, the GA will form the ensemble model from the evaluation of the cross-validation models.

2. Materials and Methods

This study has been approved by the Research Ethics Committee, National Taiwan University Hospital (NTUH) in Taiwan. Furthermore, written informed consent was received for permission by the patients. In total, a dataset of 18 patients during surgical operation was used for the evaluation. The dataset was acquired using an MP60 IntelliVue Patient Monitor (Koninklijke Philips N.V, Amsterdam, Netherlands) that is connected to a PC. More detail about the data collection can be seen on a study conducted by Liu et al. [31].

Regarding the dataset and the deep learning evaluations, the sampling rate of the PPG and ABP is 128 Hz. The window size evaluation was based on each 5-s signal, both the PPG and ABP. This phenomenon means that each 5-s PPG signal is able to predict the corresponding 5-s of the ABP signal. Initial total data were 42,498 sequences of 5-s windows of PPG and ABP. The data were manually filtered based on their signal quality due to the diathermy effect or nurse activities. Manual filtration was performed by eye by evaluating if either the PPG signal or the ABP signal was noisy. The evaluation was based on the PPG and ABP signal shapes. Finally, the abnormal sequences of these 5-s signals were discarded. This reduced the data amount by about 14% to 36,516 sequences. In this study, the range of the data was limited between 10 and 250 mmHg. Some noisy ABP signals were likely affected by the high-frequency noise. The dataset was randomly divided into 85% and 15% respectively for the training and testing data. MATLAB R2014b (The MathWorks, Inc., Natick, Massachusetts, USA) was utilized for pre-processing the data and post-processing the results. TensorFlow (Ver. 1.15.2) [32] and Keras (Ver. 2.3.1) were utilized in Google Colaboratory (Google Inc., California, USA) for the deep learning training using Python 3.6. The training was conducted for 200 epochs with a batch size of 16 with Adam optimizer [33]. The model checkpoint was also set for the training system. Further, the training data were shuffled. Finally, the cross-validation (CV) method was conducted to investigate the model regularity.

The evaluations were conducted based on mean absolute error (MAE), root mean squared error (RMSE), and Pearson's linear correlation coefficient. Furthermore, the Bland–Altman plot model was provided for comparison purposes. These evaluations are given in Equations (1-3). The Pearson's linear correlation coefficient evaluates between the MP60, as the gold standard, and the generated continuous arterial blood pressures. It also investigates the systolic blood pressure (SBP) and diastolic blood pressure (DBP) values, by taking the maximum and minimum values from the continuous signal, respectively for SBP and DBP, between the MP60 IntelliVue Patient Monitor and the models. The given error is in mmHg. The $R_{x,y}$ value is in range between 0 and 1. The model and the reference are perfectly correlated when the given $R_{x,y}$ value equals 1.

$$MAE = \frac{1}{n} \sum_{i=1}^{n} |x_i - y_i| \tag{1}$$

$$RMSE = \sqrt{\frac{1}{n} \sum_{i=1}^{n} (x_i - y_i)^2} \tag{2}$$

$$R_{x,y} = \frac{\sum_{i=1}^{n}(x_i - \bar{x})(y_i - \bar{y})}{\left[\sum_{i=1}^{n}(x_i - \bar{x})^2 \sum_{i=1}^{n}(y_i - \bar{y})^2\right]^{\frac{1}{2}}} \tag{3}$$

where x_i is the reference, y_i is the estimated result, n is the number of samples, $\bar{x}$ is the mean of the reference, and $\bar{y}$ is the mean of the predicted result.

This study evaluates two DCAE models. Basically, the autoencoder structure has the latent space between the input and the output layers. The first model is generated based on the LeNet-5 CNN model [34]. Originally, this model worked for the digit recognition system. The architecture of this model is relatively simple compared with other models. The convolution layer in this model is regularly followed by subsampling. For the classification system, there are several fully connected layers installed to the network. This study uses only the convolution layer with the subsampling from the original LeNet-5 model to form the encoder. Meanwhile, the decoder utilizes the opposite way of the encoder. The summary of the LeNet-5-based deep convolutional autoencoder (LDCAE) utilized in this study can be seen in Figure A1 in Appendix A. From this figure, it can be seen that the original 5 s of the one-dimensional PPG signal and the sampling rate of 128 Hz, with a size of 640 points, are used for the input layer. For the encoder, this study applies an increasing filter size. All convolution layers administer the rectified linear unit (ReLU) activation function, shown in Equation (4). This structure also uses the same padding. After the input layer, for the encoder, the first convolution layer starts with 16 filters and ends with 64 filters. However, the decoder works with initially 64 filters to 16 filters. The output layer is equal to the input layer. This layer is the 5-s ABP signal. This model has equal total parameters and trainable parameters, which total about sixty thousand parameters.

$$f(X) = \max(0,X) \tag{4}$$

where X is the input signal.

Another model is the deep convolutional autoencoder based on the U-Net architecture [35]. This model was originally applied for biomedical segmentation. One of the reasons behind the uniqueness of the U-Net model is the concatenating between a layer in the encoder and another layer in the decoder that has the same feature map. The detailed structure of the U-Net-based deep convolutional autoencoder (UDCAE) used in this study is shown in Figure A2 in Appendix A. In parallel with the LDCAE model, this model also has an input size of 640 data points of the PPG. The encoder and decoder structures are also very identical to the LDCAE. However, the first filter in the encoder has 32 filters and ends with 256 filters. Further, the concatenated layer filters in the decoder are formed by considering the filter from the encoder layer. The UDCAE also utilizes the ReLU activation function. This UDCAE model has an equal total number of settings and trainable parameters, which total about three hundred thousand parameters. These numbers of parameters are much bigger compared with the LDCAE structure.

Moreover, a 10-fold cross-validation (CV) system is conducted to evaluate the data generalization to the models. This CV method uses a leave-testing-out cross validation technique, meaning that the CV model shuffles only the training part and keeps the testing data outside the shuffling system. The highest average BP of the CV fold, combining the DCAE models, is selected as the best single model.

Finally, this study deploys genetic algorithm (GA) optimization, named the genetic deep convolutional autoencoder (GDCAE), to ensemble the ten CV models for each LDCAE and UDCAE. Each CV model has equally distributed weights, meaning each model will have the chance to be combined with other models. Therefore, the GA will have a total of 20 bits for each chromosome. The chromosomes are encoded in 32 bits binary format. Zero means the model is not selected and one means the model is selected. The GA is set with a single point crossover, 95% mutation rate and 2000 generations. The fitness function is given by Equation (5). This equation is a modified version of

Equation (3). Specifically, Equation (5) calculates the average Pearson's linear correlation coefficient between SBP and DBP, meaning that the weights are equally distributed.

$$\overline{R_{bp}} = \frac{1}{2}\left(\frac{\sum_{i=1}^{n}\left(x_{i,sbp} - \overline{x_{sbp}}\right)\left(y_{i,sbp} - \overline{y_{sbp}}\right)}{\left[\sum_{i=1}^{n}\left(x_{i,sbp} - \overline{x_{sbp}}\right)^2 \sum_{i=1}^{n}\left(y_{i,sbp} - \overline{y_{sbp}}\right)^2\right]^{\frac{1}{2}}} + \frac{\sum_{i=1}^{n}\left(x_{i,dbp} - \overline{x_{dbp}}\right)\left(y_{i,dbp} - \overline{y_{dbp}}\right)}{\left[\sum_{i=1}^{n}\left(x_{i,dbp} - \overline{x_{dbp}}\right)^2 \sum_{i=1}^{n}\left(y_{i,dbp} - \overline{y_{dbp}}\right)^2\right]^{\frac{1}{2}}} \right) \quad (5)$$

3. Results

This study utilizes deep convolutional autoencoder (DCAE) models to generate the continuous arterial blood pressure signal (ABP) by using single photoplethysmography (PPG). The results produced by the models are compared to investigate the better model compared to the MP60 IntelliVue Patient Monitor as the gold standard. The evaluations cover the continuous arterial blood pressure signal with systolic and diastolic blood pressures.

The training of the DCAE models can be seen in Figure 1 where UDCAE converges faster and better than the LDCAE model. Furthermore, for the testing phase, the UDCAE model also provides a preferable result compared with the LDCAE. In addition, the UDCAE model shows relatively less fluctuation.

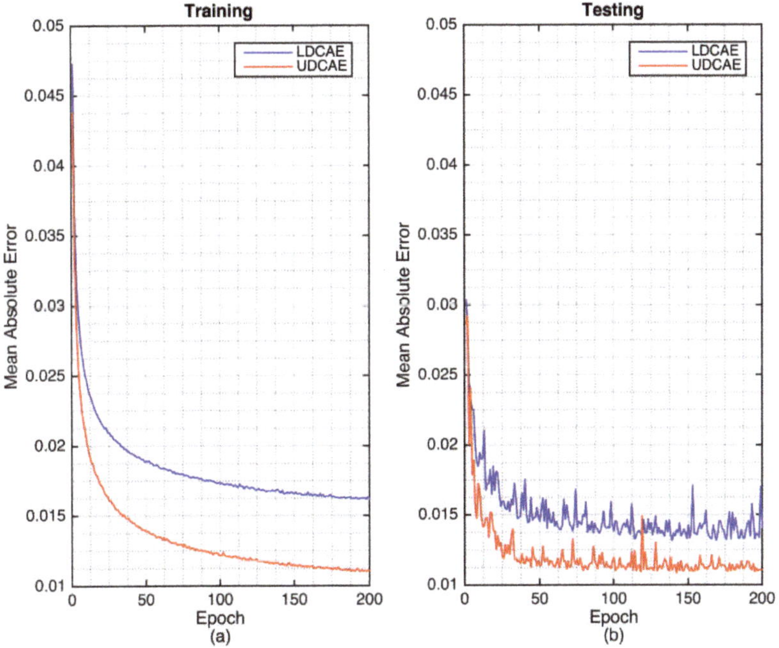

Figure 1. The training (**a**) and testing (**b**) of the LeNet-5 based deep convolution autoencoder (LDCAE) and U-Net-based deep convolutional autoencoder (UDCAE) models.

Figure 2 shows the input of the PPG signal and its corresponding output of the continuous ABP signals, generated by the DCAE-based models for the testing results. It can be seen that both models, LDCAE and UDCAE, successfully produce continuous ABP. In addition, Figure 2 also reveals that SBP and DBP can be accurately estimated. Both models display a fine estimation result in that the PPG has either a significant or non-significant second peak.

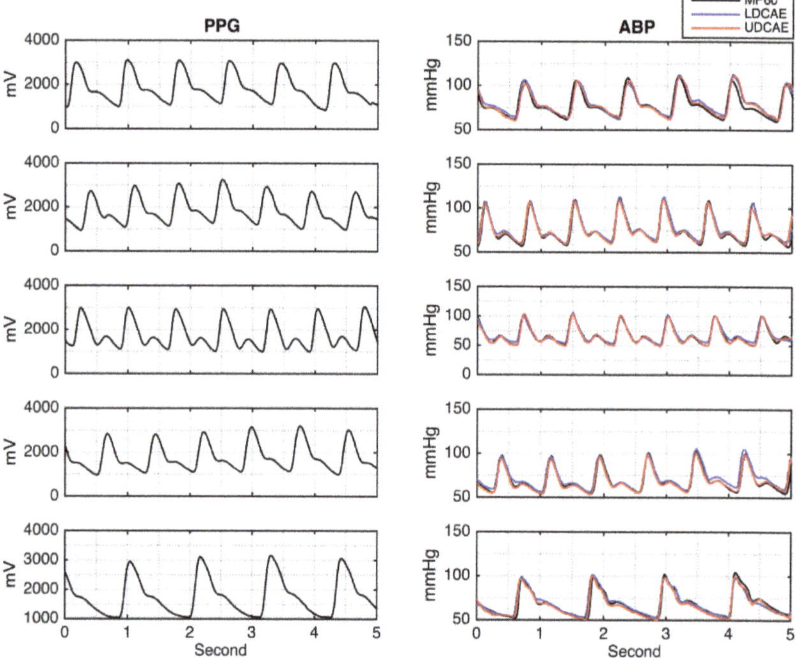

Figure 2. The photoplethysmography (PPG) input signal and arterial blood pressure (ABP) results between LDCAE and UDCAE models in comparison to MP60 IntelliVue Patient Monitor.

After performing the continuous ABP, the evaluation of SBP and DBP is further investigated. The maximum value of a 5-s segment is defined as SBP. Meanwhile, the minimum value is DBP. This approach is deployed for both the DCAE models and the MP60, as the gold standard. The evaluation of SBP and DBP can be seen on the error distribution graphs shown in Figure 3. From this figure, both LDCAE and UDCAE are compared to the MP60 IntelliVue Patient Monitor values. It can be seen that the UDCAE model produces a better outcome by delivering a higher frequency of results approaching zero than the LDCAE model.

Furthermore, to investigate the model prediction accuracy of SBP and DBP, the results are compared to the MP60 using Pearson's linear correlation coefficient, which shows heterogeneous outcomes. The UDCAE has a slightly better result in the SBP prediction. Meanwhile, the LDCAE displays insignificantly better results for the DBP estimation. The detailed evaluation is shown in Figure 4.

Another powerful approach given by the DCAE models is the ability to generate a continuous ABP signal that is not interfered by any noise since a good-quality PPG is supplied. From Figure 5, it can be seen that some signals produced by the MP60 IntelliVue Patient Monitor are relatively noisy. However, this has been overcome by the DCAE models. Moreover, the predicted SBP and DBP values are comparable, by comparing them to either the preceding or the succeeding cycles.

Cross-validation is later performed in order to evaluate the data generalization and ensemble combination. The results show that the data have very high generalization. Good generalization is given by the standard deviation of the Pearson's linear correlation for SBP, DBP and the waveform evaluations, given in Table 1. Moreover, the relatively small standard deviation of RMSE and MAE for SBP, DBP and the waveform error evaluations are shown in Table 2.

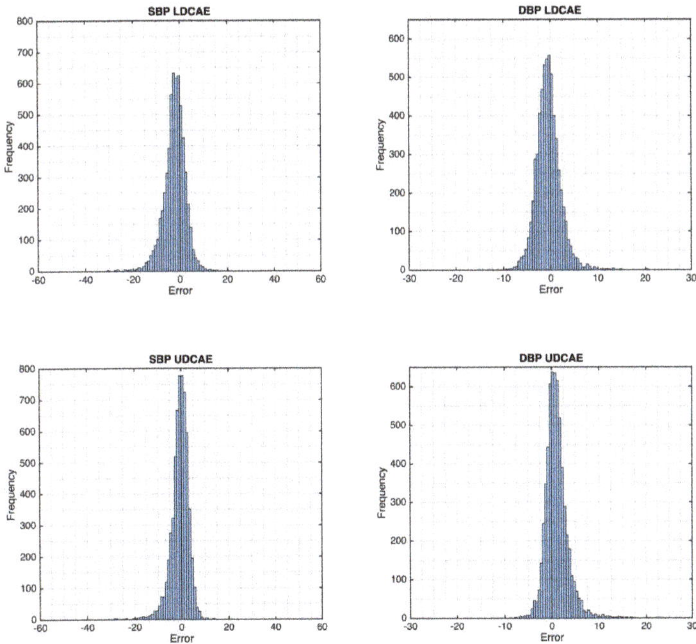

Figure 3. The error comparison between DCAE-based models and MP60 IntelliVue Patient Monitor.

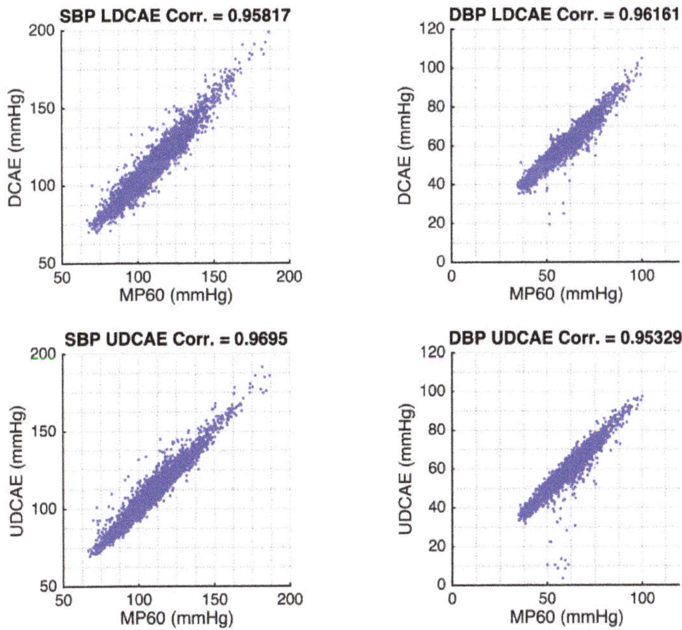

Figure 4. The Pearson's linear correlation comparison between DCAE-based models and MP60 IntelliVue Patient Monitor.

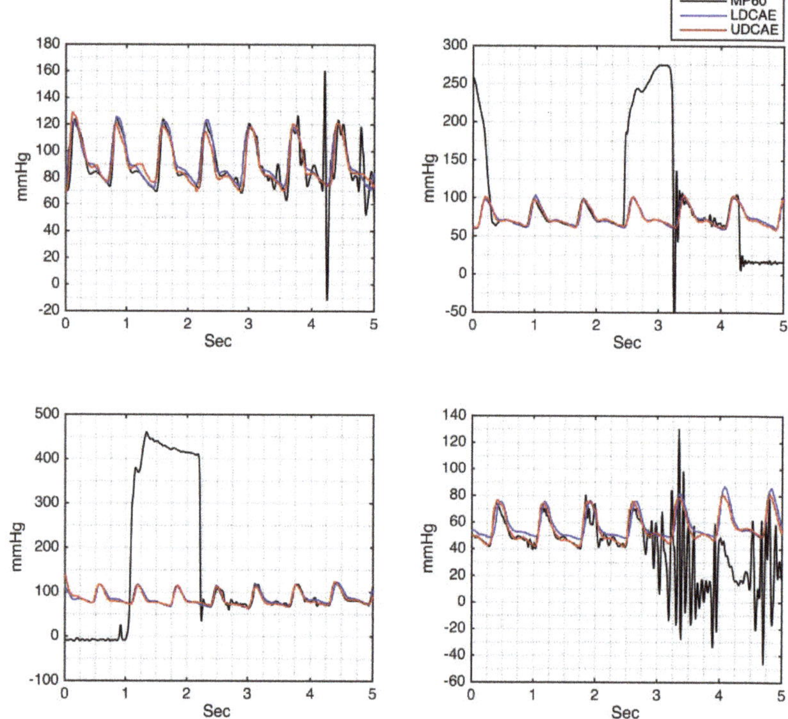

Figure 5. The comparison of the noisy MP60 ABP signal and the generated ABP signal by the LDCAE and UDCAE models.

Table 1. The Pearson's Linear Correlation Coefficient Evaluation of LDCAE and UDCAE Models from the Cross-Validation (CV) Method. Note: bold value is the best single CV model.

CV	Correlation Coefficient						Average
	SBP		DBP		Waveform		
	LDCAE	UDCAE	LDCAE	UDCAE	LDCAE	UDCAE	
1	0.956	0.958	0.958	0.953	0.968	0.974	0.9612
2	0.960	0.961	0.954	0.942	0.969	0.974	0.9600
3	0.962	0.965	0.951	0.941	0.968	0.975	0.9603
4	**0.958**	**0.969**	**0.962**	**0.953**	**0.968**	**0.976**	**0.9643**
5	0.954	0.964	0.963	0.962	0.966	0.975	0.9640
6	0.951	0.960	0.959	0.956	0.966	0.974	0.9610
7	0.956	0.957	0.947	0.951	0.967	0.973	0.9585
8	0.959	0.964	0.949	0.956	0.968	0.976	0.9620
9	0.957	0.963	0.947	0.946	0.966	0.975	0.9590
10	0.958	0.968	0.963	0.947	0.967	0.975	0.9630
Mean	0.957	0.963	0.955	0.951	0.967	0.975	
STD	0.003	0.004	0.007	0.007	0.001	0.001	

Table 2. Error Evaluations of SBP and DBP from LDCAE and UDCAE Models.

CV	SBP				DBP			
	LDCAE		UDCAE		LDCAE		UDCAE	
	RMSE	MAE	RMSE	MAE	RMSE	MAE	RMSE	MAE
1	4.69	3.44	4.62	3.26	3.10	2.22	3.06	1.82
2	4.63	3.39	4.30	3.11	3.25	2.18	3.87	2.23
3	4.91	3.72	4.81	3.42	3.09	2.04	3.38	1.92
4	5.19	3.80	3.85	2.73	2.76	2.00	3.25	1.95
5	5.11	3.64	4.12	3.01	2.70	1.86	3.01	1.96
6	6.48	4.85	5.11	3.47	2.86	2.03	3.02	1.78
7	5.02	3.61	4.48	3.07	3.40	2.14	3.25	1.99
8	4.88	3.57	4.54	3.14	3.26	2.12	3.17	1.90
9	4.63	3.39	5.18	3.65	3.27	2.08	3.30	1.77
10	6.39	4.93	5.12	3.71	2.79	1.04	3.29	1.82
Mean	5.19	3.83	4.61	3.26	3.05	1.97	3.26	1.91
STD	0.68	0.57	0.45	0.31	0.25	0.34	0.25	0.14

The selection of the best single model from the CV results is evaluated based on Pearson's linear correlation coefficient given in Table 1. It can be seen that the fourth CV model provides the highest average value between SBP and DBP, which is 0.9643. Hence, this model is selected as the best single model.

After having the CV models, both from LDACE and UDCAE, the genetic algorithm-based optimization deep convolutional autoencoder (GDCAE) is subsequently performed. The GA will work as the selector of the DCAE models that will be combined for the ensemble system. As the result, the CV models 1, 2, 3, 4, 5 and 10 are selected by the GA from the LDCAE model. Meanwhile, GA selects all the UDCAE models, except the first model. The results also show the reliability of the fourth model of the LDCAE and UDCAE systems.

The convergence of the GDCAE is shown in Figure 6. Several chromosome sizes of 4, 8, 16, 32 and 64 are investigated. The average result from the SBP and DBP of GDCAE is 0.98004. This GDCAE result is better compared with the average value of SBP and DBP from the best single CV model, 0.960 and 0.961 for the LDCAE and UDCAE models, respectively. By having this combination, the GA-optimized reconstructed signal is later performed. The results also provide some improvements in comparison with the best CV model in Pearson's linear correlation coefficient and error evaluations, which can be seen in Table 3.

Table 3. Comparison between the LDCAE, UDCAE and GDCAE Models.

Method	Correlation Coefficient			Error [mmHg]		
	Waveform	SBP	DBP	Waveform	SBP	DBP
LDCAE	R = 0.968	R = 0.958	R = 0.962	RMSE = 5.10 MAE = 3.52	RMSE = 5.19 MAE = 3.80	RMSE = 2.76 MAE = 2.00
UDACE	R = 0.976	R = 0.969	R = 0.953	RMSE = 4.25 MAE = 2.77	RMSE = 3.85 MAE = 2.73	RMSE = 3.25 MAE = 1.95
GDCAE	R = 0.984	R = 0.981	R = 0.979	RMSE = 3.46 MAE = 2.33	RMSE = 3.41 MAE = 2.54	RMSE = 2.14 MAE = 1.48

Furthermore, the Bland–Altman evaluation results can be seen in Table 4 and Figure 7. Even though the GDCAE has a slightly inferior result for the mean value to the LDCAE and UDCAE respectively for DBP and SBP, the GDCAE has lower standard deviation compared with other models. Furthermore, for GDCAE, the 95% confidence band, ± 1.96 of standard deviation of the difference, produces smaller distances compared with LDCAE and UDCAE both for SBP and DBP. Qualitative results are shown in

Figure 7, which is a good indication that the GDCAE model provides better prediction results between the 95% confidence band compared with the LDCAE and UDCAE models.

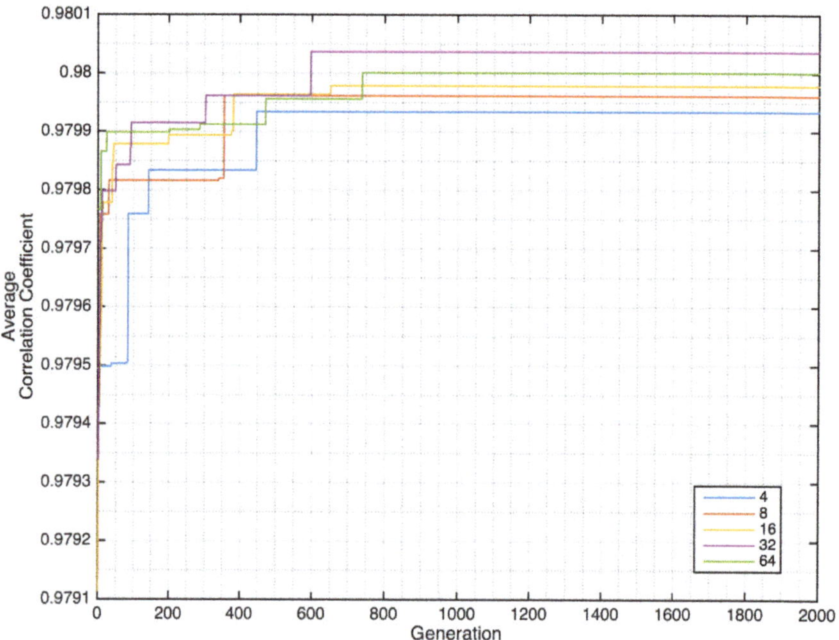

Figure 6. Genetic deep autoencoder (GDCAE) generation convergence.

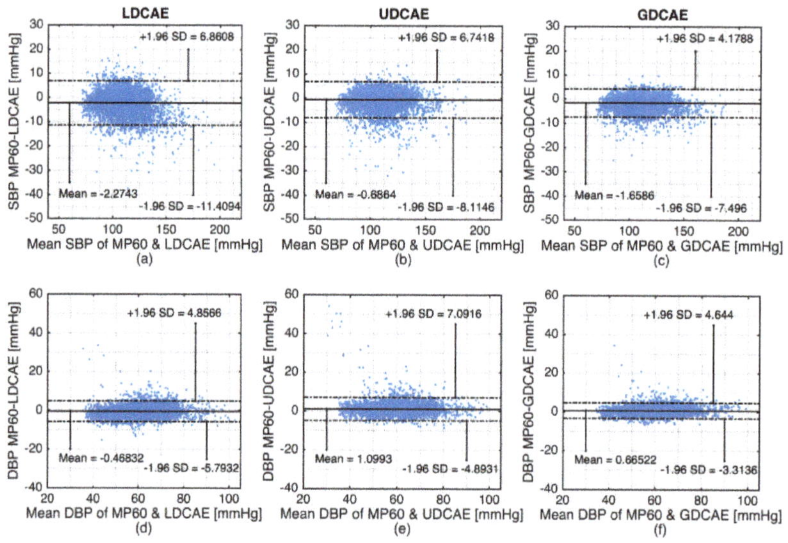

Figure 7. Bland–Altman plot. (**a**) LDCAE systolic blood pressure (SBP); (**b**) UDCAE SBP; (**c**) GDCAE SBP; (**d**) LDCAE diastolic blood pressure (DBP); (**e**) UDCAE DBP; (**f**) GDCAE DBP.

Table 4. Bland–Altman DCAE Model Comparison.

Methods	Mean [mmHg]		STD [mmHg]		−1.96 STD [mmHg]		+1.96 STD [mmHg]	
	SBP	**DBP**	**SBP**	**DBP**	**SBP**	**DBP**	**SBP**	**DBP**
LDCAE	−2.274	−0.468	4.661	2.717	−11.410	−5.793	6.862	4.857
UDCAE	−0.686	1.099	3.790	3.057	−8.114	−4.893	6.742	7.091
GDCAE	−1.659	0.665	2.978	2.030	−7.496	−3.314	4.178	4.644

4. Discussion

Initially in this study, the PPG signal is trained by using DCAE models, LeNet-5- and U-Net-based models, to generate a continuous arterial blood pressure (ABP) signal. In this step, the PPG- and MP60 IntelliVue Patient Monitor-generated continuous arterial blood pressure signals are compared. Moreover, systolic and diastolic blood pressures are evaluated by root mean squared error (RMSE), mean absolute error (MAE) and the Pearson's linear correlation coefficient between the models with the MP60 IntelliVue Patient Monitor as the gold standard. Finally, the GA-regulated DCAE based on the cross-validation results is deployed to ensemble the model and evaluate the system.

In order to investigate the quality of the proposed methods, a comparative study to the previously organized research was conducted. The comparison method included the dataset, input signal, methodology, generative system, error evaluations and linear correlations. The details of the comparative studies are given in Table 5. Sideris et al. [12] utilized the forty-two-patient dataset from MIMIC PhysioNet, originally a two hundred-patient dataset, after applying some filtering steps based on the quality of the blood pressure signal. This study also only used a single PPG signal. The overlapped window size was used in order to form either the training or testing data. Further, LSTM, one of the deep neural network methods, was applied for the prediction. One of the essential achievements from this study is the ability to generate a continuous arterial blood pressure signal. As it can be seen, the capability of LSTM is able to produce continuous arterial blood pressure by only utilizing the PPG signal. However, it did not mention specifically about the RMSE of the DBP. Nevertheless, in this study, they provided a table consisting of the tabulated RMSE result of SBP, DBP and ABP. With full respect to all the authors in this study [12], we re-evaluate the ABP and SBP results based on the corresponding table. This is conducted to recalculate the mean and standard deviation, which were found to have very identical results to their reported results. Hence, we perform the DBP calculation, in parallel to the aforementioned method for the ABP and SBP calculations. The results of DBP, for mean and standard deviation, are 1.98±1.06 mmHg. In comparison with our study, this study has slightly better results in the RMSEs of SBP and DBP error evaluations. However, in this study, the GDCAE provides a better outcome in the waveform error evaluation, which is 0.984. Moreover, our GDCAE also delivers a superior solution for the correlation coefficient for the waveform evaluation. Meanwhile, Sideris et al. [12] did not provide any information about the SBP and DBP correlation coefficient results.

Table 5. Comparative Results for Dataset and Methodology Between the Proposed Method and Previous Related Studies.

Error [mmHg]	DBP	RMSE = 1.98 STD = 1.06	RMSE = 0.73 MAE = 0.52	RMSE = 5.12	MAE = 3.33 STD = 3.42	MAE = 6.88	RMSE = 2.76 MAE = 2.00	RMSE = 3.25 MAE = 1.95	RMSE = 2.14 MAE = 1.48
	SBP	RMSE = 2.58 STD = 1.23	RMSE = 1.26 MAE = 0.93	RMSE = 7.21	MAE = 4.06 STD = 4.04	MAE = 9.43	RMSE = 5.19 MAE = 3.80	RMSE = 3.85 MAE = 2.73	RMSE = 3.41 MAE = 2.54
	Waveform	RMSE = 6.04 STD = 3.26	N/A	N/A	N/A	N/A	RMSE = 5.10 MAE = 3.52	RMSE = 4.25 MAE = 2.77	RMSE = 3.46 MAE = 2.33
Correlation Coefficient	DBP	N/A	0.998	N/A	$R^2 = 0.49$	N/A	R = 0.962	R = 0.953	R = 0.979
	SBP	N/A	0.999	N/A	$R^2 = 0.52$	N/A	R = 0.958	R = 0.969	R = 0.981
	Waveform	Mean = 0.95 STD = 0.045	N/A	N/A	N/A	N/A	R = 0.968	R = 0.976	R = 0.984
Gen. Cont. ABP		Yes	No	No	No	No	Yes		
Method		LSTM	ANN + LSTM	ARMA	CNN + Bi-GRU + Attention	Spectro temporal ResNet	LDCAE	UDCAE	GDCAE
Input Signal		PPG	ECG + PPG	PPG	ECG + PPG + BCG	PPG	PPG		
Dataset		42 subjects, MIMIC PhysioNet	39 subjects, MIMIC PhysioNet	15 subjects	15 subjects	510 subjects, MIMIC III PhysioNet	18 subjects, NTUH, Taiwan		
Studies		Sideris et al. [12]	Tanveer et al. [13]	Zadi et al. [14]	Eom et al. [15]	Slapničar et al. [16]	Proposed		

Another study related to blood pressure evaluation was conducted by Tanveer et al. [13]. This study applied multiple vital signs, which are ECG and PPG. This study used the dataset of thirty-nine patients, from originally ninety-three patients, of the MIMIC I PhysioNet database. This study had 16-s and 40-s window sizes, with 125 Hz of sampling frequency. This study also deployed the LSTM method, similar to the study performed by Sideris et al. [12], alongside the ANN. This study provided an outstanding result in the error estimation in mmHg. Based on the combination of LSTM and ANN methods, their study produced significantly small RMSEs, which are 1.26 mmHg and 0.73 mmHg, respectively for SBP and DBP. Moreover, the MAEs for SBP and DBP are respectively 0.93 mmHg and 0.52 mmHg. Identical to the error evaluation, the Pearson's linear correlation coefficient evaluation is also an exceptional finding. Nearly perfectly correlated systems are produced, which are 0.999 and 0.998 for SBP and DBP, respectively. This result is produced by the longer size, which is the 40-s window size system. However, this method has a drawback. It did not provide the information about generative continuous arterial blood pressure.

A study investigated by Zadi et al. [14] used fifteen young subjects. This study evaluated the blood pressure based on two conditions, which are normal breath and breath hold. The autoregressive moving average (ARMA) was deployed in the modeling. This study produced a relatively good result. It has RMSEs of 7.21 and 5.12 mmHg, respectively for systolic and diastolic blood pressure. However, neither correlation coefficient for waveform, SBP nor DBP was provided. Moreover, there was no available generative continuous ABP signal investigation.

Another comparative study is the finding by Eom et al. [15]. This study was conducted on fifteen subjects. It used several vital signs, which are ECG, PPG and BCG. The 5-s window size was also used in this study. The combination of CNN, bidirectional gated recurrent unit (Bi-GRU) and attention mechanism. The result showed the produced MAEs and standard deviations are 4.06 ± 4.04 and 3.33 ± 3.42 mmHg, respectively for SBP and DBP. However, this study has a disadvantage, which is no generative continuous blood pressure estimation was performed.

The latest study conducted by Slapničar et al. [16] utilizing 510 subjects using a single PPG with a ResNet-based model is used. The results showed 9.43 and 6.88 mmHg of MAE respectively for SBP and DBP. Nevertheless, there is no given information about generative continuous arterial blood pressure evaluation.

As it can be seen from the aforementioned information comparing our proposed methods to previously performed studies, our study shows assorted advantages. Our proposed methods, working based on the deep autoencoder and using only a single PPG signal, provide a leading achievement for the correlation coefficient for the waveform of the generative continuous blood pressure signal. Additionally, our proposed methods produce highly correlated results of the estimated SBP and DBP to the MP60 IntelliVue Patient Monitor, as the gold standard.

However, this study has several limitations. The number of the patients utilized in this study is relatively small. In addition, most of the utilized patient data are during surgery. This unconscious condition may reduce the noise interfering the PPG signal, especially for the motion artifact. For this reason, automatic-based filters should be applied in future work for conscious subjects. Furthermore, the algorithm to evaluate SBP and the DBP from a 5-s sliding window can be improved. This technique is selected based on the consideration that either SBP or DBP do not fluctuate significantly within five seconds. Furthermore, more advanced statistical analysis can be applied. In addition, the noisy PPG signal can contribute to the low-quality continuous ABP prediction, as it can be seen in Figure 8.

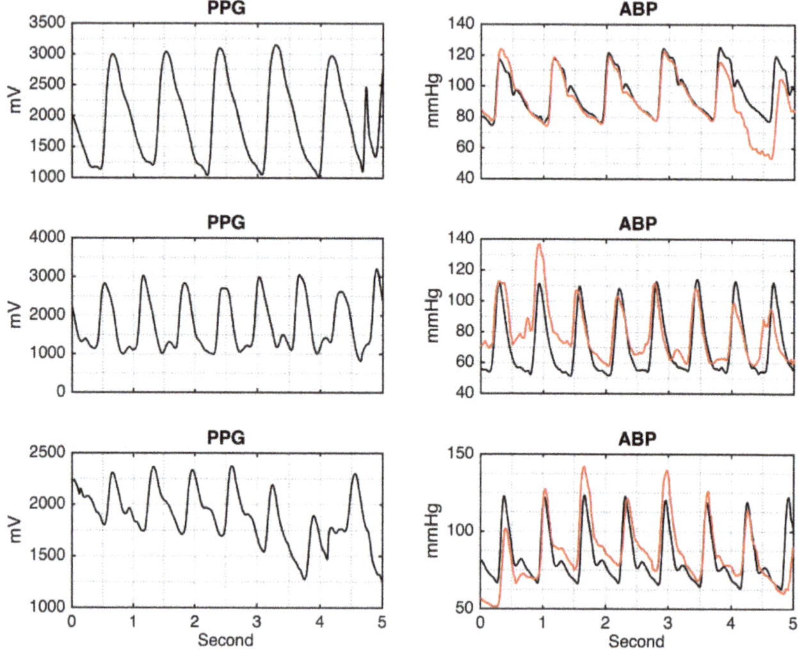

Figure 8. Low-quality generated continuous ABP result.

5. Conclusions

This study demonstrates that deep convolutional autoencoder methods with GA-based optimization have successfully evaluated the continuous arterial blood pressure system by only using a single PPG signal. In addition, supporting the previous studies, this study also shows straightforward information that the PPG is highly correlated with continuous arterial blood pressure. Hence, the SBP and DBP measurements can be precisely achieved by only using a single PPG signal.

Author Contributions: Conceptualization, M.S., Y.-T.L., M.F.A. and J.-S.S.; formal analysis, M.S.; investigation, M.S., Y.-T.L., M.F.A. and J.-S.S.; methodology, M.S., M.F.A. and J.-S.S.; software, M.S., Y.-T.L., C.-H.L., M.F.A. and J.-S.S.; supervision, Y.-T.L., S.-Z.F., M.F.A. and J.-S.S.; validation, M.S., Y.-T.L., C.-H.L., M.F.A. and J.-S.S.; visualization, M.S.; writing—original draft, M.S., Y.-T.L., B.M., M.F.A. and J.-S.S.; writing—review and editing, M.S., Y.-T.L., M.F.A. and J.-S.S. All authors have read and agreed to the published version of the manuscript.

Funding: This research received no external funding.

Acknowledgments: We are particularly grateful for the assistance given by Yi-Fang Chiu.

Conflicts of Interest: The authors declare no conflict of interest.

Appendix A

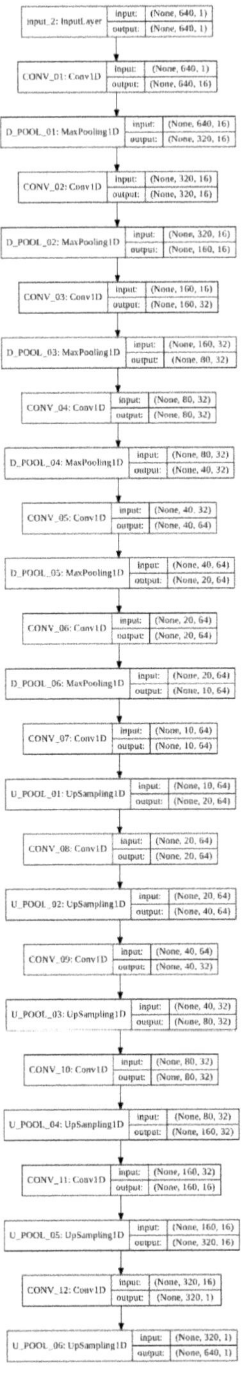

Figure A1. LeNet-5 based deep convolution autoencoder (LDCAE) structure.

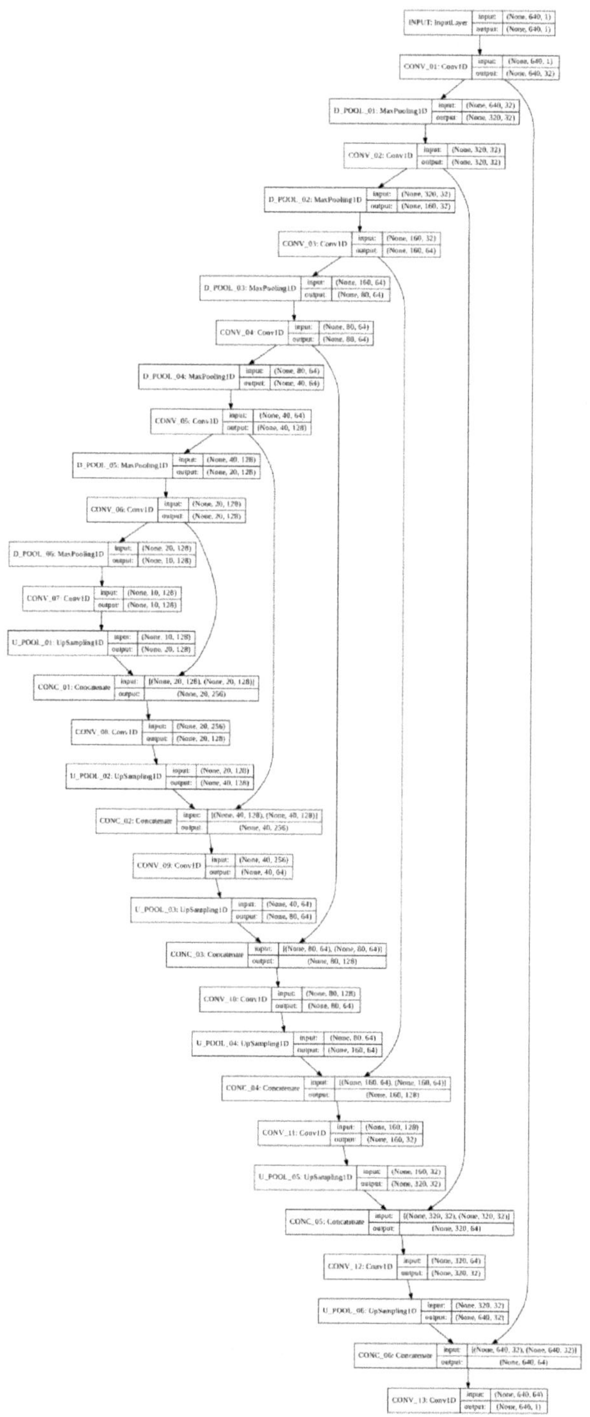

Figure A2. U-Net based deep convolution autoencoder (UDCAE) structure.

References

1. Available online: https://www.who.int/news-room/fact-sheets/detail/hypertension (accessed on 18 March 2020).
2. He, J.; Whelton, P.K. Elevated systolic blood pressure and risk of cardiovascular and renal disease: Overview of evidence from observational epidemiologic studies and randomized controlled trials. *Am. Heart J.* **1999**, *138*, S211–S219. [CrossRef]
3. Lawes, C.M.; Vander Hoorn, S.; Rodgers, A. Global burden of blood-pressure-related disease, 2001. *Lancet* **2008**, *371*, 1513–1518. [CrossRef]
4. Wong, T.Y.; McIntosh, R. Hypertensive retinopathy signs as risk indicators of cardiovascular morbidity and mortality. *Br. Med. Bull.* **2005**, *73*, 57–70. [CrossRef] [PubMed]
5. Sadrawi, M.; Shieh, J.S.; Haraikawa, K.; Chien, J.C.; Lin, C.H.; Abbod, M.F. Ensemble empirical mode decomposition applied for PPG motion artifact. In Proceedings of the 2016 IEEE EMBS Conference on Biomedical Engineering and Sciences (IECBES), Kuala Lumpur, Malaysia, 4–7 December 2016; pp. 266–269.
6. Tang, S.C.; Huang, P.W.; Hung, C.S.; Shan, S.M.; Lin, Y.H.; Shieh, J.S.; Lai, D.M.; Wu, A.Y.; Jeng, J.S. Identification of atrial fibrillation by quantitative analyses of fingertip photoplethysmogram. *Sci. Rep.* **2017**, *7*, 45644. [CrossRef] [PubMed]
7. Liang, Y.; Chen, Z.; Ward, R.; Elgendi, M. Hypertension assessment using photoplethysmography: A risk stratification approach. *J. Clin. Med.* **2019**, *8*, 12. [CrossRef] [PubMed]
8. Phillips, J.P.; Hickey, M.; Kyriacou, P.A. Evaluation of electrical and optical plethysmography sensors for noninvasive monitoring of hemoglobin concentration. *Sensors* **2012**, *12*, 1816–1826. [CrossRef] [PubMed]
9. Perpetuini, D.; Chiarelli, A.M.; Cardone, D.; Rinella, S.; Massimino, S.; Bianco, F.; Bucciarelli, V.; Vinciguerra, V.; Fallica, G.; Perciavalle, V.; et al. Photoplethysmographic Prediction of the Ankle-Brachial Pressure Index through a Machine Learning Approach. *Appl. Sci.* **2020**, *10*, 2137. [CrossRef]
10. Wei, H.C.; Ta, N.; Hu, W.R.; Xiao, M.X.; Tang, X.J.; Haryadi, B.; Liou, J.J.; Wu, H.T. Digital Volume Pulse Measured at the Fingertip as an Indicator of Diabetic Peripheral Neuropathy in the Aged and Diabetic. *Entropy* **2019**, *21*, 1229. [CrossRef]
11. Sadrawi, M.; Shieh, J.S.; Fan, S.Z.; Lin, C.H.; Haraikawa, K.; Chien, J.C.; Abbod, M.F. Intermittent blood pressure prediction via multiscale entropy and ensemble artificial neural networks. In Proceedings of the 2016 IEEE EMBS Conference on Biomedical Engineering and Sciences (IECBES), Kuala Lumpur, Malaysia, 4–7 December 2016; pp. 356–359.
12. Sideris, C.; Kalantarian, H.; Nemati, E.; Sarrafzadeh, M. Building continuous arterial blood pressure prediction models using recurrent networks. In Proceedings of the 2016 IEEE International Conference on Smart Computing (SMARTCOMP), St. Louis, MO, USA, 18–20 May 2016; pp. 1–5.
13. Tanveer, M.S.; Hasan, M.K. Cuffless blood pressure estimation from electrocardiogram and photoplethysmogram using waveform based ANN-LSTM network. *Biomed. Signal Process. Control.* **2019**, *51*, 382–392. [CrossRef]
14. Zadi, A.S.; Alex, R.; Zhang, R.; Watenpaugh, D.E.; Behbehani, K. Arterial blood pressure feature estimation using photoplethysmography. *Comput. Biol. Med.* **2018**, *102*, 104–111. [CrossRef]
15. Eom, H.; Lee, D.; Han, S.; Hariyani, Y.S.; Lim, Y.; Sohn, I.; Park, K.; Park, C. End-to-End Deep Learning Architecture for Continuous Blood Pressure Estimation Using Attention Mechanism. *Sensors* **2020**, *20*, 2338. [CrossRef] [PubMed]
16. Slapničar, G.; Mlakar, N.; Luštrek, M. Blood pressure estimation from photoplethysmogram using a spectro-temporal deep neural network. *Sensors* **2019**, *19*, 3420.
17. Sadrawi, M.; Yunus, J.; Khalil, M.; Sofyan, S.E.; Abbod, M.F.; Shieh, J.S. Computational fluid dynamics based fuzzy control optimization of heat exchanger via genetic algorithm. In Proceedings of the 2019 IEEE International Conference on Cybernetics and Computational Intelligence (CyberneticsCom), Banda Aceh, Indonesia, 22–24 August 2019; pp. 1–6.
18. Silitonga, A.S.; Shamsuddin, A.H.; Mahlia, T.M.I.; Milano, J.; Kusumo, F.; Siswantoro, J.; Dharma, S.; Sebayang, A.H.; Masjuki, H.H.; Ong, H.C. Biodiesel synthesis from Ceiba pentandra oil by microwave irradiation-assisted transesterification: ELM modeling and optimization. *Renew. Energy* **2020**, *146*, 1278–1291. [CrossRef]

19. Song, C.; Lee, S.; Gu, B.; Chang, I.; Cho, G.Y.; Baek, J.D.; Cha, S.W. A Study of Anode-Supported Solid Oxide Fuel Cell Modeling and Optimization Using Neural Network and Multi-Armed Bandit Algorithm. *Energies* **2020**, *13*, 1621. [CrossRef]

20. Zaidan, M.A.; Harrison, R.F.; Mills, A.R.; Fleming, P.J. Bayesian hierarchical models for aerospace gas turbine engine prognostics. *Expert Syst. Appl.* **2015**, *42*, 539–553. [CrossRef]

21. Sadrawi, M.; Fan, S.Z.; Abbod, M.F.; Jen, K.K.; Shieh, J.S. Computational depth of anesthesia via multiple vital signs based on artificial neural networks. *Biomed Res. Int.* **2015**, *2015*, 13. [CrossRef]

22. Sadrawi, M.; Lin, C.H.; Lin, Y.T.; Hsieh, Y.; Kuo, C.C.; Chien, J.C.; Haraikawa, K.; Abbod, M.F.; Shieh, J.S. Arrhythmia evaluation in wearable ECG devices. *Sensors* **2017**, *17*, 2445. [CrossRef]

23. Liao, Y.H.; Shih, C.H.; Abbod, M.F.; Shieh, J.S.; Hsiao, Y.J. Development of an E-nose system using machine learning methods to predict ventilator-associated pneumonia. *Microsyst. Technol.* **2020**, 1–11. [CrossRef]

24. Liao, Y.H.; Wang, Z.C.; Zhang, F.G.; Abbod, M.F.; Shih, C.H.; Shieh, J.S. Machine Learning Methods Applied to Predict Ventilator-Associated Pneumonia with Pseudomonas aeruginosa Infection via Sensor Array of Electronic Nose in Intensive Care Unit. *Sensors* **2019**, *19*, 1866. [CrossRef]

25. Zhou, Z.H.; Wu, J.; Tang, W. Ensembling neural networks: Many could be better than all. *Artif. Intell.* **2002**, *137*, 239–263. [CrossRef]

26. Sadrawi, M.; Sun, W.Z.; Ma, M.H.M.; Yeh, Y.T.; Abbod, M.F.; Shieh, J.S. Ensemble genetic fuzzy neuro model applied for the emergency medical service via unbalanced data evaluation. *Symmetry* **2018**, *10*, 71. [CrossRef]

27. LeCun, Y.; Bengio, Y.; Hinton, G. Deep learning. *Nature* **2015**, *521*, 436–444. [CrossRef]

28. Hannun, A.Y.; Rajpurkar, P.; Haghpanahi, M.; Tison, G.H.; Bourn, C.; Turakhia, M.P.; Ng, A.Y. Cardiologist-level arrhythmia detection and classification in ambulatory electrocardiograms using a deep neural network. *Nat. Med.* **2019**, *25*, 65. [CrossRef] [PubMed]

29. Acharya, U.R.; Oh, S.L.; Hagiwara, Y.; Tan, J.H.; Adeli, H. Deep convolutional neural network for the automated detection and diagnosis of seizure using EEG signals. *Comput. Biol. Med.* **2018**, *100*, 270–278. [CrossRef] [PubMed]

30. Liu, Q.; Cai, J.; Fan, S.Z.; Abbod, M.F.; Shieh, J.S.; Kung, Y.; Lin, L. Spectrum analysis of eeg signals using cnn to model patient's consciousness level based on anesthesiologists' experience. *IEEE Access* **2019**, *7*, 53731–53742. [CrossRef]

31. Liu, Q.; Ma, L.; Fan, S.Z.; Abbod, M.F.; Lu, C.W.; Lin, T.Y.; Jen, K.K.; Wu, S.J.; Shieh, J.S. Design and evaluation of a real time physiological signals acquisition system implemented in multi-operating rooms for anesthesia. *J. Med. Syst.* **2018**, *42*, 148. [CrossRef]

32. Abadi, M.; Barham, P.; Chen, J.; Chen, Z.; Davis, A.; Dean, J.; Devin, M.; Ghemawat, S.; Irving, G.; Isard, M.; et al. Tensorflow: A system for large-scale machine learning. In Proceedings of the 12th {USENIX} Symposium on Operating Systems Design and Implementation ({OSDI} 16), Savannah, GA, USA, 2–4 November 2016; pp. 265–283.

33. Kingma, D.P.; Ba, J. Adam: A method for stochastic optimization. *Arxiv Prepr.* **2014**, arXiv:1412.6980.

34. LeCun, Y.; Bottou, L.; Bengio, Y.; Haffner, P. Gradient-based learning applied to document recognition. *Proc. IEEE* **1998**, *86*, 2278–2324. [CrossRef]

35. Ronneberger, O.; Fischer, P.; Brox, T. U-net: Convolutional networks for biomedical image segmentation. In Proceedings of the 18th International Conference, Munich, Germany, 5–9 October 2015.

Article

Motion Artifact Reduction in Wearable Photoplethysmography Based on Multi-Channel Sensors with Multiple Wavelengths

Jongshill Lee [1,†], Minseong Kim [1,†], Hoon-Ki Park [2,*] and In Young Kim [1,*]

[1] Department of Biomedical Engineering, Hanyang University, Seoul 04763, Korea; netlee@hanyang.ac.kr (J.L.); minseong5905@hanyang.ac.kr (M.K.)
[2] Department of Family Medicine, Hanyang University, Seoul 04763, Korea
* Correspondence: hoonkp@hanyang.ac.kr (H.-K.P.); iykim@hanyang.ac.kr (I.Y.K.)
† These authors contributed equally to this work.

Received: 12 February 2020; Accepted: 7 March 2020; Published: 9 March 2020

Abstract: Photoplethysmography (PPG) is an easy and convenient method by which to measure heart rate (HR). However, PPG signals that optically measure volumetric changes in blood are not robust to motion artifacts. In this paper, we develop a PPG measuring system based on multi-channel sensors with multiple wavelengths and propose a motion artifact reduction algorithm using independent component analysis (ICA). We also propose a truncated singular value decomposition for 12-channel PPG signals, which contain direction and depth information measured using the developed multi-channel PPG measurement system. The performance of the proposed method is evaluated against the R-peaks of an electrocardiogram in terms of sensitivity (Se), positive predictive value (PPV), and failed detection rate (FDR). The experimental results show that Se, PPV, and FDR were 99%, 99.55%, and 0.45% for walking, 96.28%, 99.24%, and 0.77% for fast walking, and 82.49%, 99.83%, and 0.17% for running, respectively. The evaluation shows that the proposed method is effective in reducing errors in HR estimation from PPG signals with motion artifacts in intensive motion situations such as fast walking and running.

Keywords: photoplethysmography; motion artifact; independent component analysis; multi-wavelength

1. Introduction

Photoplethysmography (PPG) is an optical method used to detect volume changes in blood in the peripheral circulation. PPG can determine these volume changes from the surface of the skin, and is a low-cost and noninvasive method. This technique provides useful information related to cardiovascular systems such as heart rate, oxygen saturation, blood pressure [1], and cardiac output [2,3], and is also used to determine stress levels by analyzing the response of the autonomic nervous system based on pulse rate variability (PRV) [4].

Recently, wearable PPG sensors have attracted attention because they can continuously measure and monitor heart rate (HR), and numerous devices in the form of bands or watches (e.g., Apple watch, Fitbit, and Samsung Gear) are being used to monitor instantaneous heart rate using PPG. While these PPG-based devices have the advantages of being lightweight, portable, and easy to use, the distortion of the signal due to motion artifacts in the PPG signal is a challenge to overcome. Currently, these devices are only used for general wellness purposes because they are accurate only in limited conditions such as resting or walking slowly.

PPG uses a sensor composed of light-emitting and light-receiving elements. When light is irradiated to the body tissue by the light-emitting element, it is transmitted, reflected, and scattered by the Beer–Lambert law in the tissues, blood vessels, and blood of the body and detected by the light-receiving element [5,6]. In general, a green, red, or infrared light-emitting diode (LED) is used as a light-emitting unit, and a photo diode (PD) is used for a light-receiving unit [7]. LEDs used for PPG measurements generally have a wavelength of 400–1000 nm. Short wavelengths do not reveal much cardiac activity and blood vessel information due to low skin penetration depths, but are less affected by motion artifacts due to the shorter light path. In the case of long wavelengths, the penetration depth of the skin is deep, which can clearly indicate activity of the heart and blood vessels such as the dicrotic notch, but these are affected by motion artifacts because of the long light path [8].

The normal frequency range for PPG signals is 0.5 to 5 Hz, while for motion artifacts it is 0.01 to 10 Hz [9–11]. Therefore, it is not easy to obtain a clean signal by applying a general filter to a PPG signal contaminated by motion artifacts. In order to solve this problem, adaptive filters or moving average filters are commonly used in the industrial field. However, satisfactory performance in removing or reducing motion artifacts has not yet been achieved, and various signal processing methods for reducing motion artifacts of PPG signals have been proposed.

Poh M.-Z. et al. developed an earlobe-wearable PPG measuring device and presented a method of removing motion artifacts by applying adaptive noise cancellation (ANC) using an accelerometer [12]. The correlation between the heart rate calculated via electrocardiogram (ECG) and the heart rate measured via PPG was shown as a performance evaluation. The results showed a correlation coefficient of 0.75 ($p < 0.001$) when ANC was applied while running. Due to the size of the earlobe-wearable measurement system, it is difficult to apply in everyday life and it may be inconvenient when measuring because the attachment method uses neodymium magnets. In addition, the ANC method requires an additional sensor, such as an accelerometer, because it must provide a reference signal for motion artifacts.

In a recent study, Zhang Y. et al. proposed the use of optical signals rather than accelerometers as the motion reference for the cancellation of motion artifacts [13]. The proposed framework uses the infrared (IR) PPG signal as the motion reference and the green PPG signal for HR estimation. This approach helps to reduce burden on additional hardware such as accelerometers and the computational complexity.

Reddy K. A. et al. proposed the CFSA (Cycle-by-cycle Fourier Series Analysis) method using Fourier series analysis for each cycle using the autocorrelation of PPG signals [14]. The results show that randomly applied Gaussian noise is removed. However, due to the limitation that CFSA can only be applied to periodic signals, it is difficult to apply to situations where loss of periodicity occurs due to distortion caused by motion artifacts during the actual PPG measurement.

In order to improve the performance of the noise reduction algorithm, methods using a multi-channel PPG system have been proposed. Warren K. et al. measured six-channel PPG signals at the forehead using six red and infrared LEDs [15]. They proposed an algorithm that selects the channel with the least influence of noise by quantifying the amount of motion artifacts for each channel during exercise. As a result, it was possible to automatically select an accurate channel from the measured multi-channel PPG signals. However, since the algorithm does not include noise reduction, and selects the signal with the lowest noise level from the measured signals, it is difficult to apply when motion artifacts exist in all channels.

It is necessary to design a filter suitable for multiple channels, and various algorithms such as independent component analysis (ICA), principal component analysis (PCA), and singular value decomposition (SVD) have been actively studied [9,16–19]. PCA is used to find an orthogonal linear transformation that maximizes the variance of variables, whereas ICA is used to find the linear transformation of the basis vectors that are statistically independent and non-Gaussian. Unlike PCA, the major feature of ICA is that the basis vectors are neither orthogonal nor ranked in order. The PPG signal and the motion artifacts are independent components of the detected signal, so ICA or PCA can be used to separate the cleaned PPG signal from these artifacts. However, in most studies, the number of PPG channels used for verifying a multi-channel signal processing algorithm is limited, and the relationship between the multi-channel signal and the algorithm is not clear.

In this study, we develop a multi-channel PPG measurement to consider the effects of motion artifacts on the direction of the sensor module and the change of penetration depth in the skin according to the wavelength. Further, we propose a multi-channel motion artifact reduction algorithm based on the signals obtained through this system. Using a multi-channel PPG system, 12-channel PPG signals for three wavelengths are acquired in four directions (up, down, left, and right). We present a method by which to reduce motion artifacts through applying ICA and a truncated SVD to 12 channels of PPG signals. We extract the independent components using an ICA of three channels of PPG signals measured in each module, then select the most pulsatile components. Using PCA, the statistical method by which the basis vectors are ranked in order, we can obtain the cleaned PPG signal. PCA can be implemented with powerful, robust techniques such as singular value decomposition (SVD) [20]. Ultimately, we implement PCA with SVD.

2. Materials and Methods

Our research focuses on reducing motion artifacts using multi-channel PPGs. In this section, we introduce our in-house-built wearable multi-channel PPG measurement system and describe the proposed motion artifact reduction algorithm using a multi-sensor module. We also describe in detail the experimental protocol and data acquisition using the measurement system.

2.1. Wearable Multi-Channel PPG System

We developed a wearable multi-channel PPG system consisting of the main system, inertial measurement units (IMUs), and PPG sensors, as shown in Figure 1a. The sensor module connected to the main system has sensors arranged in four directions perpendicular to its center, with each sensor consisting of a green, red, and infrared LED and one PD (Figure 1b). In addition, the sensor module includes a nine-axis IMU to detect movement.

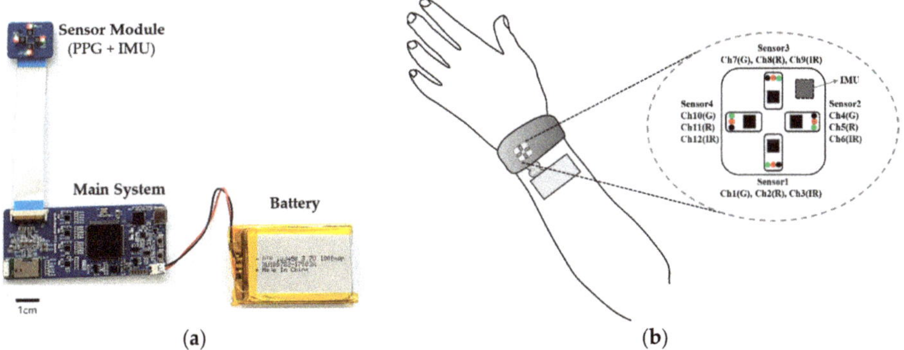

Figure 1. In-house-built wearable multi-channel PPG acquisition system: (**a**) wearable PPG hardware system consisting of a main system, sensor module (PPG and IMU), and battery; (**b**) the direction and wavelength (G: green; R: red; IR: infrared) of the sensor for each channel when the wearable system is worn on the wrist.

The system architecture of the in-house-built wearable multi-channel PPG measurement system is shown in Figure 2. The main system consists of an ARM Cortex TM-M4-based microcontroller (STM32F407VGT, STMicroelectronics, Geneva, Swiss), an analog front-end (AFE4900, Texas Instruments, Dallas, TX, USA), and a Bluetooth module (PAN1321i, Panasonic, Osaka, Japan). The sensor module was designed and implemented using four SFH7050 sensors (OSRAM, Munich, Germany) and one motion sensor (MPU9250, InvenSense, San Jose, CA, USA). SFH7050 is a sensor for heart rate monitoring or oximetry. It is an integrated sensor that contains three LEDs (green, red, and IR) of different wavelengths.

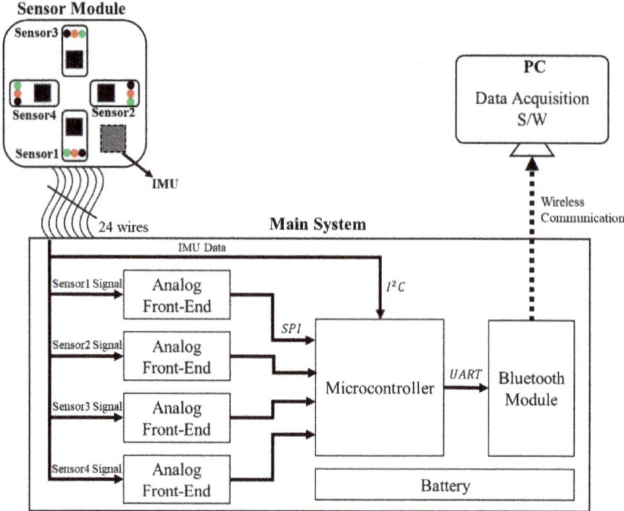

Figure 2. System diagram of the wearable multi-channel PPG acquisition system including PC application software (S/W). The solid line indicates a wired connection (USART (Universal Asynchronous Receiver/Transmitter), SPI (Serial Peripheral Interface) and I2C (Inter-Integrated Circuit)) while the dashed line indicates a wireless connection.

The readings of the PPG sensor are acquired via the analog front-end under the control of the microcontroller, then the amplified digital signals are sent to the microcontroller using serial peripheral interface (SPI) communication. Motion sensor data are transmitted to the microcontroller via I2C communication. The main system includes a Bluetooth module so that the data can be transferred to a PC via wireless communication. Additionally, we have implemented in-house data acquisition software for wireless transmission and storage to the PC.

The PPG signal of each channel was acquired at a 100-Hz sampling rate with a 24-bit high resolution, while the three-axis acceleration data were acquired at 16 bits and 100 Hz. In this paper, we use the vector sum magnitude of three-axis acceleration to determine the degree of motion.

2.2. Data Acquisition

The subjects were seven healthy males and one female without cardiovascular disease between the ages of 20 and 30 years (mean age: 27.1 years). This study was approved by Hanyang University IRB (IRB Approved no. HYI-17-048-4) and informed consent was received from all subjects before the experiment.

Since ECG is characterized by robustness to motion artifacts, most studies use ECG as the ground truth [21,22]. Therefore, in this study, the experiment was conducted using the polar chest strap electrode with high reliability in ECG measurement. In addition, the electrode was wetted with water before the experiment to minimize contact noise between the skin and the electrode. The R-peak could be easily extracted from the measured ECG through the Pan–Tompkins algorithm.

PPG and ECG were measured in walking, fast walking, and running environments that may occur in everyday life. The developed multi-channel PPG measurement system was worn on the subject's wrist to measure 12-channel PPG and acceleration signals. The ECG signal to be used as a reference for the heart rate was obtained at a sampling rate of 300 Hz using the developed prototypes of ECG acquisition systems [23] based on ADS1298 (Texas Instruments, Dallas, TX, USA) and the chest strap (Polar Pro Strap, Polar Electro, Ltd., Kempele, Finland). In order to minimize the time delay difference between the multi-channel PPG and ECG systems, two systems can simultaneously perform Analog-to-Digital Conversion (ADC) and transmit measured data based on trigger signals transmitted from the in-house program. Figure 3 shows a photograph of a subject wearing an ECG and multi-channel PPG measurement system. The ECG system for measuring the signal to be used as a reference for HR is mounted on the chest strap, and the main system of the multi-channel PPG measuring system is fixed with a stretchable band on the arm. The sensor module is also fixed with a wrist support band.

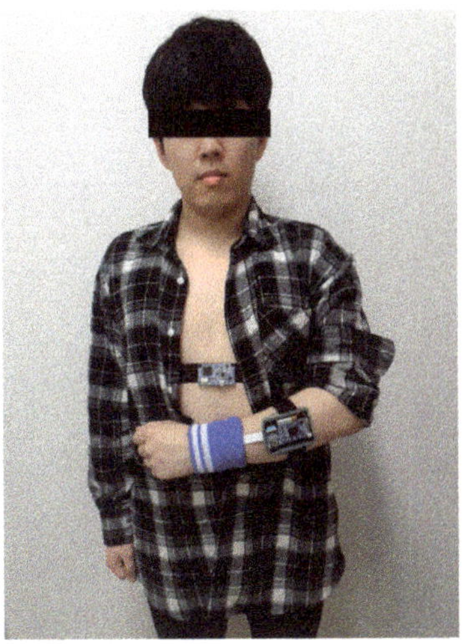

Figure 3. The picture of the subject being equipped with an ECG and multi-channel PPG measurement system.

Experimental protocols consisted of resting, walking, fast walking, and running sessions. The measured signals were normalized using PPG measured during the first 1-min resting session. In order to minimize the effect between sessions, each session had a 1-min rest period. In the walking (about 1 m/s) session, the subjects walked for 2 min at the typical pace of their daily activities. Next, 2 min of fast walking (about 1.8 m/s) and 2 min of running (about 2.2 m/s) were performed. The experiment was carried out by the subject reciprocating a distance of 20 m in the corridor of the building. Subjects were allowed to move along both sides close to the wall instead of the center of the corridor for a smooth turn of the subject at both turning points. In order to maintain the subject's speed at each session, the observer induced the subject to reach a fixed time at the start of the round trip. By conducting this protocol two times for each subject, a total of 16 datasets were obtained (one additional dataset whose ECG was damaged by the lead fault of the electrode was excluded). The number of beats for walking, fast walking, and running were 2757, 2913, and 3563, respectively, for a total of 9233 beats.

2.3. Motion Artifact Reduction Algorithm Based on Multi-Sensors

The penetration depth of radiation in human skin is known to increase with increasing wavelengths [24]. When measuring PPG in different directions and locations on the wrist, different signals are detected for the same movement due to the diversity of the directions and distribution of blood vessels in the skin of the wrist. Therefore, in the presence of motion artifacts caused by movement, multi-channel PPG signals measured by sensors with multiple wavelengths and in various locations contain information about blood vessel characteristics under the skin and movement in various directions and depths.

In this study, we extracted three independent component signals via preprocessing and independent component analysis for three channels of PPG signals with three different wavelengths (green: solid line arrow; red: dashed line arrow; infrared: dotted line arrow) for each of the four sensors, as shown in Figure 4. Among these signals, the signal with the component pulsating the most was selected, and a reconstructed PPG signal was obtained by applying the truncated SVD to a total of four pulsating component signals selected one by one.

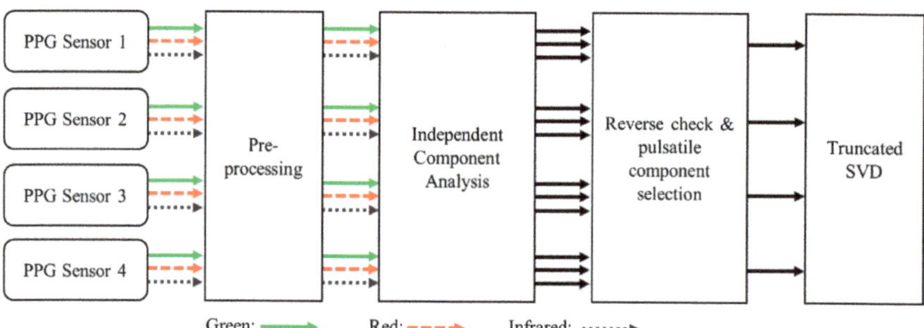

Figure 4. Block diagram of signal flows for each sensor and proposed algorithm.

Figure 5 shows the detailed proposed algorithm. The raw signals measured by the sensors contained various noise components such as power line interference, baseline drift, and ambient noise. In order to remove or reduce these noises, digital filtering was performed using a 3-order Butterworth band pass filter with cutoff frequencies of 0.5 and 5 Hz.

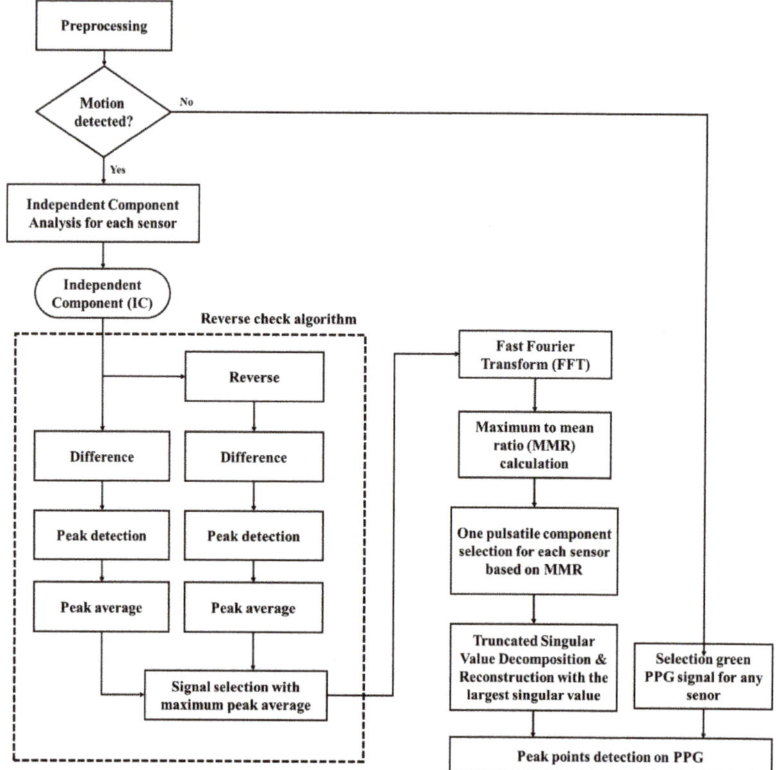

Figure 5. Flowchart of the proposed multi-sensor-based motion artifact reduction algorithm. (MMR is defined as Maximum to mean ratio).

In little or no motion environments, all channel signals can be measured with high quality. Therefore, heart rate can only be obtained by detecting peak points in PPG signals without any special processing. In this case, we used a signal with a green wavelength that is known to be measured well on the wrist. To quantify the motion, we used the three-axis accelerometer of the IMU mounted on the sensor module. High-pass filtering was performed to remove the gravity component from the acceleration, and the vector sum (acc_{sum}) was calculated for three axes of the gravity-free accelerometer, as shown in Equation (1). The threshold of the presence or absence of motion was set to the vector sum value to distinguish resting and walking.

$$acc_{sum} = \sqrt{(acc_x^2 + acc_y^2 + acc_z^2)} \tag{1}$$

In the presence of movement, PPG signals are a mixture of pulsation and motion artifact components. In this multi-PPG system, the PPG measured by sensors facing different directions showed the change of blood volume of various depths at each location, so the pulsatile component of each sensor can be extracted through ICA algorithms. However, because the output components of the ICA appear in random order and the signals can be reversed, a reverse-check and pulsatile component selection algorithm is required. The reverse-check algorithm (shown as a dotted box in Figure 5) performs differential and peak detection on each independent component (IC), calculates the average of the peak values, and applies the same process to the inverted signal. Due to the morphological characteristics of the PPG, the IC signal with the largest average peak value among the inverted and non-inverted signals becomes the correct PPG. Therefore, the IC with a large average value is the output.

The number of outputs of the ICA is the same as the number of input channels, and the output signals are randomly output regardless of the order of the input signals. The pulsation component is periodic because it is a change of blood volume that occurs with every beat of the heart. Therefore, by analyzing the periodicity for each IC signal, it is possible to select an IC with a pulsation component. We applied fast Fourier transform (FFT) to IC signals of each sensor module and selected one pulsatile component based on the maximum to mean ratio (MMR) in power spectral density.

As shown in Figure 4, the four components selected based on the MMR were pulsatile components measured at different locations. The truncated SVD [25] was applied to reduce the motion artifacts remaining on the pulsatile components extracted from four sensor modules with different directions, and to obtain the PPG signal that best reflected the change in blood volume. By applying the truncated SVD, the PPG is reflected in the largest singular value. Therefore, reconstruction using only this singular value yields a pure PPG signal with motion artifacts removed.

Figures 6–8 show an example of applying the proposed motion artifact reduction algorithm to a subject's data. Figure 6 shows the 12-channel PPG signals measured in the running state most affected by motion artifacts. The x-axis represents the time elapsed after 70 s, including 1 min of resting and 10 s of running.

Figures 7 and 8 show the results of applying the algorithm to each step shown in Figure 5. Figure 7 shows the 12 ICs after passing the ICA. Figure 8a shows four ICs selected through pulsatile component selection for each sensor module, and Figure 8b shows the results of applying the truncated SVD.

Figure 6 shows a 12-channel PPG signal measured from sensors placed in four directions while running at about 8 km/h. The four rows represent signals measured by the four sensor modules, and each column is a frequency-specific (green, red, and infrared) signal that can reflect information of the same depth. It can be seen that all channels are contaminated by motion artifacts. The R-peak time points extracted from the ECG are marked with the symbol (▼), and purple dotted lines represent the heart rate. The Pan–Tompkins algorithm was used to detect R-peaks from ECG data [26]. In this paper, all algorithms (including the motion artifact reduction algorithm) were implemented in Matlab 2019 (MathWorks Inc, Natick, MA, USA).

Figure 7 shows the 12 independent components (ICs) applying ICA to the 12-channel input PPG signals shown in Figure 6. This represents the same number of ICs as the number of input signals and includes pulsatile, motion artifacts, and other noise components for the signals acquired from each sensor. Each component appears in random order and some may appear as inverted signals. Each sensor module has three ICs for PPG signals from three wavelengths, which are the signals shown in each row of Figure 7.

In Figure 7, the most periodic signals for the IC signals and their inverted signals for each sensor module are IC1, IC1, and IC3 for the sensor modules 1, 2, and 3, respectively; the signals for sensor 4 are unclear. The results of applying the reverse-check and pulsatile component selection algorithm to select these pulsatile components are shown in Figure 8a. One pulsatile signal was selected for each sensor. In this example, the signals selected for each of the sensor modules 1, 2, and 3 are the reversed signals of IC1 and the reversed signal of IC1 and IC2. After applying the truncated SVD to the four selected ICs, reconstruction using only the largest singular value yielded a pure PPG signal with motion artifacts reduced, as shown in Figure 8b.

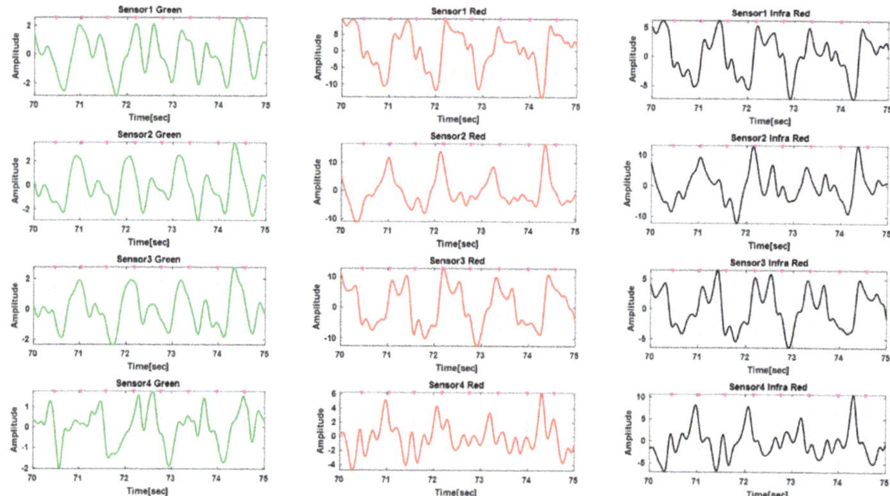

Figure 6. Twelve-channel PPG signals measured while running at about 8 km/h.

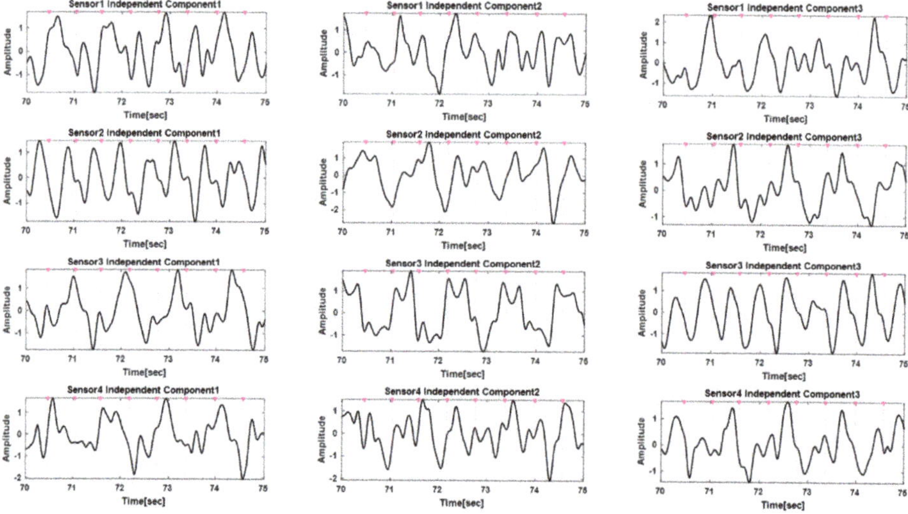

Figure 7. ICA components of 12-channel PPG signals.

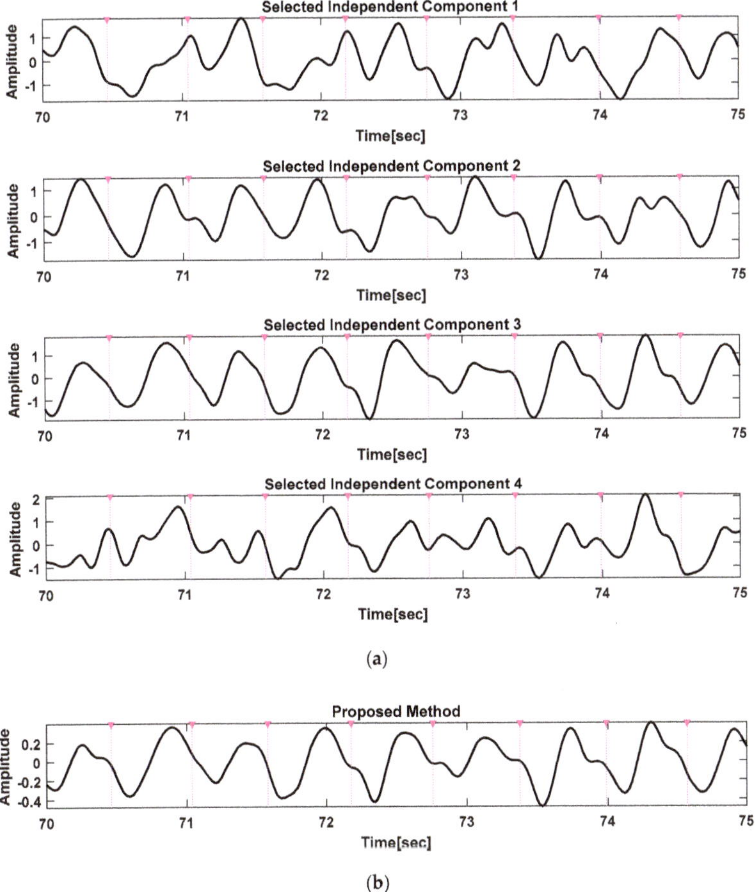

Figure 8. Selected independent components and motion artifact reduced PPG signals. (**a**) Selected ICs applying the reverse-check and pulsatile component selection algorithm, (**b**) reconstructed PPG signal based on the truncated SVD.

3. Experimental Results

In order to evaluate the proposed algorithm, the motion artifact reduction algorithm was applied to the signal measured by the multi-channel PPG measurement system, and the indexes for evaluating the performance of the algorithm were calculated by comparing the ECG with the reference signal. Sensitivity (Se) as Equation (2), positive predictive value (PPV) as Equation (3), and failed detection rate (FDR) as Equation (4) were used as performance indexes [27]. In these equations, TP (True Positive) is the number of peaks detected, FN (False Negative) is the number of peaks non-detected, and FP (False Positive) is the number of artifacts or noise classified as peaks. Algorithms based on the best signal selection [15], ICA [28], or SVD [25] were compared with the proposed method and evaluated using performance indexes.

$$Se = \frac{TP}{TP + FN} \times 100\% \tag{2}$$

$$PPV = \frac{TP}{TP + FP} \times 100\% \tag{3}$$

$$\text{FDR} = \frac{FP}{TP} \times 100\% \tag{4}$$

Figure 9 shows the results of applying the proposed method and algorithms based on the best signal selection, ICA, or SVD to the measured data at about 8 km/h. The R-peak time points extracted from the ECG as a reference for the heartbeat are indicated by the (▼) symbol and purple dotted lines. Compared with other methods, the proposed algorithm is clearest in terms of the heart rate extraction of PPG signals and had a low FDR.

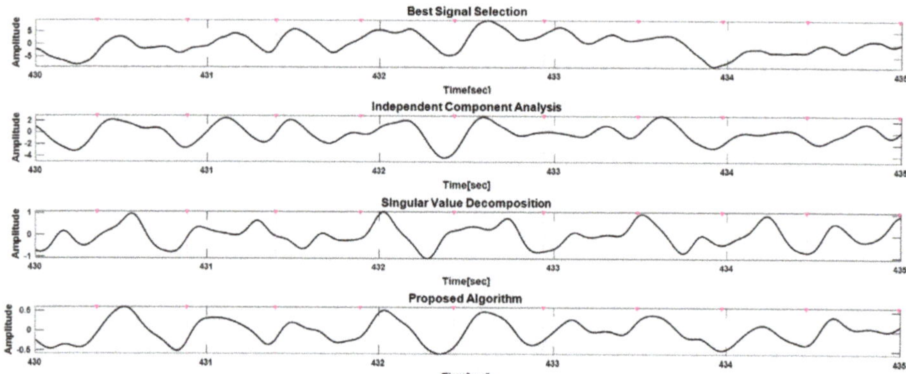

Figure 9. Example of applying the proposed method and algorithms based on best signal selection, ICA, and SVD during running at about 8 km/h.

Table 1 shows Se, PPV, and FDR as performance indexes for peak detection on 12-channel data measured during walking, fast walking, and running. As the intensity of the motion increases, Se decreases and PPV changes less. From the results, it can be seen that as the motion artifacts became more severe, the false negative (FN) for beat detection increased a lot while the false positive (FP) tended to fall slightly. In terms of wavelength, the green wavelength channels Ch1, Ch4, Ch7, and Ch10 had high sensitivity.

Table 1. Comparison of the results of Se, PPV, and FDR for the 12-channel signals acquired during walking, fast walking, and running conditions.

Conditions	Performance Parameters	Channels					
		Ch1	Ch2	Ch3	Ch4	Ch5	Ch6
	Se (%)	93.66	91.14	92.22	97.07	88.33	91.27
	PPV (%)	97.63	92.19	94.03	99.10	92.77	93.99
	FDR (%)	2.43	8.47	6.35	0.91	7.79	6.39
Walking		Ch7	Ch8	Ch9	Ch10	Ch11	Ch12
	Se (%)	96.59	88.46	90.36	96.48	90.39	92.10
	PPV (%)	99.47	92.96	95.96	98.30	92.68	94.80
	FDR (%)	0.53	7.57	4.21	1.73	7.90	5.49
		Ch1	Ch2	Ch3	Ch4	Ch5	Ch6
	Se (%)	84.17	80.47	82.80	93.69	77.07	79.73
Fast walking	PPV (%)	98.23	94.10	95.70	99.08	94.07	95.48
	FDR (%)	1.80	6.27	4.49	0.93	6.30	4.73
		Ch7	Ch8	Ch9	Ch10	Ch11	Ch12
	Se (%)	91.03	78.95	82.74	90.05	80.50	84.64
	PPV (%)	99.43	93.60	95.21	98.58	94.29	95.04
	FDR (%)	0.57	6.84	5.03	1.44	6.06	5.22

Table 1. *Cont.*

Conditions	Performance Parameters	Channels					
		Ch1	Ch2	Ch3	Ch4	Ch5	Ch6
Running	Se (%)	72.48	67.67	68.68	76.53	66.54	68.21
	PPV (%)	99.42	99.12	99.41	99.60	99.43	99.17
	FDR (%)	0.58	0.89	0.59	0.40	0.57	0.84
		Ch7	Ch8	Ch9	Ch10	Ch11	Ch12
	Se (%)	77.28	67.53	69.36	76.60	66.14	66.94
	PPV (%)	99.57	99.05	99.29	99.69	98.75	99.04
	FDR (%)	0.43	0.96	0.72	0.31	1.27	0.97

Table 2 shows the performance parameters Se, PPV, and FDR from algorithms based on the best selection method, SVD, ICA, and proposed method for the data measured in walking (2757 beats), fast walking (2913 beats), and running (3563 beats). As the movement increased with walking, fast walking, and running, the detection rate of heart beats decreased. The performance of the proposed method for walking was 99%, 99.55%, and 0.45% for Se, PPV, and FDR, respectively. The performance of the proposed method was higher than that of the other algorithms, and the SVD-based algorithm had the lowest accuracy. For the fast walking and running conditions, the proposed method showed the best performance, as well as for these two conditions with a lot of motion.

Table 2. Comparison of the results of Se, PPV, and FDR from algorithms based on the best signal selection method, SVD, ICA, or the proposed method for signals acquired under the walking, fast walking, and running conditions.

Conditions	Performance	Best Signal Selection	SVD	ICA	Proposed
Walking	Se (%)	97.07	91.30	98.65	99.00
	PPV (%)	99.10	92.89	99.03	99.55
	FDR (%)	0.91	7.65	0.98	0.45
Fast walking	Se (%)	93.69	82.33	95.42	96.28
	PPV (%)	99.08	93.71	99.06	99.24
	FDR (%)	0.93	6.71	0.95	0.77
Running	Se (%)	77.28	68.00	79.98	82.49
	PPV (%)	99.57	99.09	99.81	99.83
	FDR (%)	0.43	0.92	0.19	0.17

4. Discussion

ECG is a representative signal for calculating heart rate that measures the bio-potential generated by electrical signals that control the expansion and contraction of the heart. Another signal is that of PPG, a light-based technology to sense volumetric changes in blood in peripheral circulation as controlled by the heart's pumping action. ECG produces an electrical signal that is robust in the presence of motion artifacts and has the advantage of stably extracting heartbeats even compromised by motion. However, ECG signals are obtained by measuring a weak electrical potential difference between two points, and thus cannot be measured on a single arm. In the case of using both hands with more electrical potential difference, the user must intentionally make contact with the electrode. In contrast, PPG has an advantage over ECG in terms of user convenience and wearability, as it can take measurements in any location with a high concentration of blood vessels.

Over the past decade, there HR monitors and wearable fitness equipment have been made commercially available. Many people use HR monitors to inform their training and to access aerobic fitness.

In the clinical field, physicians and trainers often refer to physiological and behavioral data such as energy consumption, step counts, sleep/wake information, and HR obtained from patients' wearable devices. Energy consumption is calculated using HR and accelerometer motion information. If accurate HR information is provided, more accurate energy consumption can be estimated. Accurate HR monitoring is an essential component of a systematic exercise prescription because a target HR is set to guide patient-specific exercise intensity.

As another field of the clinical application of HR, heart rate variability (HRV) is widely used to investigate the state of autonomic nervous systems and related diseases. HRV is calculated using beat-by-beat HRs, and accurate HR detection is required because HR errors lead to false ANS (Autonomic nervous system) analysis.

Currently, many devices are being used to monitor heart rate in the form of bands or watches (e.g., Apple watch, Fitbit, and Samsung Gear), and these PPG devices have the advantages of being lightweight, portable, and easy to use. However, the PPG signal is very vulnerable to motion artifacts. It is well known that the depth of penetration of light into human skin increases with decreasing wavelength [29]. Therefore, the wavelength of the LED and the direction of the PPG sensor module are important factors for analyzing the influence of motion artifacts in the PPG signal. In this study, we developed a 12-channel PPG measurement system with up, down, left, and right directions using three wavelengths (green, red, and infrared) for each direction. In addition, we proposed a multi-channel PPG motion artifact reduction algorithm based on independent component analysis and truncated singular value decomposition. In this study, PPG signals were measured at multiple skin penetration depths according to LED wavelength. In addition, PPG signals for each position and direction of blood vessels in the skin were measured through sensor modules with up, down, left and right directions. In order to apply independent component analysis to multi-channel PPG signals, signal inversion and pulsation component detection ability were confirmed by comparing MMR in the power spectral density through FFT. Furthermore, truncated singular value decomposition was applied to the detected pulsating components to reduce motion artifacts. When the proposed algorithm was applied to the signal measured during running using the developed system, a sensitivity of 82.49%, a positive predictive value of 99.83%, and a false detection rate of 0.17% were obtained. These results are due to the motion artifact reduction algorithm applied to the multi-channel PPG measurement system using various wavelengths and directions of the sensor modules. As the wavelengths increased, the change in blood volume at a deeper depth could be measured. Therefore, each PPG has a different amount of information, and when applied to the ICA algorithm, a pulsatile component can be obtained. In addition, the pulsatile components acquired from the sensor modules in different directions are represented as a single PPG signal that shows the change in blood volume without being affected by motion through the truncated SVD.

The proposed algorithm extracts the independent component signals from PPG signals with different wavelengths measured for each sensor module, as shown in Figure 4. In addition, in order to compare the effect of the wavelength and the direction of the sensor, we applied the proposed algorithm for PPG signals with the same wavelength in each senor module. This approach did not provide clear results as compared with the use of ICA-based algorithms for each sensor module. This means that signals measured by LEDs of the same wavelength in each sensor are measured through different positions and paths of light, so that information on motion artifacts is inconsistent when applying independent component analysis. Therefore, applying independent component analysis to the signals measured by the same sensor module rather than the same wavelength signals can increase the motion artifact reduction performance.

The limitation of this study is that the protocols for PPG measurement were limited to walking, fast walking, and running; therefore, more detailed protocols need to be set up to remove the noise caused by various kinds of movements in real life. Additional experiments and research are needed to develop real-world applications. As further work, we plan to quantitatively analyze the effects of motion artifacts on PPG signals at various speeds on the treadmill. In addition, long-term measurements are required to analyze the various features in the signal measured during physical activity. We will try to diversify the protocols and measure data in various environments through long-term experiments.

The proposed method is effective for monitoring heart rate. Furthermore, if the beat-to-beat interval can be obtained precisely like the R-peak of ECG in the presence of motion artifacts, it can be applied to various fields, such as emotion or stress, by analyzing the autonomic nervous system based on pulse rate variability.

Author Contributions: Conceptualization, J.L. and M.K.; methodology, J.L. and M.K.; formal analysis, J.L. and M.K.; writing—original draft preparation, J.L. and M.K.; writing—review and editing, I.Y.K. and H.-K.P.; supervision, I.Y.K. and H.-K.P.; funding acquisition, H.-K.P. All authors have read and agreed to the published version of the manuscript.

Funding: This work was supported by the National Research Foundation of Korea (NRF) grant funded by the Korea government (MSIT) (No. 2018R1A5A7025522).

Conflicts of Interest: The authors declare no conflict of interest.

References

1. Sharma, M.; Barbosa, K.; Ho, V.; Griggs, D.; Ghirmai, T.; Krishnan, S.K.; Hsiai, T.K.; Chiao, J.C.; Cao, H. Cuff-Less and Continuous Blood Pressure Monitoring: A Methodological Review. *Technologies* **2017**, *5*, 21. [CrossRef]

2. Wang, L.; Pickwell-Macpherson, E.; Liang, Y.P.; Zhang, Y.T. Noninvasive cardiac output estimation using a novel photoplethysmogram index. *Conf. Proc. IEEE Eng. Med. Biol. Soc.* **2009**, *2009*, 1746–1749. [PubMed]

3. Wang, L.; Poon, C.C.Y.; Zhang, Y.T. The non-invasive and continuous estimation of cardiac output using a photoplethysmogram and electrocardiogram during incremental exercise. *Physiol. Meas.* **2010**, *31*, 715–726. [CrossRef] [PubMed]

4. Mohan, P.M.; Nagarajan, V.; Das, S.R. Stress Measurement from Wearable Photoplethysmographic Sensor Using Heart Rate Variability Data. In Proceedings of the 2016 International Conference on Communication and Signal Processing (ICCSP), Melmaruvathur, India, 6–8 April 2016; pp. 1141–1144.

5. Karlen, W.; Raman, S.; Ansermino, J.M.; Dumont, G.A. Multiparameter respiratory rate estimation from the photoplethysmogram. *IEEE Trans. Bio-Med. Eng.* **2013**, *60*, 1946–1953. [CrossRef]

6. Paul, B.; Manuel, M.P.; Alex, Z.C. Design and Development of Non Invasive Glucose Measurement System. In Proceedings of the 2012 1st International Symposium on Physics and Technology of Sensors (ISPTS-1), Pune, India, 7–10 March 2012; pp. 43–46.

7. Boulnois, J.-L. Photophysical processes in recent medical laser developments: A review. *Lasers Med. Sci.* **1986**, *1*, 47–66. [CrossRef]

8. Clarke, G.W.; Chan, A.D.; Adler, A. Effects of motion artifact on the blood oxygen saturation estimate in pulse oximetry. In Proceedings of the 2014 IEEE International Symposium on Medical Measurements and Applications (MeMeA), Lisbon, Portugal, 11–12 June 2014; pp. 1–4.

9. Rojano, J.F.; Isaza, C.V. Singular value decomposition of the time-frequency distribution of PPG signals for motion artifact reduction. *Int. J. Signal Process. Syst.* **2016**, *4*, 475–482. [CrossRef]

10. Bagha, S.; Shaw, L. A real time analysis of PPG signal for measurement of SpO2 and pulse rate. *Int. J. Comput. Appl.* **2011**, *36*, 45–50.

11. Park, H.; Nam, J.; Lee, J. Design of filter to reject motion artifacts of PPG signal using multiwave optical source. *J. Korea Soc. Comput. Inf.* **2014**, *19*, 101–107.

12. Poh, M.-Z.; Swenson, N.C.; Picard, R.W. Motion-tolerant magnetic earring sensor and wireless earpiece for wearable photoplethysmography. *IEEE Trans. Inf. Technol. Biomed.* **2010**, *14*, 786–794. [CrossRef] [PubMed]

13. Zhang, Y.; Song, S.; Vullings, R.; Biswas, D.; Simões-Capela, N.; Van Helleputte, N.; Van Hoof, C.; Groenendaal, W. Motion artifact reduction for wrist-worn photoplethysmograph sensors based on different wavelengths. *Sensors* **2019**, *19*, 673. [CrossRef]

14. Reddy, K.A.; George, B.; Kumar, V.J. Use of fourier series analysis for motion artifact reduction and data compression of photoplethysmographic signals. *IEEE Trans. Instrum. Meas.* **2008**, *58*, 1706–1711. [CrossRef]

15. Warren, K.; Harvey, J.; Chon, K.; Mendelson, Y. Improving pulse rate measurements during random motion using a wearable multichannel reflectance photoplethysmograph. *Sensors* **2016**, *16*, 342. [CrossRef]

16. Ram, M.R.; Madhav, K.V.; Krishna, E.H.; Reddy, K.N.; Reddy, K.A. Use of multi-scale principal component analysis for motion artifact reduction of PPG signals. In Proceedings of the 2011 IEEE Recent Advances in Intelligent Computational Systems, Trivandrum, India, 22–24 September 2011; pp. 425–430.

17. Ram, M.R.; Madhav, K.V.; Krishna, E.H.; Komalla, N.R.; Sivani, K.; Reddy, K.A. ICA-based improved DTCWT technique for MA reduction in PPG signals with restored respiratory information. *IEEE Trans. Instrum. Meas.* **2013**, *62*, 2639–2651. [CrossRef]

18. Lee, H.-W.; Lee, J.-W.; Jung, W.-G.; Lee, G.-K. The periodic moving average filter for removing motion artifacts from PPG signals. *Int. J. Control Autom. Syst.* **2007**, *5*, 701–706.

19. Ram, M.R.; Madhav, K.V.; Krishna, E.H.; Komalla, N.R.; Reddy, K.A. A Novel Approach for Motion Artifact Reduction in PPG Signals Based on AS-LMS Adaptive Filter. *IEEE Trans. Instrum. Meas.* **2012**, *61*, 1445–1457. [CrossRef]

20. Chawla, M. PCA and ICA processing methods for removal of artifacts and noise in electrocardiograms: A survey and comparison. *Appl. Soft Comput.* **2011**, *11*, 2216–2226. [CrossRef]

21. Gilgen-Ammann, R.; Schweizer, T.; Wyss, T. RR interval signal quality of a heart rate monitor and an ECG Holter at rest and during exercise. *Eur. J. Appl. Physiol.* **2019**, *119*, 1525–1532. [CrossRef]

22. Vogelaere, P.; De Meyer, F.; Duquet, W.; Vandevelde, P. Sport Tester PE 3000 vs Holter ECG for the measurement of heart rate frequency. *Sci. Sports* **1986**, *1*, 321–329. [CrossRef]

23. Ahn, H.J.; You, S.M.; Cho, K.; Park, H.K.; Kim, I.Y. Multi-modal Wearable Device for Cardiac Arrest Detection. *J. Biomed. Eng. Res.* **2017**, *38*, 330–335.

24. Spigulis, J.; Gailite, L.; Lihachev, A.; Erts, R. Simultaneous recording of skin blood pulsations at different vascular depths by multiwavelength photoplethysmography. *Appl. Opt.* **2007**, *46*, 1754–1759. [CrossRef]

25. Lee, H.; Chung, H.; Ko, H.; Lee, J. Wearable multichannel photoplethysmography framework for heart rate monitoring during intensive exercise. *IEEE Sens. J.* **2018**, *18*, 2983–2993. [CrossRef]

26. Pan, J.; Tompkins, W.J. A real-time QRS detection algorithm. *IEEE Trans. Bio-Med. Eng.* **1985**, 230–236. [CrossRef] [PubMed]

27. Lin, S.-T.; Chen, W.-H.; Lin, Y.-H. A pulse rate detection method for mouse application based on multi-PPG sensors. *Sensors* **2017**, *17*, 1628.

28. Kim, B.S.; Yoo, S.K. Motion artifact reduction in photoplethysmography using independent component analysis. *IEEE Trans. Bio-Med. Eng.* **2006**, *53*, 566–568. [CrossRef] [PubMed]

29. Vizbara, V. Comparison of green, blue and infrared light in wrist and forehead photoplethysmography. *Biomed. Eng.* **2013**, *17*, 78–81.

Article

Retrieval and Timing Performance of Chewing-Based Eating Event Detection in Wearable Sensors

Rui Zhang * and Oliver Amft

Chair of Digital Health, Friedrich-Alexander Universität Erlangen-Nürnberg (FAU), Henkestraße 91, 91052 Erlangen, Germany; oliver.amft@fau.de
* Correspondence: rui.rui.zhang@fau.de; Tel.: +49-9131-852-3604

Received: 20 December 2019; Accepted: 17 January 2020; Published: 20 January 2020

Abstract: We present an eating detection algorithm for wearable sensors based on first detecting chewing cycles and subsequently estimating eating phases. We term the corresponding algorithm class as a bottom-up approach. We evaluated the algorithm using electromyographic (EMG) recordings from diet-monitoring eyeglasses in free-living and compared the bottom-up approach against two top-down algorithms. We show that the F1 score was no longer the primary relevant evaluation metric when retrieval rates exceeded approx. 90%. Instead, detection timing errors provided more important insight into detection performance. In 122 hours of free-living EMG data from 10 participants, a total of 44 eating occasions were detected, with a maximum F1 score of 99.2%. Average detection timing errors of the bottom-up algorithm were 2.4 ± 0.4 s and 4.3 ± 0.4 s for the start and end of eating occasions, respectively. Our bottom-up algorithm has the potential to work with different wearable sensors that provide chewing cycle data. We suggest that the research community report timing errors (e.g., using the metrics described in this work).

Keywords: automated dietary monitoring; eating detection; eating timing error analysis; biomedical signal processing; smart eyeglasses; wearable health monitoring

---

1. Introduction

Eating occasion detection is at the core of automated dietary monitoring (ADM) in humans, targeting healthy diet management [1,2]. We regard intake to consume food pieces with dietary activities including ingestion, chewing, and swallowing [3] as an eating occasion if all dietary activities start and end in a given temporal relation. Meals or snacks are typical examples of eating occasions. Eating occasions thus have a start and end denoting the timing of intake beginning and intake completion. For solid and semi-solid food, chewing (i.e., the cyclic opening and closing of the jaw) is typically the longest activity within eating occasions [3]. We therefore consider chewing as representative of eating occasions, denoted as *eating events* in this work.

Recording chewing to interpret eating has been attempted in a variety of approaches intended for free-living ADM (see Section 2), as accurate eating event timing detection is essential for diet management. For example, users could be reminded to check vital parameters such as glucose level when the initial moment of an eating event is detected. Similarly, users could be asked to confirm food details or take a photo of leftovers immediately after an eating event ends. In both examples it is important that timing errors of the eating event detection are minimal. Hence, timing errors determine whether an eating event detection approach is suitable across the ADM application spectrum.

Detecting dietary activities, including eating events, in wearable or ambient sensor data is a complex pattern analysis and modelling problem due to the inter- and intra-individual variability in free-living behaviour patterns. Approaches to eating event detection and analysis can be categorised as top-down or bottom-up sensor data processing: In the top-down approach, eating events are detected by applying sliding windows to the sensor time series and applying feature pattern models. If necessary, further information details such as chewing cycles, intake gestures, etc. could be derived using the detected eating events. Conversely, in a bottom-up approach, individual dietary activities are modelled and the result is subsequently used to detect eating events. The early abstraction in bottom-up processing may help to deal with varying dietary activity patterns. Furthermore, bottom-up processing fits into hierarchical data processing schemes of resource-constrained wearable and IoT systems, where instead of raw data, derived parameters or events are communicated between system components.

This investigation proposes a bottom-up eating detection algorithm and compares it with two top-down algorithms. The bottom-up eating detection algorithm first detects individual chewing cycles. Retrieved chewing cycles are then used to detect eating events and estimate start and end of eating occasions. In contrast, top-down algorithms apply sliding windows over the sensor time series to detect eating events. The bottom-up algorithm proposed here is potentially agnostic to the particular sensor used, as long as chewing cycle information is acquired. In particular, the following contributions are made:

1. We present a bottom-up algorithm for eating event detection based on chewing time-series data. The algorithm works based on chewing cycle information and has only four parameters.
2. We evaluate and compare bottom-up and top-down eating event detection algorithms in data of a free-living study, where participants continuously wore unobtrusive diet monitoring eyeglasses. The diet eyeglasses recorded electromyographic (EMG) data of the temporalis muscles. We analysed retrieval performance as well as start and end timing errors of detected eating events.
3. We describe and analyse a procedure to derive eating event reference data in a free-living context. Our approach combines participant self-reports with a mostly unobtrusive chewing reference measurement. The analysis confirms that our reference estimation approach reached a timing resolution of less than one second in free-living behaviour data.

2. Related Work

ADM has received increasing research interest over the last decade, where eating event detection based on data from various body-worn and ambient sensors has been frequently considered. Most investigations that considered quantitative performance for eating event detection focused on detection accuracy or retrieval metrics. In this investigation, we highlight that timing errors are critical for detection performance and investigate timing errors specifically.

Eating event detection has often been approached by top-down data processing. For example, Dong et al. used a wrist motion sensor to detect eating, reporting 81% accuracy in 449 hours of free-living data [4]. Thomaz et al. also used a wrist-worn three-axis accelerometer to monitor eating in free-living conditions [5]. The random forest classifier yielded 66% precision and 88% recall for one day of data and intra-individual analysis. Bi et al. implemented a headband carrying a bone-conducting acoustic sensor and reported eating detection performance of over 90% [6]. Farooq et al. used accelerometer-equipped eyeglasses to detect food intake in the lab and in short-term free-living [7]. The highest F1 score of 87.9% $\pm$ 13.8% (mean $\pm$ standard deviation) was achieved with a 20 s sliding window using a k-nearest neighbour classifier. Studies involving multiple sensor modalities are a recent trend in eating event detection applications. Wahl et al. implemented an eyeglasses prototype equipped with an inertial measurement unit (IMU), an ambient light sensor, and a photoplethysmogram (PPG)sensor for the recognition of nine daily activities, including eating [8]. The classification reached an average accuracy of 77%. Merck et al. realised a multi-device monitoring system involving in-ear audio, head motion, and wrist motion sensors, which could recognise eating with 92% precision and 89% recall [9]. Papapanagiotou et al. proposed an ear-worn eating monitoring system based on PPG, audio and accelerometer, achieving an accuracy up to 93.8% and class-weighted accuracy up to 89.2% in eating detection [10]. Bedri et al. used an ear-worn system for chewing instance detection. An F1 score of over 80% and accuracy of over 93% was reported [11]. Timing error for eating start was 65.4 s. The authors did not report the timing error at eating ends. Doulah et al. investigated the effect of the temporalresolution of eating microstructure analysis, including the duration of eating events [12]. The analysis did not yield insight into start and end time estimates for eating events. In our prior investigation of top-down eating detection based on free-living EMG recordings, a one-class support vector machine (ocSVM) yielded an F1 score of 95%. Timing error analysis showed 21.8 ± 29.9 s for eating start and 14.7 ± 7.1 s for eating end [13].

For the bottom-up data processing approach, dietary activities that characterise eating are modelled, and eating is subsequently derived from these activities. Chewing has frequently been investigated as a basis for subsequent eating analysis. Amft et al. investigated chewing detection for ADM using an ear-plug acoustic sensor, capturing vibration patterns during chewing [14]. Bedri et al. proposed earwear using proximity sensors for the detection of tiny deformations of the outer ear during chewing [15]. Eating could be detected with 95.3% accuracy with a user-dependent classification. Zhang et al. was the first to use smart eyeglasses to detect chewing, analysing EMG electrode positions in eyeglasses frames and the effect of hair on the EMG signal [16]. EMG electrodes were embedded into the eyeglasses' temples, and chewing cycles were detected with a precision and recall of 80%. In subsequent work [17], a refined version of the eyeglasses was used for eating detection, yielding an accuracy of above 95% in natural, free-living data. Furthermore, it was demonstrated that soft foods such as banana provide identifiable EMG signatures. Chung et al. incorporated a force-sensitive load cell in eyeglasses hinges to monitor temple movement during chewing, head movement, talking, and winking. A classification of these activities yielded an F1 score of 94% [18]. Farooq et al. attached a strain sensor at the temporalis muscle area to obtain chewing cycle information [19]. With additional accelerometer data, the authors reported an F1 score of 99.85% for recognising eating from other physical activities in laboratory recordings.

So far, timing performance has been rarely reported, partly because methods to derive eating reference in free-living studies were missing. Here, we evaluated three algorithms in free-living EMG recordings with a realistic ratio of eating vs. non-eating time. All algorithms can be used with one or more sensors and in multimodal configurations. In particular, the bottom-up algorithm builds on chewing cycle information extracted from sensor data, and thus can be applied with other sensors besides EMG by adapting the chewing cycle extraction. Our current work focuses in particular on the analysis of timing errors.

3. Eating Event Detection Algorithms

We propose a bottom-up eating event detection algorithm and compare it to two top-down algorithms. As input for all algorithms we consider a multi-source sensor data stream of chewing cycle measurements, corresponding to a random process $X_n(t)$, where n indexes the random variables (e.g., sensor channels or features) and t is the time index. For example, the sensor could be an EMG monitor measuring the temporalis muscle contraction or acoustic transducers measuring vibration patterns due to food fracture. An overview of the algorithm pipelines for all algorithms considered is shown in Figure 1. Below, we formally describe the algorithms.

3.1. Bottom-Up Algorithm

The idea of this algorithm is to estimate eating events from the density of chewing cycles, where a relatively high frequency of chewing cycles indicates eating. After pre-processing multi-source sensor signals $X_n(t)$, chewing cycle onsets C_n were detected. Subsequently, a sliding window of length w_0, was applied around each retrieved onset of C_n (i.e., with a step size of one onset). Then, the sliding window moved to the next detected onset. At every onset, we calculated chewing cycle frequency f_n as the number of detected onsets per time interval w_0. A chewing segment start $t_{n,\,start}$ was detected as the first onset in C_n at the signal start or an onset after a preceding detected chewing segment, where f_n equalled or exceeded θ_0. The end of a chewing segment $t_{n,\,end}$ was determined as the onset in C_n where f_n equalled θ_0 and the $(\theta_0 - 1)$-th subsequent f_n equalled 1. Detection results of n sensor sources were combined and post-processed by eliminating gaps between adjacent groups of chewing segments. The details of each step are described below.

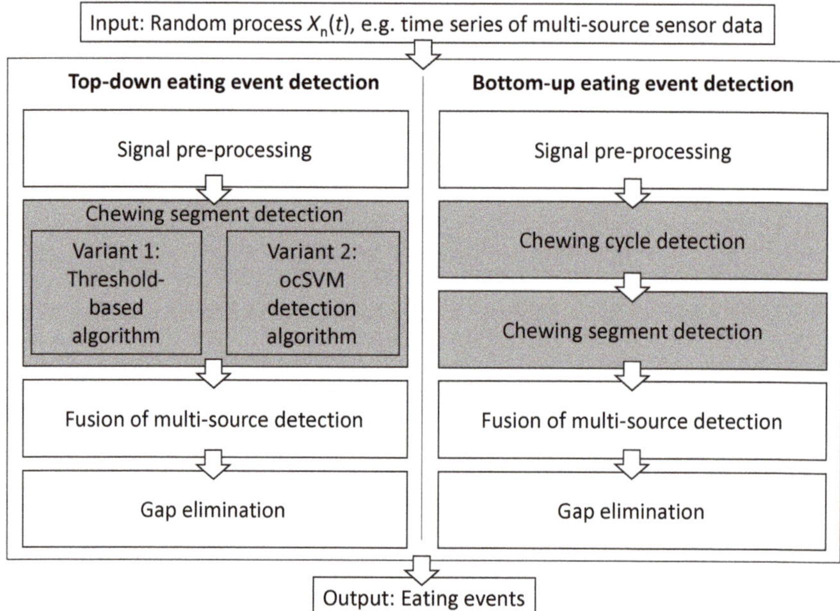

Figure 1. Overview of the top-down and bottom-up eating event detection algorithms investigated in this work. White processing blocks indicate functions shared by the algorithms. Shaded processing blocks are specific functions for each algorithm. Both top-down algorithms follow the same detection pipeline with different implementations of the "Chewing segment detection" block. ocSVM: one-class support vector machine.

3.1.1. Signal Pre-Processing

Pre-processing steps vary depending on the type of sensors used. It is likely that the human body acts as an antenna and picks up power line noise. Thus, we applied a notch filter to raw signal $X_n(t)$ to eliminate potential power line interference at frequency f_{nf}. In this study, we used dual-channel smart eyeglasses EMG data sampled at 256 Hz per channel. Hence, $X_n(t)(n = 1, 2)$ represents EMG data in this case. The notch filter frequency was set to $f_{nf} = 50$ Hz. Baseline wander and motion artifacts were removed using a high-pass filter with a cut-off frequency of $f_{hpf} = 20$ Hz—a typical value for EMG signal processing. The resulting data $X_{n,\,hpf}$ were rectified for detection. The pre-processed and rectified data were abbreviated as X_n. The pseudo code is in Algorithm block 1.

Algorithm block 1 : Signal pre-processing.

Input: Multi-source free-living data $X_n(t)$

Parameter: Notch filter band-stop frequency f_{nf}, high-pass filter cut-off frequency f_{hpf}

Output: Pre-processed data X_n

1: $X_{n,nf} = \text{NotchFilt}(X_n(t),\ f_{nf})$

2: $X_{n,hpf} = \text{HighPassFilt}(X_{n,\,nf},\ f_{hpf})$

3: $X_n = |X_{n,\,hpf}|$

3.1.2. Chewing Cycle Detection

Chewing cycle detection was performed by adapting the EMG onset detection principle initially proposed by Abbink et al. [20]. Every chewing cycle has an onset time corresponding to the moment when the muscle contraction starts, and an offset time corresponding to the contraction end. Hence, the number of onsets should represent the number of chewing cycles. First, a sliding window of size w was applied to X_n. The value of w should be no larger than the duration of a typical chewing cycle. Here we used $0.4\,$s (100 samples for the EMG signal), as chewing cycle frequency typically ranges between 0.94 and $2.17\,$Hz [21]. We derived a conditional summation of sensor samples within the window: For samples 0 to $w/2$ within the current window starting at i_0, we derived $index_1 = \sum_{i=0}^{w/2} 1$ if $X_n[i_0 + i] < \theta_C$. For samples in the second half-window, $index_2 = \sum_{i=w/2+1}^{w} 1$ if $X_n[i_0 + i] > \theta_C$ was summed. Finally, $index = index_1 + index_2$ was derived. Parameter θ_C was set to $\mu + 3 \times \sigma$, where μ was the mean and σ the standard deviation derived from baseline noise of X_n. Both μ and σ were estimated across training data of all participants. The amplitude of the baseline noise was assumed to be Gaussian distributed and threshold θ_C was set to cover 99% of the confidence interval. As the window with size w was slid with a step size of one sample, an $index$ in range $[0, w]$ was obtained for each sample, forming a new time series I_n per signal source n. To determine chewing onsets, we derived points of I_n that exceeded $\theta_P \times w$, with θ_P in the range $[0, 1]$. Considering the chewing frequency, the temporal distance between neighbouring detection points of I_n should be larger than $t_{\text{interval}} = 1/3\,$s. Detected chewing cycle onsets were sequentially saved in a list C_n. The pseudo code is shown in Algorithm block 2.

Algorithm block 2 : Chewing cycle detection.

Input: Pre-processed data X_n

Parameter: EMG burst threshold θ_C, sliding window size w, peak threshold θ_P, peak interval t_{interval}

Output: A list of detected chewing cycle onsets C_n

1:   $index = 0, I_n \leftarrow \varnothing, C_n \leftarrow \varnothing$

2:   **for** $(i = 1,\ i < w/2,\ i++)$ **do**

3:       **if** $X_n[i] < \theta_C$ **then**

4:         $index+ = 1$

5:   **for** $(i = w/2,\ i < w,\ i++)$ **do**

6:       **if** $X_n[i] > \theta_C$ **then**

7:         $index+ = 1$

8:   **for** $(i = w/2+1,\ i < \text{length}(X_n) - w/2,\ i++)$ **do**

9:       **if** $X_n[i - w/2 - 1] < \theta_C$ **then**

10:         $index- = 1$

11:       **if** $X_n[i - 1] < \theta_C$ **then**

12:         $index+ = 1$

13:       **if** $X_n[i - 1] > \theta_C$ **then**

14:         $index- = 1$

15:       **if** $X_n[i + w/2] > \theta_C$ **then**

16:         $index+ = 1$

17:       $I_n.\text{append}(index)$

18: **for** $(i = 0,\ i < length(I_n) - 2,\ i++)$ **do**

19:       **if** $I_i < I_{i+1}$ and $I_{i+1} > I_{i+2}$ and $I_{i+1} > \theta_P$ **then**

20:         $C_n.append(i+1)$

21:         $i+ = t_{\text{interval}}$

3.1.3. Chewing Segment Detection

We applied a sliding window of size w_0 to C_n, with the start of the window located at the first chewing cycle onset $C_n[0]$, and subsequently slid to the adjacent onset until reaching the end of C_n. With the window starting at $C_n[j]$, the chewing cycles in the window were counted and noted as the jth chewing cycle frequency $f_n[j]$. We applied a criterion $f_n[j] \geq \theta_0$ to confirm that onset $C_n[j]$ belonged to a chewing segment. Correspondingly, the first onset in C_n that also satisfied the criterion $f_n[j_{start}] \geq \theta_0$ was considered as the start of the first chewing segment $t_{n,\,start}[0]$. An onset with $f_n[j_{end}] = \theta_0$ and $f_n[j_{end} + \theta_0 - 1] = 1$ indicated that $C_n[j_{end} + \theta_0 - 1]$ was the only onset in the latest window, that is, the final onset/end of the k-th estimated chewing segment, denoted as $t_{n,\,end}[k]$. The next onset after $t_{n,\,end}[k]$ that satisfied the criterion $f_n[j] \geq \theta_0$ was considered as the $(k+1)$-th chewing segment start $t_{n,\,start}[k+1]$. The pseudo code is shown in Algorithm block 3.

Algorithm block 3 : Chewing segment detection.

Input: List of detected chewing cycle onsets C_n

Parameter: Sliding window size w_0, chewing cycle frequency threshold θ_0

Output: Detected chewing segment starts and ends $(t_{n,\,start}, t_{n,\,end})$ from each signal source n

1:   $t_{n,\,start} \leftarrow \varnothing, t_{n,\,end} \leftarrow \varnothing$

2:   **function** FIND_START_AND_END(C_n, j, θ_0, w_0)

3:      $f_{n,\,end}$ = onset count in interval $[C_n[j], C_n[j] + w_0]$

4:      **if** $f_{n,\,end} == 1$ **then**

5:         $t_{n,\,end}$.append($C_n[j + \theta_0 - 1]$)

6:         **for** $(i = j + \theta_0,\ i < \text{length}(C_n),\ i++)$ **do**

7:             $f_{n,\,start}$ = onset count in interval $[C_n[i], C_n[i] + w_0]$

8:             **if** $f_{n,\,start} >= \theta_0$ **then**

9:                 $t_{n,\,start}$.append($C_n[i]$)

10:                  break

11:        **return** i

12:      **else**

13:         FIND_START_AND_END(C_n, $j_0 + f_{n,\,end} + \theta_0 - 1$, θ_0, w_0)

14: **for** $(j = 1,\ j < \text{length}(C_n),\ j++)$ **do**

15:      $f_n[j]$ = onset count in interval $[C_n[j], C_n[j] + w_0]$

16:      $f_n[j+1]$ = onset count in interval $[C_n[j+1], C_n[j+1] + w_0]$

17:      **if** $t_{n,\,start} == \varnothing$ and $f_n[j] >= \theta_0$ **then**

18:         $t_{n,\,start}$.append($C_n[j]$)

19:      **if** $f_n[j] >= \theta_0$ and $f_n[j+1] < \theta_0$ **then**

20:         $step$ = FINDSTARTEND(C_n, j, θ_0, w_0)

21:         $j += step + \theta_0 - 1$

3.1.4. Fusion of Multi-Source Detection

The fusion of N sensor or feature channels was made by taking the union of source-specific chewing segments:

$$T_{\text{merge}} = \bigcup_{n=1}^{N} \bigcup_{k=1}^{K_n} [t_{n,\,start}[k],\, t_{n,\,end}[k]], \tag{1}$$

where T_{merge} was a list of the merged chewing segments of N sources, and K_n was the number of chewing segments in Channel n. All detected segments were collected chronologically regardless of any overlapping among sources. For the evaluation data used in this investigation, bilateral EMG channels yielded two lists of chewing segments. Hence, $N = 2$.

3.1.5. Gap Elimination

In free-living, eating is often accompanied by interrupts (e.g., conversations). Thus, an eating event is usually represented by several chewing segments in T_{merge}, where the gaps indicate interrupts without chewing cycles. Depending on the detection application and choice of eating event definition, it is reasonable to combine temporally close segments into one final eating event. We denote the start and end of the k-th segment $T_{\text{seg}}[k]$ in T_{merge} as $start[k]$ and $end[k]$ respectively, and the gap between $T_{\text{seg}}[k]$ and $T_{\text{seg}}[k+1]$ as $T_{\text{gap}}[k]$. We generated a new list $T_{\text{concatenated}}$ by removing all gaps that were smaller than t_{gap}:

$$T_{\text{concatenated}} = \bigcup_{k \in S} (T_{\text{seg}}[k] \cup T_{\text{gap}}[k] \cup T_{\text{seg}}[k+1]), \tag{2}$$

where

$$S = \{k \mid start[k+1] - end[k] < t_{\text{gap}}\}. \tag{3}$$

An estimated eating event start $\hat{T}_{\text{start}}[q]$ and end $\hat{T}_{\text{end}}[q]$ with ($q = 1, 2, ..., Q$) were thus obtained as the start and end of every segment in $T_{\text{concatenated}}$, where Q was the number of segments (i.e., detected eating events) in $T_{\text{concatenated}}$. In the present investigation, t_{gap} was set to 5 min.

3.2. Top-Down Algorithms

Two top-down algorithm variants were considered with different chewing segment detection blocks (see Figure 1): Threshold-based top-down and ocSVM top-down. Several blocks of the top-down and bottom-up pipelines were identical, including signal pre-processing (Section 3.1.1), fusion of multi-source detection (Section 3.1.4), and gap elimination (Section 3.1.5). Here we concentrate on the individual variants of the chewing segment detection.

3.2.1. Threshold-Based Top-Down Algorithm

A sliding window of size w_1 and step size s_1 was applied to X_n. We computed the chewing intensity feature F in each sliding window and applied threshold θ_1. If $F > \theta_1$, the window was reported as chewing. For the present investigation, we considered EMG readings as time series containing chewing information and extracted EMG work as chewing intensity feature F. EMG work was defined as the summation of rectified EMG samples within the sliding window. For the EMG data, s_1 was 256 samples (1 s). The pseudo code is shown in Algorithm block 4.

Algorithm block 4 : Chewing segment detection.

Input: Preprocessed signals X_n

Parameter: Sliding window size w_1, window step size s_1, chewing intensity feature threshold θ_1

Output: Detected eating starts/ends from each signal source n: $t_{n,\,\text{start}}$ and $t_{n,\,\text{end}}$

1: $t_{n,\,\text{start}} \leftarrow \varnothing$, $t_{n,\,\text{end}} \leftarrow \varnothing$

2: **for** $(i = s_1,\ i < \text{length}(X_n) - w_1,\ i{+}{=}\,s_1)$ **do**

3:     extract F_{previous} from $X_n[i - s_1 : i + w_1 - s_1]$

4:     extract F_{current} from $X_n[i : i + w_1]$

5:     extract F_{next} from $X_n[i + s_1 : i + w_1 + s_1]$

6:     **if** $F_{\text{previous}} < \theta_1$ and $F_{\text{current}} > \theta_1$ **then**

7:       $t_{n,\,\text{start}}.\text{append}(i)$

8:     **if** $F_{\text{current}} > \theta_1$ and $F_{\text{next}} < \theta_1$ **then**

9:       $t_{n,\,\text{end}}.\text{append}(i + s_1)$

3.2.2. ocSVM Top-Down Algorithm

We applied a non-overlapping sliding window of size w_2 to the EMG data. An ocSVM model was trained based on the windows to detect chewing segments using the same features as described in [13]. The radial basis function (RBF) was used as the kernel. The hyper-parameters γ and v were varied, where γ weighted the non-support vectors' influence on the hyper plane, and v was an upper bound on the fraction of margin errors as well as a lower bound of the fraction of support vectors relative to the number of training samples. The ocSVM predicted the class of each sliding window as either eating or non-eating.

4. Evaluation Methodology

We evaluated the algorithms using a free-living dataset collected from smart eyeglasses with integrated EMG electrodes. Details of the eyeglasses design and data collection process can be found in [17]. Here we summarise the relevant data collection procedures, as well as evaluation methods.

4.1. Participants and Recording Protocol

The dataset was collected from a group of 10 participants (6 male, 4 female, average age of 25.1 years, average BMI of 23.8 kg/m^2) each wearing the smart eyeglasses for one day of regular activity without script or specific protocol. The study was approved by the Ethical Committee of FAU Erlangen-Nürnberg. All participants were healthy and consented to participate after having received oral and written study information.

Each participant received a pair of 3D-printed smart eyeglasses mechanically fitted to their head using a personalisation procedure similar to [22], ensuring that the effect of hair, loss of contact between skin and electrodes, or movement was minimal. In each temple of the eyeglasses frame, dry stainless-steel electrodes of 3 mm × 20 mm (EL-DRY-STEEL-5-20, BITalino, Lisbon, Portugal) were integrated, yielding a two-channel EMG recording system on each side of the head. The EMG electrode pairs were positioned to capture activity of the temporalis muscle. A reference EMG channel was recorded from the right temporalis muscle via gel electrodes attached to the skin at the corresponding forehead region. All EMG channels were acquired with an EMG recorder (ACTIWAVE, CamNtech, Cambridgeshire, United Kingdom) at a sampling rate of 256 Hz per channel.

Participants were suggested to wear the eyeglasses during one entire recording day (i.e., attaching the system right after getting up and ending before going to bed at night). Recordings were conducted in free-living conditions without dietary constraints. Participants chose their diets and conducted other daily activities at their choice. Participants were asked to log activities in a paper-based 24-h activity journal with 1 min resolution, including any food intake as well as start and end times of eating events. As Figure 2 show:

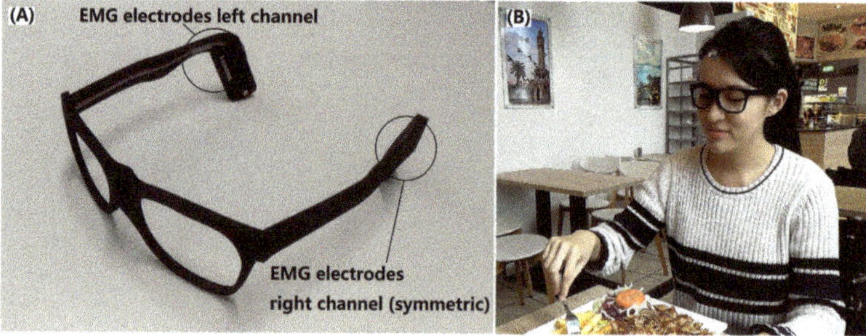

Figure 2. Illustration of the EMG eyeglasses and study: (**A**) Eyeglasses frame with electromyographic (EMG) electrodes symmetrically integrated on the temples. (**B**) Study participant wearing the EMG eyeglasses. Reference EMG electrodes were attached to the skin at the right forehead temporalis muscle position.

4.2. Data Corpus

By the end of the recording, we collected a total of 122.3 h of free-living data including 44 eating events ranging from 54 s to 35.8 min, which summed up to 429 min of eating for all participants combined. Eating took up 5.8% of the whole dataset. Participants took off eyeglasses for a total time of 12 min during the recordings, which corresponds to 0.16% of the total recordings. Known activities reported by participants in the activity journal included cooking, eating, walking, transportation, attending lectures, performing office work, having conversations, doing housework, brushing teeth, playing video games, going to the cinema, and engaging in physical exercise. Through visual inspection we observed various artefacts in the data corpus including, for example, suspected teeth grinding [17].

4.3. Free-Living Eating/Non-Eating Reference Construction

Obtaining accurate reference information on eating events in unsupervised free-living studies is particularly challenging. Here, we propose a combination of participant activity journal and EMG reference recordings. All eating events were annotated using a custom Matlab annotation software. Our annotation process comprised two steps: coarse manual annotation using the activity journal and fine-tuning through reference EMG recordings. Coarse manual annotation was realised by searching the journal for the participant-logged start time $T_{start}[i]$ and end time $T_{end}[i]$ of each annotated eating event, indexed i. As manual journaling is often imprecise in identifying event times, a fine-tuning step was used to adjust coarse eating event times: Start and end times $T_S[i]$ and $T_E[i]$ of eating event i were adjusted by visually searching the reference EMG data for chewing cycle patterns in the neighbourhood of approx. ± 1 min (journal resolution) around the coarse annotations $T_{start}[i]$ and $T_{end}[i]$. Since each chewing cycle had a duration of around 1/3 s, the fine-tuned eating event labels $T_S[i]$ and $T_E[i]$ resulted in a chew-accurate eating/non-eating reference with resolution of approximately 1/3 s. The derived start and end times were considered as eating/non-eating reference for algorithm evaluation. The eating/non-eating reference construction is illustrated in Figure 3.

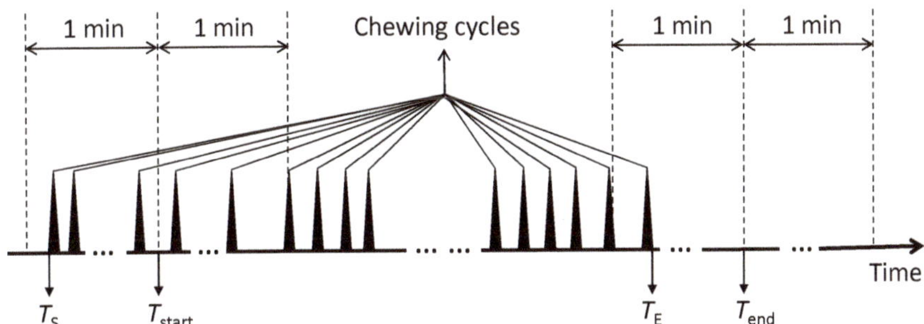

Figure 3. Illustration of the free-living eating/non-eating reference construction. T_{start} and T_{end} are start and end times of an eating event obtained from the participant journal, while T_S and T_E are the corrected start and end times derived by searching the EMG reference ± 1 min around T_{start} and T_{end}. The eating/non-eating reference construction is described in Section 4.3.

Type 1 errors (false positives) could occur in the eating/non-eating reference if an activity journal entry could not be matched to any chewing-like pattern in the reference EMG signal. We inspected all entries in the participant journal and compared them to the reference EMG signal. In the present dataset, all participant-annotated events could be matched to the EMG reference.

Type 2 errors (false negatives) could occur in the eating/non-eating reference if participants omitted annotations. To amend potential omissions from the activity journal, we first inspected the entire reference EMG data for chewing-like signal patterns that did not correspond to any entry in the journal. For each chewing-like pattern found, we inspected the activity journal to obtain insight into the participant's momentary context. We observed that concise activations in the EMG reference occurred occasionally without corresponding eating annotations (e.g., during a lecture). Yet, EMG activations were typically short (i.e., less than five consecutive activations with lower EMG work compared to confirmed chewing). Given a non-eating context and the clear non-chewing signal patterns, we attributed the activations to teeth grinding. Jaw motion during speaking does not involve profound temporalis muscle activation, as there is hardly any teeth clenching and thus substantially lower EMG work than during chewing [16]. In addition, non-chewing muscle activity is typically non-periodic, thus observable and distinguishable during time series inspection. Overall, we did not find Type 2 errors in the dataset, supporting our eating/non-eating reference construction approach for free-living recordings.

4.4. Evaluation Metrics

A grid search over the window length parameters w_i and thresholds θ_i with $i = 0, 1, 2$, and $\theta_2 = (\gamma, \nu)$ representing the combination of the ocSVM hyper-parameters was performed to investigate optimal parameter combinations. To evaluate the eating event detection algorithms, we derived the overlap between retrieved eating events and any eating/non-eating reference label. The precision and recall of each algorithm were calculated according to: $Recall = \frac{T_{tp}}{T_{gt}}$ and $Precision = \frac{T_{tp}}{T_{ret}}$, where T_{gt} was the summed duration of all P eating events according to the constructed eating/non-eating reference labels, calculated as:

$$T_{gt} = \sum_{p=1}^{P} (T_{end}[p] - T_{start}[p]), \tag{4}$$

while T_{ret} was the summed duration of all Q detected eating events by the algorithm:

$$T_{ret} = \sum_{q=1}^{Q} (\hat{T}_{end}[q] - \hat{T}_{start}[q]), \tag{5}$$

and T_{tp} was the summed overlap duration between retrieved eating events and the eating/non-eating reference:

$$T_{tp} = \sum_{p=1}^{P} \sum_{q=1}^{Q} (\min(T_{end}[p], \hat{T}_{end}[q]) - \max(T_{start}[p], \hat{T}_{start}[q])), \tag{6}$$

given the following premise:

$$\min(T_{end}[p], \hat{T}_{end}[q]) - \max(T_{start}[p], \hat{T}_{start}[q]) > 0. \tag{7}$$

$\hat{T}_{end}[q]$ and $\hat{T}_{start}[q]$ were the start and end time points of the qth retrieved eating event, Q was the number of retrieved eating events, and P was the number of eating events in the eating/non-eating reference. All times were computed at a resolution of 1 sample ($1/256\,$s). Finally, the F1 score was calculated as the harmonic mean of precision and recall.

The evaluation was performed using leave-one-participant-out (LOPO) cross-validation. In each evaluation fold, the EMG data were split into a training set of nine participants and a test set of one participant. This process was repeated 10 times until every participant's data were in the test set once. Training data were used in a grid search to estimate performance under different parameter combinations. Optimal parameter combinations were chosen according to the training data performance and applied with the test data to estimate algorithm performance. The test results of all folds were averaged to obtain the total algorithm performance. For the bottom-up algorithm, w_0, θ_0, and θ_P were analysed. For the threshold-based top-down algorithm, w_1 and θ_1 were analysed, and for the ocSVM top-down algorithm, w_2, γ, and ν were analysed.

4.5. Detection Timing Errors

We further investigated the detection timing error of every algorithm. The average start and end timing errors of the algorithms were calculated as follows:

$$\Delta \overline{T}_S = \frac{\sum_{q=1}^{Q} \min \left(\left| \hat{T}_S[q] - T_S[p] \right| \Big|_{p=1,2,\ldots,P} \right)}{Q}, \tag{8}$$

and

$$\Delta \overline{T}_E = \frac{\sum_{q=1}^{Q} \min \left(\left| \hat{T}_E[q] - T_E[p] \right| \Big|_{p=1,2,\ldots,P} \right)}{Q}. \tag{9}$$

$\Delta \overline{T}_S$ and $\Delta \overline{T}_E$ were the average absolute detection errors at the start and end of eating events.

To investigate retrieval performance in detail and identify the algorithms' behaviour, different optimisation objectives were analysed. Using the grid search over the parameter space, the best performance point according to maximal F1 score (termed P_X), minimal start timing error $\Delta \overline{T}_S$ (termed P_S), and minimal end timing error $\Delta \overline{T}_E$ (termed P_E) were derived.

5. Results

Algorithm detection performances according to the test data are shown in Figure 4 for varying parameter combinations. The threshold-based top-down algorithm could not reach meaningful F1 scores, indicating that detecting eating events is not a trivial task. The performance map of the ocSVM algorithm shows a periodic landscape due to the variation of parameters γ and ν. The best performance of the bottom-up algorithm was achieved with $\theta_P = 0.7$. The bottom-up algorithm had a smooth landscape across the parameters. For all algorithms, the three performance points (P_X, P_S, P_E), did not coincide at the same parameter settings. To illustrate the performance points quantitatively, they are summarised in Table 1. The bottom-up algorithm yielded comparable performance values across all performance points (P_X, P_S, P_E). At best, the bottom-up algorithm reached an F1 score of 99.2%, yielding a start/end error ($\Delta \overline{T}_S$ and $\Delta \overline{T}_E$) of $2.4 \pm 0.4\,$s and $4.3 \pm 0.4\,$s, respectively. The results show that the bottom-up algorithm outperformed the top-down algorithms.

Table 1. Performance comparison among algorithms using optimal parameter settings for each performance point (P_X, P_S, P_E). For timing metrics, mean performance $\pm$ std. dev. are shown. For example, the bottom-up algorithm reached an F1 score of 99.2% at best, where the start/end error was $2.4 \pm 0.4\,$s and $4.3 \pm 0.4\,$s, respectively.

Metric		Performance Points		
		P_X	P_S	P_E
F1 score (%)	Threshold-based top-down	36.7	0.03	0.001
	ocSVM top-down	95.1	90.9	93.2
	Bottom-up	99.2	97.8	97.7
$\Delta \overline{T}_S$ (s)	Threshold-based top-down	152.4 ± 21.7	10.1 ± 3.0	185.9 ± 35.9
	ocSVM top-down	30.0 ± 36.4	18.8 ± 27.9	53.2 ± 61.7
	Bottom-up	3.0 ± 0.6	2.4 ± 0.4	4.8 ± 2.9
$\Delta \overline{T}_E$ (s)	Threshold-based top-down	177.4 ± 12.1	265.8 ± 86.5	63.0 ± 11.9
	ocSVM top-down	25.9 ± 39.4	26.9 ± 38.3	15.2 ± 19.0
	Bottom-up	4.9 ± 0.3	6.4 ± 0.5	4.3 ± 0.4

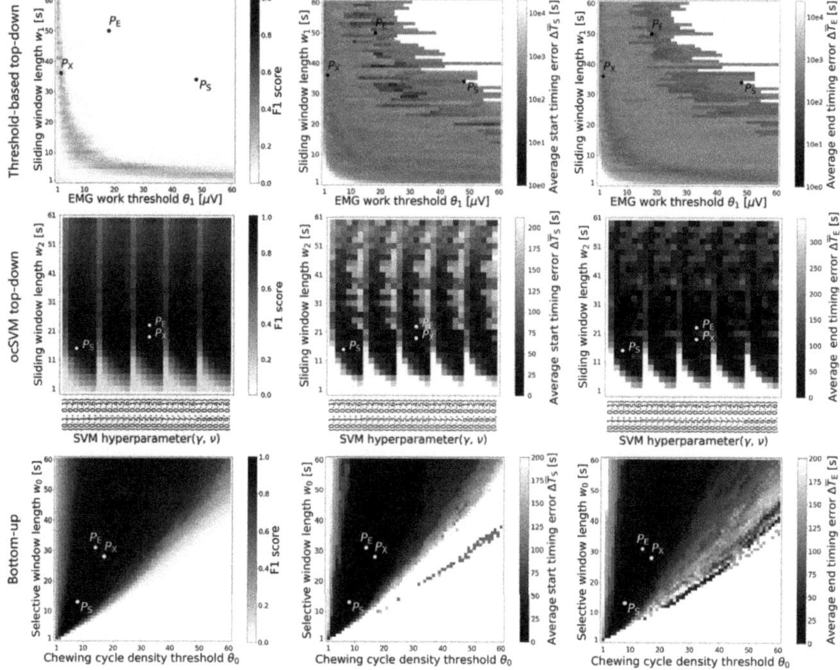

Figure 4. F1 score, average start and end timing errors for test data and each eating event detection algorithm using grid search over the parameter space. The highest F1 score location was denoted as P_X, while P_S and P_E indicate the minimal start timing error $\Delta \overline{T}_S$ and minimal end timing error $\Delta \overline{T}_E$, respectively. The bottom-up algorithm performance was obtained with fixed peak detection threshold $\theta_P = 0.7$.

Figure 5 shows the effect of varying the peak detection threshold θ_P of the bottom-up algorithm, indicating robust retrieval and timing performance (P_X, P_S, P_E) for a parameter range of $0.65 < \theta_P < 0.8$. The best retrieval and timing performances were achieved at $\theta_P = 0.7$.

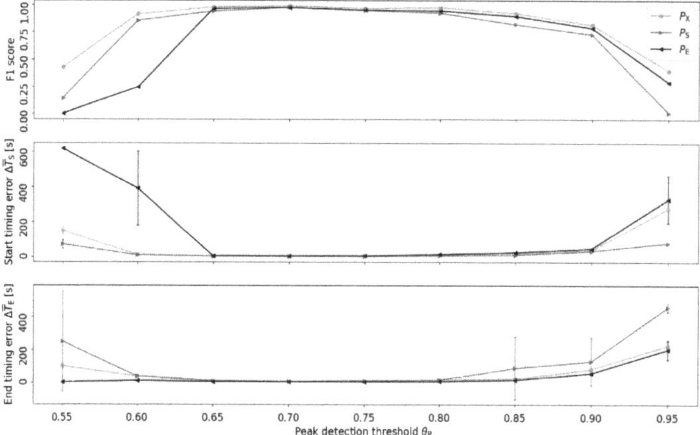

Figure 5. Retrieval and timing performance of the bottom-up algorithm at different peak detection thresholds θ_P. In the timing error diagrams, caps on vertical line ends indicate the standard deviation.

Figure 6 illustrates retrieved eating events as point pairs across F1 scores, where the line ends represent average start and end timing errors ($\Delta\overline{T}_S$ and $\Delta\overline{T}_E$). $\Delta\overline{T}_S$ and $\Delta\overline{T}_E$ were obtained by varying the algorithm parameters and averaging the individual timing errors obtained for specific retrieval performances. For the bottom-up algorithm, the graph shows the performance obtained by varying sliding window size w_0 and chewing cycle frequency threshold θ_0 at fixed peak detection threshold $\theta_P = 0.7$. There was no parameter combination for the threshold-based top-down algorithm that yielded an F1 score above 40%. In contrast, bottom-up and ocSVM top-down algorithms provided retrieval performances of up to 99% and 95% respectively. With increasing F1 score, timing errors tended to decline. It can be derived from Figure 6 that the relation between start and end timing errors varied between algorithms. For the bottom-up algorithm and F1 score >80%, the start timing error $\Delta\overline{T}_S$ became smaller than the end timing error $\Delta\overline{T}_E$.

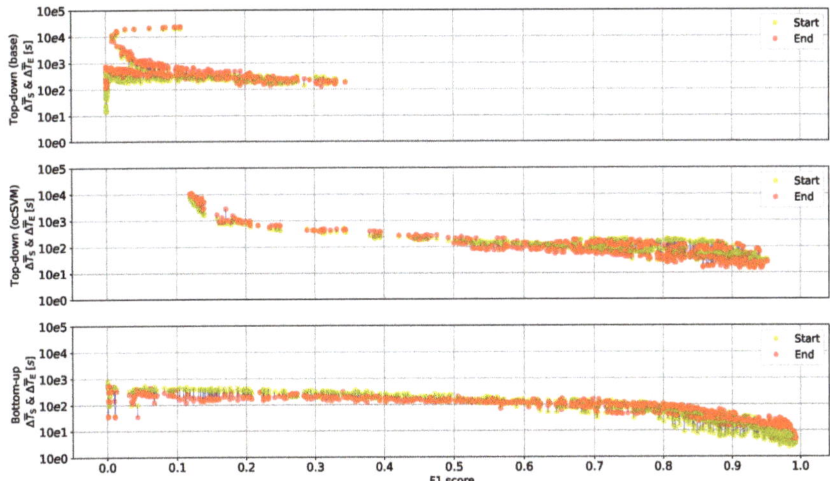

Figure 6. Relation of retrieval and timing performance of all three algorithms. $\Delta\overline{T}_S$ and $\Delta\overline{T}_E$ were obtained by varying algorithm parameters. Blue lines link average start and end timing errors of all eating events at a given algorithm parameter set. With increasing F1 score, timing errors declined. Note that timing error analysis could be performed only for eating events retrieved by an algorithm. The bottom-up algorithm ($\theta_P = 0.7$) achieved the highest F1 score at smallest timing errors among all algorithms investigated. Point pairs were down-sampled for visualisation.

Figure 7 shows examples of the detected eating event starts and ends. The bottom-up algorithm yielded similar detected labels to the eating/non-eating reference whereas the ocSVM top-down algorithm incurred larger timing errors for some eating event instances.

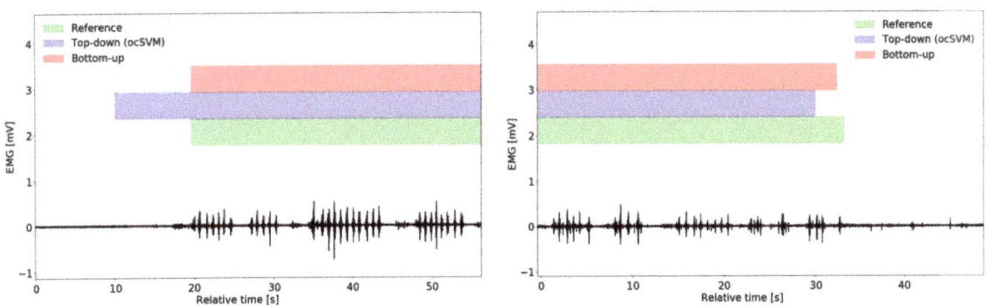

Figure 7. *Cont.*

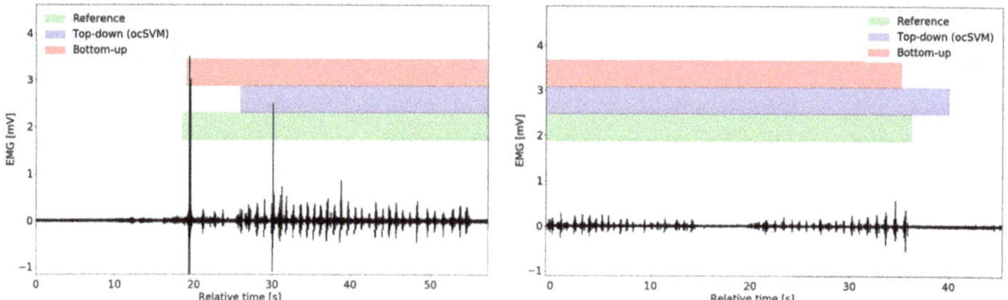

Figure 7. Examples of data situations with the corresponding retrieval results of bottom-up and ocSVM top-down algorithms obtained at each algorithm's performance point P_S (left column) and P_E (right column). As the diagrams illustrate, the ocSVM algorithm may anticipate or delay eating events' starts, as ocSVM deploys a time-domain sliding windowing with a given step size, whereas the bottom-up algorithm did not.

6. Discussion

The F1 score describes the algorithm's retrieval performance by retrieved and missed eating instances, while timing errors reveal the accuracy of estimated event timing. Considering the varying eating durations in a free-living context, the two metrics are not necessarily similar in their sensitivity, thus we argue here that both are relevant metrics for evaluation. Among the few investigations on event timing in ADM, Dong et al. [4] reported event start-timing errors of 0.6 minutes, and end errors of 1.5 min. The authors determined intake from bites using arm motion, while the present investigation was based on chewing. Bedri et al. [11] evaluated eating event detection using a metric called delay, measuring the time from the beginning of an eating event until it was recognised. The average delay reported was 65.4 s. In contrast to the investigation of Bedri et al. [11], we also evaluated the timing error at the end of eating events. Our bottom-up algorithm yielded average start/end timing errors of 2.4 s and 4.3 s.

We believe that the bottom-up method is practically useful for eating event start and end detection, as well as, for example, sending reminders, sampling user responses, and gathering environmental variables. Study participants did not complain or reject wearing the eyeglasses for one day. Hence, the combination of the bottom-up algorithm and smart eyeglasses could be adopted in unconstrained free-living applications. In contrast to several previous investigations of eating detection that require the training of many parameters, our bottom-up approach requires that only four parameters be set (w_0, θ_0, θ_P, and t_{gap}). Our analysis indicates that performance was unaffected by parameter changes across a wide value range (i.e., shown as a smooth performance space in Figure 4). Pattern learning may work reliably when trained on sufficient data with proper features. Considering the variability in free-living behaviour and the unbalanced distribution of eating and non-eating times, substantial training data is needed to implement any learning method and therefore a minimal number of free parameters is key. The bottom-up method outperformed our top-down methods, with a higher F1 score and lower detection timing errors. We attribute the higher performance yielded in the present investigation to the expert knowledge incorporated in the bottom-up approach.

In both top-down and bottom-up methods, the sliding window size w_i influenced the algorithm performance. In top-down methods, a small sliding window of length w_i contained fewer data samples, which usually led to less representative features. Thus, the lowest timing errors were typically not achieved with smallest sliding window sizes (e.g., $w_i < 10\,s$). Similarly, in the bottom-up method, both window size w_0 and the second parameter θ_0 influenced the detection performance. Hence, a small window size w_0 did not always give the best performance.

The timing errors of top-down methods were highly dependent on the combination of sliding window length and window step size. Large sliding window sizes included more dietary activity information, but usually failed in accurately detecting the starts and ends of eating events as the window was filled with both eating and non-eating data. Figure 7 shows impressively that the ocSVM top-down algorithm indeed incurred larger timing errors due to the larger sliding window size. In our previous investigation [13], we adopted window overlaps and majority voting on windows with differing results. We observed that retrieval performances differed marginally when comparing overlapping and non-overlapping windowing approaches. The bottom-up algorithm was not affected by the window parameterisation problem, as the window step size is determined by distance of neighbouring chewing onsets. Thus, eating and non-eating rarely coincided in one window.

The bottom-up algorithm is based on chewing cycle detection, which decouples the eating event detection from the sensor type. The detection leverages event frequency information (i.e., chewing cycle frequencies), which can be obtained with different chewing monitoring approaches. We expect that the algorithms could be applied with various sensors or sources that provide chewing cycle information, including acoustics [1], ear canal deformation [15], strain on head skin [19], eyeglasses temple motion [18], etc.

The present investigation analysed relevant free parameters of the proposed algorithms to determine their stability. For example, the sweep of the peak detection threshold θ_P showed desirable performance trends (Figure 5) allowing us to set θ_P to a proper range—approximately $[0.65, 0.8]$. In addition, the pipeline block "gap elimination" used the parameter $t_{gap} = 5$ min to merge temporally close eating detections. The parameter t_{gap} supports our informal definition of eating events as temporally linked sequences of dietary activities during one meal or snack [3] and was set based on experience. Varying t_{gap} means to change the representation of eating occasions (i.e., meals and snacks), which is outside of the scope of this investigation.

While this investigation focuses on the retrieval performance, the computational complexity of the algorithms is an important consideration for wearable resource-limited systems. In a detection, the computational complexity is $\mathcal{O}(n)$ for the threshold-based top-down algorithm, and $\mathcal{O}(n_{sv} \times n)$ for the ocSVM top-down algorithm. Here, n is the input data dimension and n_{sv} is the number of support vectors of the ocSVM model. The complexity of the bottom-up algorithm is decided by the chewing cycle detection method. For the proposed bottom-up algorithm, the corresponding complexity is $\mathcal{O}(n)$. With a proper chewing cycle detection approach, the bottom-up algorithm is suitable to execute, for example, on wearables at a minimal computational cost. The delay due to processing was not addressed in this investigation. However, with the low complexity of all algorithms, processing delay is expected to have a negligible effect compared to the algorithm timing errors.

This investigation was supported by a new method to obtain reference data on eating times in a free-living context, where we combined the participants' activity journals with reference EMG measurements. While the activity journals yielded rather coarse timing, they provided us with context information on the users' behaviour. The reference EMG measurement complemented the journal with accurate timing resolution of individual chewing cycles. However, adherence to journals is known to decline quickly over several days of measurement [23]. Hence, it is reasonable to assume that journals alone would be too inaccurate. We avoided video recordings to retrieve eating/non-eating reference due to privacy concerns and the potential impact of cameras on natural, free-living behaviour.

One limitation of our study is that only young healthy participants were involved. For other populations, the eating structure could vary, which could generate different eating durations. However, our present investigation already showed that eating events ranging from short snacks of 54 s to 35.8 min meals could be recognised. Other populations may benefit from different pre-processing steps or other sensors to apply the discussed bottom-up algorithm. We are planning longer-term studies in the future.

7. Conclusions

We proposed a bottom-up eating event detection algorithm that uses chewing cycle information as input and compared it to two top-down algorithms, including threshold-based and ocSVM algorithms.

Evaluation of the algorithms was performed using free-living data with smart eyeglasses recording EMG data bilaterally from the temporalis muscles. Our results indicate that the F1 score became less meaningful at high retrieval rates above 0.9. The analysis of timing errors revealed substantial differences of several tens to hundreds of seconds on average between top-down and bottom-up algorithms. The grid search analysis showed smooth performance transitions during parameter variation for the bottom-up algorithm. We conclude that timing error analysis is an important component in performance estimation, besides a relevant retrieval metric, as the F1 score. We suggest that the research community report timing errors (e.g., using the metrics described in this work). The bottom-up algorithm yielded the overall best results with the lowest timing errors of $2.4 \pm 0.4\,\text{s}$ for eating start and $4.3 \pm 0.4\,\text{s}$ for eating end. The bottom-up algorithm is thus suitable for eating event detection.

Author Contributions: R.Z. and O.A. devised the methodology. R.Z. performed data curation and implemented the algorithms. O.A. provided feedback throughout the implementation phase. R.Z. and O.A. prepared the manuscript. All authors have read and agreed to the published version of the manuscript.

Funding: This research did not receive external funding.

Acknowledgments: The present work was performed in partial fulfilment of the requirements for obtaining the degree "Dr. rer. biol. hum." We are thankful to the participants for the time and effort spent in the study.

Conflicts of Interest: The authors declare no conflict of interest.

References

1. Amft, O.; Stäger, M.; Lukowicz, P.; Tröster, G. Analysis of Chewing Sounds for Dietary Monitoring. In Proceedings of the 7th International Conference on Ubiquitous Computing (UbiComp 2005), Tokyo, Japan, 11–14 September 2005; pp. 56–72.

2. Amft, O.; Junker, H.; Tröster, G. Detection of eating and drinking arm gestures using inertial body-worn sensors. In Proceedings of the Ninth International Symposium on Wearable Computers (ISWC 2005), Osaka, Japan, 18–21 October 2005; pp. 160–163.

3. Schiboni, G.; Amft, O. Automatic Dietary Monitoring Using Wearable Accessories. In *Seamless Healthcare Monitoring*; Springer: Cham, Switzerland, 2018; pp. 369–412.

4. Dong, Y.; Scisco, J.; Wilson, M.; Muth, E.; Hoover, A. Detecting periods of eating during free-living by tracking wrist motion. *IEEE J. Biomed. Health Inform.* **2014**, *18*, 1253–1260. [CrossRef] [PubMed]

5. Thomaz, E.; Essa, I.; Abowd, G.D. A Practical Approach for Recognizing Eating Moments with Wrist-mounted Inertial Sensing. In Proceedings of the 2015 ACM International Joint Conference on Pervasive and Ubiquitous Computing (UbiComp'15), Osaka, Japan, 7–11 September 2015; pp. 1029–1040.

6. Bi, S.; Wang, T.; Davenport, E.; Peterson, R.; Halter, R.; Sorber, J.; Kotz, D. Toward a Wearable Sensor for Eating Detection. In Proceedings of the 2017 Workshop on Wearable Systems and Applications (WearSys'17), Niagara Falls, NY, USA, 19–23 June 2017; pp. 17–22.

7. Farooq, M.; Sazonov, E. Accelerometer-Based Detection of Food Intake in Free-Living Individuals. *IEEE Sens. J.* **2018**, *18*, 3752–3758. [CrossRef] [PubMed]

8. Wahl, F.; Freund, M.; Amft, O. WISEglass—Multi-purpose context-aware smart eyeglasses. In Proceedings of the 2015 ACM International Symposium on Wearable Computers (ISWC 2015), Osaka, Japan, 7–11 September 2015; pp. 159–160.

9. Merck, C.; Maher, C.; Mirtchouk, M.; Zheng, M.; Huang, Y.; Kleinberg, S. Multimodality Sensing for Eating Recognition. In Proceedings of the 10th EAI International Conference on Pervasive Computing Technologies for Healthcare (PervasiveHealth'16), Cancun, Mexico, 16–19 May 2016; pp. 130–137.

10. Papapanagiotou, V.; Diou, C.; Zhou, L.; Boer, J.v.d.; Mars, M.; Delopoulos, A. A Novel Chewing Detection System Based on PPG, Audio, and Accelerometry. *IEEE J. Biomed. Health Inform.* **2017**, *21*, 607–618. [CrossRef] [PubMed]

11. Bedri, A.; Li, R.; Haynes, M.; Kosaraju, R.P.; Grover, I.; Prioleau, T.; Beh, M.Y.; Goel, M.; Starner, T.; Abowd, G. EarBit: Using Wearable Sensors to Detect Eating Episodes in Unconstrained Environments. *Proc. ACM Interact. Mob. Wearable Ubiquitous Technol.* **2017**, *1*, 37:1–37:20. [CrossRef] [PubMed]

12. Doulah, A.; Farooq, M.; Yang, X.; Parton, J.; McCrory, M.A.; Higgins, J.A.; Sazonov, E. Meal Microstructure Characterization from Sensor-Based Food Intake Detection. *Front. Nutr.* **2017**, *4*, 31. [CrossRef] [PubMed]

13. Zhang, R.; Amft, O. Free-living eating event spotting using EMG-monitoring eyeglasses. In Proceedings of the 2018 IEEE EMBS International Conference on Biomedical Health Informatics (BHI '18), Las Vegas, NV, USA, 4–7 March 2018; pp. 128–132.

14. Amft, O. A Wearable Earpad Sensor for Chewing Monitoring. In Proceedings of the IEEE Sensors Conference (Sensors 2010), Waikoloa, HI, USA, 1–4 November 2010; pp. 222–227.

15. Bedri, A.; Verlekar, A.; Thomaz, E.; Avva, V.; Starner, T. A Wearable System for Detecting Eating Activities with Proximity Sensors in the Outer Ear. In Proceedings of the 2015 ACM International Symposium on Wearable Computers (ISWC'15), Osaka, Japan, 9–11 September 2015; pp. 91–92.

16. Zhang, R.; Bernhart, S.; Amft, O. Diet eyeglasses: Recognising food chewing using EMG and smart eyeglasses. In Proceedings of the International Conference on Wearable and Implantable Body Sensor Networks (BSN'16), San Francisco, CA, USA, 14–17 June 2016; pp. 7–12.

17. Zhang, R.; Amft, O. Monitoring chewing and eating in free-living using smart eyeglasses. *IEEE J. Biomed. Health Inform.* **2018**, *22*, 23–32. [CrossRef] [PubMed]

18. Chung, J.; Chung, J.; Oh, W.; Yoo, Y.; Lee, W.G.; Bang, H. A glasses-type wearable device for monitoring the patterns of food intake and facial activity. *Sci. Rep.* **2017**, *7*, 41690. [CrossRef] [PubMed]

19. Farooq, M.; Sazonov, E. A Novel Wearable Device for Food Intake and Physical Activity Recognition. *Sensors* **2016**, *16*, 1067. [CrossRef] [PubMed]

20. Abbink, J.H.; Bilt, A.v.d.; Glas, H.W.v.d. Detection of onset and termination of muscle activity in surface electromyograms. *J. Oral Rehabil.* **1998**, *25*, 365–369. [CrossRef] [PubMed]

21. Po, J.; Kieser, J.; Gallo, L.; Tésenyi, A.; Herbison, P.; Farella, M. Time-Frequency Analysis of Chewing Activity in the Natural Environment. *J. Dental Res.* **2011**, *90*, 1206–1210. [CrossRef] [PubMed]

22. Wahl, F.; Zhang, R.; Freund, M.; Amft, O. Personalizing 3D-printed smart eyeglasses to augment daily life. *IEEE Comput.* **2017**, *50*, 26–35. [CrossRef]

23. Witschi, J.C. Short-Term Dietary Recall and Recording Methods. In *Nutritional Epidemiology*; Willett, W., Ed.; Oxford University Press: Oxford, UK, 1990; Volume 4, pp. 52–68.

Article

Determining the Online Measurable Input Variables in Human Joint Moment Intelligent Prediction Based on the Hill Muscle Model

Baoping Xiong [1,2], Nianyin Zeng [3,*], Yurong Li [4], Min Du [1,5,*], Meilan Huang [1], Wuxiang Shi [1], Guojun Mao [2] and Yuan Yang [6,*]

[1] College of Physics and Information Engineering, Fuzhou University, Fuzhou City 350116, Fujian Province, China; xiongbp@fjut.edu.cn (B.X.); hml19940515@icloud.com (M.H.); shiwuxiang3@163.com (W.S.)
[2] Department of Mathematics and Physics, Fujian University of Technology, Fuzhou City 350118, Fujian Province, China; maximmao@hotmail.com
[3] Department of Instrumental and Electrical Engineering, Xiamen University, Fujian 361005, China
[4] Fujian Key Laboratory of Medical Instrumentation & Pharmaceutical Technology, Fuzhou University, Fuzhou City 350116, Fujian Province, China; liyurong@fzu.edu.cn
[5] Fujian provincial key laboratory of eco-industrial green technology, Wuyi University, Wuyishan City 354300, Fujian Province, China
[6] Department of Physical Therapy and Human Movement Sciences, Northwestern University, Chicago, IL 60208, USA
*   Correspondence: zny@xmu.edu.cn (N.Z.); dm_dj90@163.com (M.D.); yuan.yang@northwestern.edu (Y.Y.)

Received: 15 January 2020; Accepted: 19 February 2020; Published: 21 February 2020

Abstract: *Introduction*: Human joint moment is a critical parameter to rehabilitation assessment and human-robot interaction, which can be predicted using an artificial neural network (ANN) model. However, challenge remains as lack of an effective approach to determining the input variables for the ANN model in joint moment prediction, which determines the number of input sensors and the complexity of prediction. *Methods*: To address this research gap, this study develops a mathematical model based on the Hill muscle model to determining the online input variables of the ANN for the prediction of joint moments. In this method, the muscle activation, muscle-tendon moment velocity and length in the Hill muscle model and muscle-tendon moment arm are translated to the online measurable variables, i.e., muscle electromyography (EMG), joint angles and angular velocities of the muscle span. To test the predictive ability of these input variables, an ANN model is designed and trained to predict joint moments. The ANN model with the online measurable input variables is tested on the experimental data collected from ten healthy subjects running with the speeds of 2, 3, 4 and 5 m/s on a treadmill. The variance accounted for (VAF) between the predicted and inverse dynamics moment is used to evaluate the prediction accuracy. *Results*: The results suggested that the method can predict joint moments with a higher accuracy (mean VAF = 89.67±5.56 %) than those obtained by using other joint angles and angular velocities as inputs (mean VAF = 86.27±6.6%) evaluated by jack-knife cross-validation. *Conclusions*: The proposed method provides us with a powerful tool to predict joint moment based on online measurable variables, which establishes the theoretical basis for optimizing the input sensors and detection complexity of the prediction system. It may facilitate the research on exoskeleton robot control and real-time gait analysis in motor rehabilitation.

Keywords: artificial neural network; joint moment prediction; extreme learning machine; Hill muscle model; online input variables

1. Introduction

Human joint moment prediction is crucial to rehabilitation evaluation [1–3], athlete training evaluation [4–6], prosthesis and orthosis design [7–9], intramedullary device design [10–12] and human-robot interaction [13–21]. The precise prediction of joint moment can be fulfilled by the use of instrumented implants [22] which measures the relevant parameters of joint load in real time. However, this approach is not always feasible since only few people (likely those suffering from musculoskeletal deficits) have implants.

Although computational models can serve as alternative methods for joint moment prediction when the implants are not available, they face a challenge of eliminating the measurement error. This is due to the individual differences in the anatomical and functional characteristics of the musculoskeletal system [22]. Furthermore, the joint moment is not easily measured in real time. Previous studies [23–26] indicated that this challenge may be addressed by using the artificial neural network (ANN) model, because of its excellent adaptive ability to individual characteristics [27,28]. For example, Uchiyama et al. [29], used an ANN model to predict the elbow joint moment with the inputs of EMG signals, elbow and shoulder joint angles, while Luh et al. [30], and Song and Tong [31] utilized an ANN model with EMG signals, elbow joint angle and angular velocity for the same purpose. Hahn [32] intelligently predicted the isokinetic knee extensor and flexor moment with the inputs of EMG signals, gender, age, height and body mass. Ardestani et al. [33], combined the EMG signals and ground reaction force (GRFs) with ANN model to study the lower limbs' joint moment. Recently, Xiong et al. [34], used the optimized EMG signals and joint angles as the inputs of ANN model to calculate the lower extremity joint moment.

As listed above, different studies used different input variables in their ANN models to predict joint moments. However, the number of input variables determines the number of sensors and the complexity of the system. It is yet to develop a mathematical model to determine the optimal online measurable input variables. This model will provide a theoretical basis for designing a system with few sensors and high accurate of joint moment prediction. Therefore, the purpose of this study is to introduce a novel method for determining the online measurable input variables for human joint moment intelligent prediction.

In this method, musculoskeletal geometry [35,36] comprised of Hill muscle models [37,38] are utilized for representing the muscle mechanical response. Furthermore, the input variables to predict joint moment based on the Hill muscle model includes four time-varying variables: the muscle activation, muscle-tendon moment arm, velocity and length are found [39], that generally cannot be measured online in vivo. Thus, a surrogate model is built for each tested muscle to convert these four input variables to the online measurable variables, i.e., muscles EMG, the muscle actuates joints' angles and angular velocities.

To test the predictive ability of the online measurable input variables, a commonly used ANN model, i.e., Extreme Learning Machine (ELM), is designed and trained to predict joint moments. The ELM is a feedforward ANN [40], which has a much lower computational cost than traditional machine learning algorithms, especially for the single hidden layer mode [41–43]. The method is tested on the experimental data of ten healthy male subjects running at different speeds, i.e., 2, 3, 4 and 5 m/s on a treadmill. The ELM predictions are validated against inverse dynamics and compared with those obtained by jack-knife cross-validation with other online measurable variables as inputs [29–31,34].

2. Materials and Methods

2.1. Experimental Data

The lower limbs' kinematics and dynamics experimental data of ten healthy male subjects (height 1.77 ± 0.04 m, age 29 ± 5 years, mass 70.9 ± 7.0 kg) was obtained from an open database (https://simtk.org/projects/nmbl_running; accessed on, 18 October 2019). In the experiment, the motion data, EMG signals and ground reaction force were measured, while the subjects ran at different speeds of 2, 3, 4 and 5 m / s on the treadmill. At least six gait cycles were recorded for each speed. The EMG signals included gluteus medius, rectus femoris, gluteus maximus, vastus lateralis, biceps femoris long head, vastus medialis, tibialis anterior, soleus, gastrocnemius medialis and gastrocnemius lateralis. All the EMG signals were rectified, filtered and normalized. The motion and force data were filtered accordingly. A complete description of these data can be found in [44].

After obtaining the experimental data, all the ten subjects' moment of ankle plantar-dorsiflexion, knee flexion-extension, hip adduction-abduction and hip flexion-extension are firstly calculated by using the inverse dynamics method [45] with opensim software, then the moment, force, motion and EMG signals are resampled to obtain 101 time points of each gait cycle. All the inverse dynamics moment will be used as the target value of the ANN model's training samples.

2.2. Determination of Online Measurable Variables

In order to obtain the online measurable input variables, the Hill muscle model [37,38] and musculoskeletal geometry [35] is used to establish a mathematical model of input-output relation for joint moment prediction. The data processing pipeline is shown as Figure 1.

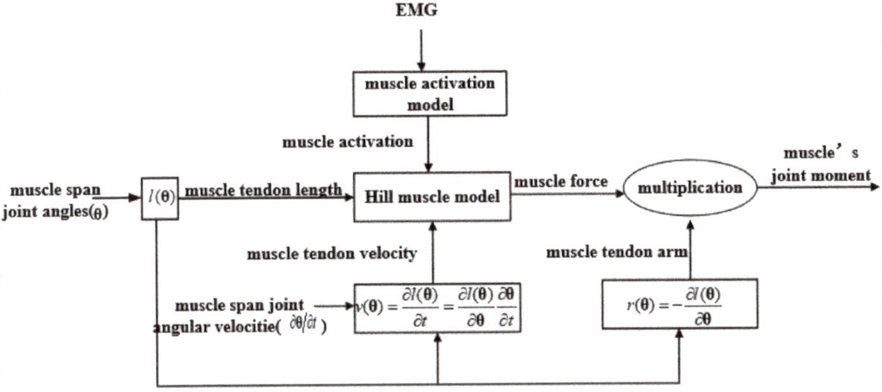

Figure 1. Data processing pipeline of the method based on Hill muscle model, where $l(\theta)$ is a polynomial function of the muscle spans joint angles.

In the Hill muscle model, the muscle moment about the spanned joint [46] is indicated by:

$$M = r{\cdot}F_o^M{\cdot}[a(emg(t-d)){\cdot}f_l(\frac{l-l_s^T}{l_o^M\cos\phi}){\cdot}f_v(\frac{v}{10{\cdot}l_o^M}) + f_p(\frac{l-l_s^T}{l_o^M\cos\phi}))]\cos(\phi) \tag{1}$$

where M and r are the muscle moment and moment arm about the joint it actuates, F_O^M is muscle's peak isometric force, $a()$ is the muscle's activation which can be calculated as a function of EMG data, t is the time, d is the electromechanical delay, v and l are muscle-tendon velocity and length, ϕ is pennation angle of the muscle, l_o^M is the optimal fiber length and l_s^T is the tendon slack length. The relationship of muscle-tendon length, muscle fiber length, tendon length, pennation angle can be seen in Figure 2. $f_v()$, $f_l()$ and $f_P()$ represent muscle force-velocity, active force-length and passive force-length curve. F_o^M, d, ϕ, l_s^T and l_o^M are assumed to remain constant for the individual. l, v and r are time variables that can be calculated as polynomial functions of joint angles and angular velocities with the same constant coefficients [47,48]. When θ is the muscle spans joint angles, those time variables can be expressed as follows:

$$l(t) = l(\theta) \tag{2}$$

$$v(t) = \frac{\partial l(t)}{\partial t} = \frac{\partial l(\theta)}{\partial t} = \frac{\partial l(\theta)}{\partial \theta}\frac{\partial \theta}{\partial t} = v(\theta, \dot{\theta}) \tag{3}$$

$$r(t) = -\frac{\partial l(\theta)}{\partial \theta} = r(\theta) \tag{4}$$

where $\theta(t)$ and $\dot{\theta}(t)$ are the muscle spans joint angles and angular velocities; $l(\theta)$ is muscle-tendon length which is polynomial functions of the muscle spans joint angles; $v(\theta, \dot{\theta})$ is muscle-tendon velocity which is the first derivative of $l(\theta)$ with respect to time t; $r(\theta)$ is muscle-tendon moment arm which is the first derivative of $l(\theta)$ with respect to θ. The sign of the variable is used to determine the direction of the moment.

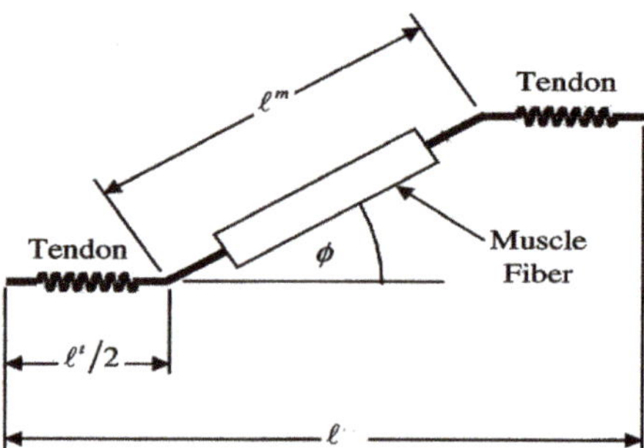

Figure 2. A diagram of muscle-tendon unit that shows the relationship of muscle-tendon length, muscle fiber length, tendon length, pennation angle. Where l is the muscle-tendon length, l^m is the muscle fiber length, l^t is the tendon length, ϕ is the pennation angle.

From Equations (1)–(4), the muscle moment about the spanned joint can be calculated as a function of the muscle's EMG signal, and the muscle actuates joints' angle and angular velocity (Figure 1):

$$M(emg, \theta, \dot{\theta}) = r(\theta) \cdot F_o^M \cdot [a(emg(t-d)) \cdot f_l(l(\theta)) \cdot f_v(v(\theta, \dot{\theta})) + f_p(l(\theta))]\cos(a) \tag{5}$$

where d is an electromechanical delay, and its value is generally 10-100ms [49]. From Equations (1)–(5) the j-th joint moment is represented by the following equation:

$$M^j = \sum_{i=1}^{m} M(emg(i), \theta(i), \dot{\theta}(i)) \tag{6}$$

where m is the number of muscles associated with the joint moment.

It can be seen from Equation (6) that the online measurable input variables for the human joint moment prediction are joint moment-associated muscles' EMG signals, and their muscles actuates joints' angles and angular velocities.

2.3. The Designed ANN

To confirm the predictive effect of the online measurable input variables, the ELM is designed and trained as the ANN model to predict joint moments, which is a feedforward ANN algorithm [40]. It can be seen from Equation (6) that different joint moments correspond to different inputs which is not suitable to use the multi-output ANN model, so the ELM only has one output neuron. Its structure is generally shown as Figure 3, which is divided into an input layer, a hidden layer and an output layer. Its expression is provided as follows:

$$O = \beta g(W \cdot X + b) \tag{7}$$

where X is the input, O is output, $W = [W_1, W_2, \cdots, W_L]$ is the matrix of input-to-hidden-layer weights, $\beta = [\beta_1, \beta_2, \cdots, \beta_L]$ is the matrix of hidden-to-output-layer weights, $b = [b_1, b_2, \cdots b_L]$ is the matrix threshold of the hidden node and $g()$ is the activation function. The distinguishing feature of ELM from the traditional feedforward neural network is that W and b are randomly selected and does not need to be adjusted during the training process, and β are calculated in the training process [45]. The feature makes the process of determining network parameters without iterations, reduces the adjustment time of network parameters, and greatly improves the learning speed. The ELM is widely used in regression analysis and classification [41,50].

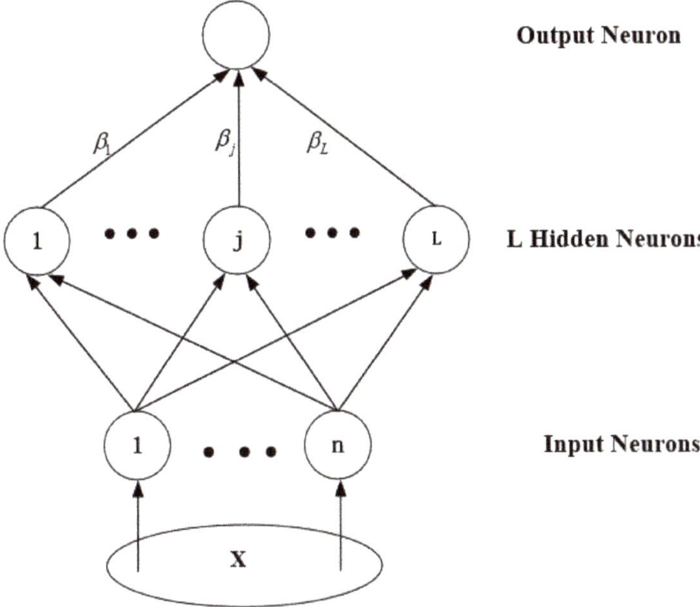

Figure 3. Structure of the designed ELM.

The ELM is trained to predict four DOFs' moment in the right leg: ankle plantar-dorsiflexion (Ankle PDF), knee flexion-extension (Knee FE), hip adduction-abduction (Hip AA) and hip flexion-extension (Hip FE), and the inverse dynamics moment is used as the target value of the training sample. It can be seen from Table 1 with Equation (6) that the input variables of Hip FE's joint moment prediction contains the EMG signals of four muscles and three joint angles and angular velocities. There are 10 input variables in total.

Table 1. The list of EMG signal sources and their muscle actuates.

EMG Signal Source	Actuates
Gluteus maximus	Hip AA, Hip FE,
Gluteus medius	Hip AA, Hip FE
Biceps femoris long head	Knee FE, Hip AA, Hip FE
Rectus femoris	Knee FE, Hip AA, Hip FE
Vastus medialis	Knee FE
Vastus lateralis	Knee FE
Gastrocnemius lateral	Knee FE, Ankle PDF, Ankle IE
Gastrocnemius medial	Knee FE, Ankle PDF, Ankle IE
Tibialis anterior	Ankle PDF, Ankle IE
Soleus	Ankle PDF, Ankle IE

2.4. Prediction Evaluation

Considering that Equation (6) is obtained under the assumption that F_o^M (muscle's peak isometric force), d (the electromechanical delay), ϕ (pennation angle of the muscle), l_s^T (the tendon slack length) and l_o^M (the optimal fiber length) are remain constant for the individual, which is not suitable for training multiple subjects at a time, so per ELM only trains one joint moment of a subject. A generic three-layer ELM is designed and trained using two strategies for evaluating the generalization ability of the method at two different levels: (1) training with all four speeds (level 1) and (2) training only with the three low speeds (2, 3 and 4 m/s) (level 2). During the supervised training, the inverse dynamics moment is used as the target value of the training samples. The variance accounted for (VAF) [51] is used to evaluate the accuracy of the ELM, its expression is as follows:

$$\text{VAF} = [1 - \frac{\text{var}(\hat{y} - y)}{\text{var}(y)}] \times 100\% \tag{8}$$

where y is the inverse dynamics moment and $\hat{y}$ is predicted joint moment. For each speed, six gait cycles ($6 \times 101 = 606$) are selected for training and testing. Since a complete gait cycle data may contain all gait features at the current speed, training and testing must take the whole gait cycle as input or it is easy to cause feature loss to make the prediction result unstable. Therefore, the data set is smaller, a greater percentage of 30% as testing data set and 70% as training data set must be used to train and test the ELM, so four ($6 \times 0.7 = 4.2$) gait cycles ($4 \times 101 = 404$ time points) data are randomly selected from each tested speed for training, and the remaining two ($6 \times 3 = 1.8$) gait cycles (202 time points) for testing. Then, in order to set the appropriate number of neurons in the hidden layer for better prediction effect, an experiment is done to observe the relationship between the number of neurons in the hidden layer and the prediction accuracy. In the experiment, four gait cycles data are selected from each speed for training, and two gait cycles for testing. The ten subjects' average predicted accuracy evaluated by the VAF (%) are shown as Figure 4.

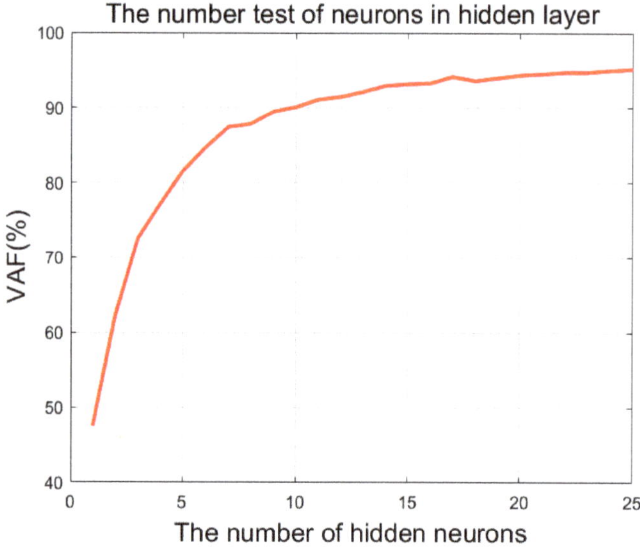

Figure 4. The ten subject's average predict accuracy evaluated by the variance accounted for (%) with the increase of neurons.

It can be seen from Figure 4 that the value of VAF increased rapidly with the increase of neurons at the beginning, but the value of VAF slowed down when the number of neurons exceeded 10. Considering the structural complexity of ELM and the time cost for training, the number of neurons in the hidden layer is set to 20.

3. Results

When training with all four speeds (level 1), the trained ANN model is used to predict the lower limbs' joint moment of all subjects at different speeds. Joint moment prediction of a typical subject at each speed are shown in Figure 5. As shown, the general pattern of lower limb joint moment can be predicted well at each speed. Comparing with inverse dynamics moment, there only have some difference in minimum and maximum values of waveforms (cross-correlation coefficient > 0.987). The VAF of the predicted joint moment for Ankle PDF, Knee FE, Hip FE and Hip AA at level 1, with the mean VAF (± standard deviation) of 97.15 ± 0.99%, 94.23 ± 2.99%, 95.39 ± 3.62% and 95.01 ± 7.46% as shown in Table 2.

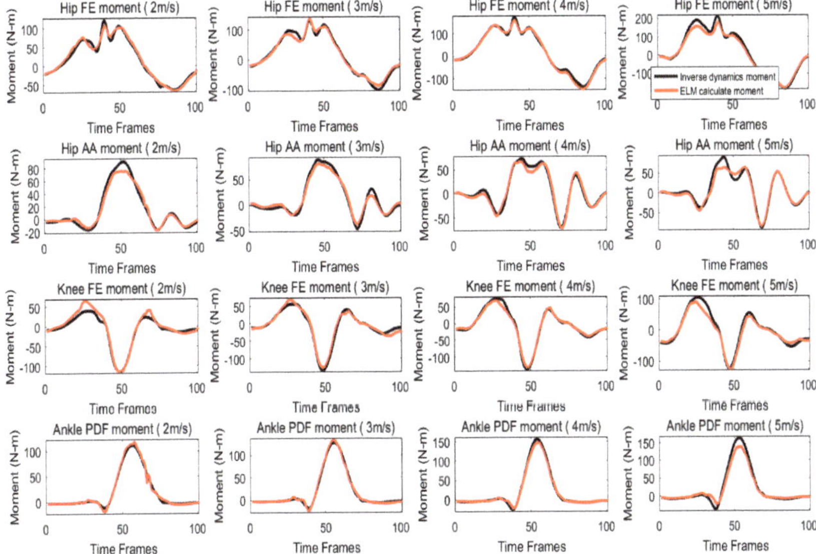

Figure 5. Joint moment prediction of a typical subject at each speed when all four speeds are used for training (level 1).

Table 2. Joint moment prediction performances for level 1, evaluated by VAF (%).

Participants	Hip FE	Hip AA	Knee FE	Ankle PDF
subject 1	97	94.50	96.47	98.11
subject 2	96.98	95.80	96.90	97.61
subject 3	94.85	87.02	86.69	73.89
subject 4	97.69	96.17	98.20	98.27
subject 5	96.86	92.15	95.12	96.94
subject 6	96.37	93.58	94.65	96.40
subject 7	97.78	96.74	95.46	96.65
subject 8	97.88	96.54	97.62	98.42
subject 9	98.15	96.46	98.02	96.22
subject 10	97.94	93.37	95.73	97.62
mean	97.15	94.23	95.39	95.01
Std	0.99	2.99	3.62	7.46

When training with the three low speeds (level 2), the trained ANN model is also used to predict the lower limbs' joint moment of all objects at different speeds. Joint moment prediction of a typical subject at each speed are shown in Figure 6. As shown, the errors between the predicted and inverse dynamics moment were slightly increased, when compared to the corresponding errors at level 1 (cross-correlation coefficient > 0.984), especially the speed of 5m/s. The VAF of the predicted joint moment for Ankle PDF, Knee FE, Hip FE and Hip AA at level 1, with the mean VAF (± standard deviation) of 94.31 ± 7.13, 93.04 ± 3.62, 92.08 ± 2.93% and 89.95 ± 2.31% as shown in Table 3.

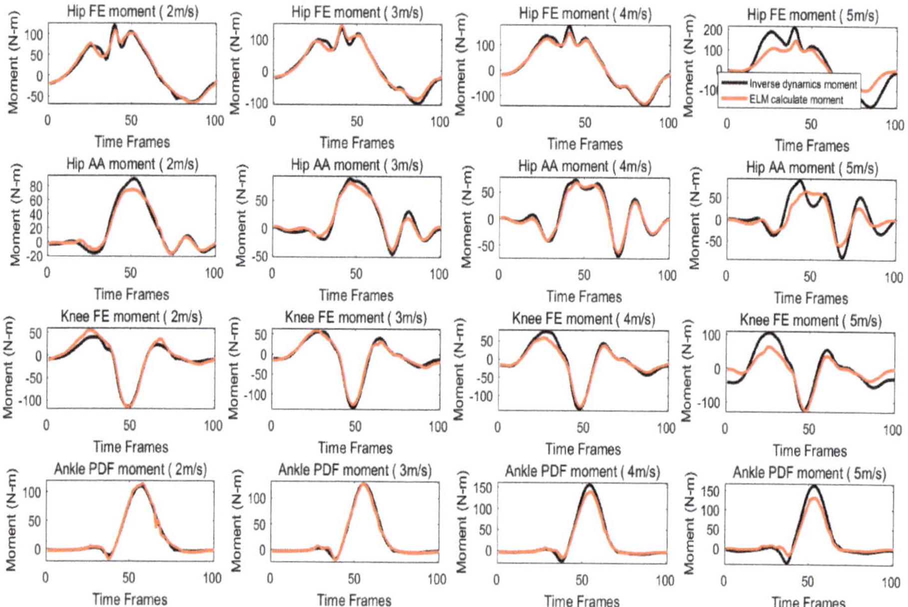

Figure 6. Joint moment prediction of a typical subject at each speed when only the three low speeds are used for training (level 2).

Table 3. Joint moment prediction performances for level 2, evaluated by VAF (%).

Participants	Hip FE	Hip AA	Knee FE	Ankle PDF
subject 1	88.31	94.06	93.04	97.50
subject 2	88.09	94.26	93.48	96.82
subject 3	89.80	86.52	84.55	74.20
subject 4	92.07	94.84	97.58	98.06
subject 5	85.36	89.84	92.40	95.92
subject 6	89.17	90.44	92.73	95.66
subject 7	92.14	94.56	92.68	95.85
subject 8	92.81	94	96.20	97.72
subject 9	91.67	93.43	96.49	94.89
subject 10	90.08	88.81	91.32	96.51
mean	89.95	92.08	93.04	94.31
Std	2.31	2.93	3.62	7.13

In order to examine generalizability over multiple conditions, a more exhaustive validation of the test result data is conducted using jack-knife cross-validation [52] which all cross-validation subsets consist of only one data set each. In the jack-knife cross-validation, six gait cycles at each speed are taken as one data set, and there are four data sets in total. In each test, three data sets are selected as training sets and one data set as test set, and their average VAF of ten subjects 'predicted joint moment for Ankle PDF, Knee FE, Hip FE and Hip AA are shown in Table 4. As shown in Table 4, the obtained results have little difference from level 2.

Table 4. Joint moment prediction performances for jack-knife cross-validation, evaluated by VAF (%).

Participants	Hip FE	Hip AA	Knee FE	Ankle PDF
mean	81.07	91.88	92.68	93.09
Std	16.37	14.10	13.67	13.42

Furthermore, the method (EAV) is compared with other combination of inputs using jack-knife cross-validation by VAF (Figure 7).

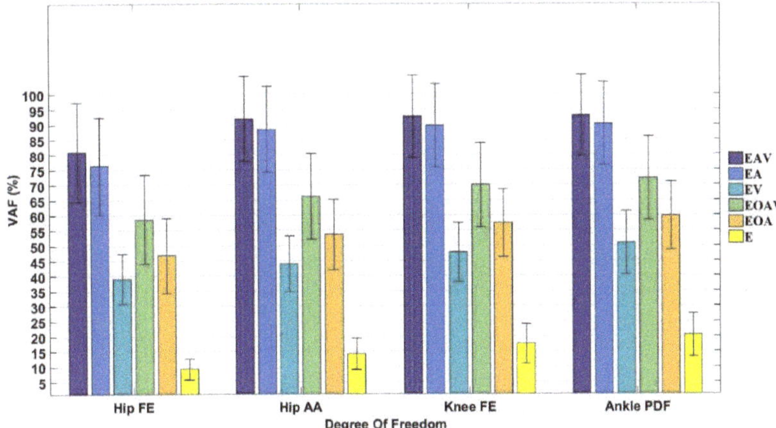

Figure 7. Comparison of performance by jack-knife cross-validation for several combination of inputs: EAV = relevant muscles' EMG, and their muscles actuate joints' Angles and angular Velocities; EA = relevant muscles' EMG and their muscles actuate joints' Angles; EV = relevant muscles' EMG and their muscles actuate joints' Angles; EJAV= relevant muscles' EMG, the Joint's Angle and Angular velocity; EJA = relevant muscles' EMG and the Joint's Angular velocity; E = relevant muscles' EMG signals.

They are five different inputs as following: (1) Relevant EMG signals and their muscles actuate joints' Angles (EA); (2) Relevant EMG signals and their muscles actuate joints' angular Velocities (EV); (3) Relevant muscles' EMG signals, the Joint's Angle and angular Velocity (EJAV); (4) Relevant muscles' EMG signals and the Joint's Angle (EJA); (5) Relevant muscles' EMG signals as inputs (E). The relevant muscles' EMG signals means that the joint moment-associated muscles' EMG signals. Take EAV ($\overline{VAF}$= 89.67 ± 5.56%) as reference and compare with the above inputs respectively, It can be seen that the $\overline{VAF}$ of the moment predicted by the EA ($\overline{VAF}$= 86.21 ± 6.60%), EV($\overline{VAF}$= 45.48 ± 5.08%), EJAV($\overline{VAF}$= 66.80 ± 5.91%), EJA($\overline{VAF}$= 54.41 ± 5.70%), and E($\overline{VAF}$ = 15.39 ± 4.81%) are almost reduced by 3.85%, 49.27%, 25.50%, 39.31% and 82.83%.

4. Discussion and Conclusions

This study demonstrated that the ELM with the online measurable input variables could be used as a real-time surrogate model to predict joint moments under different gait speeds. Compared with the previous studies [29–33,53–55], this research extends our knowledge by establishing the mathematical model of input-output relation in the human joint moment prediction based on the Hill muscle model. The online measurable input variables are obtained for the ANN model. It does not need ground reaction force and marker trajectories which increases the number of input sensors and the complexity of prediction. The novel method has high prediction accuracy with $\overline{VAF}$ = 96.07 ± 3.484%. Thus, the proposed method is suitable for online rehabilitation assessment and human-robot interaction which need to obtain joint moment in real time.

It can be seen from Equations (1)–(6) that the muscles actuate joints are very limited, while inertial magnetic measurement systems are good at measuring the limited joints' angles and angular velocities [56], so unlike previous computational models, such as inverse dynamics [57,58] and EMG-driven models [39,46,59], the method can online predict joint moment without essential 3D motion capture and complicated calculation, which make the hospitals and laboratories to predict joint moments without site requirements, even in a free state. It can also adapt to the individual differences in the process of training, and does not need the musculoskeletal model or the scaling of specific objects, thereby reducing the error caused by individual differences. Furthermore, the training time is less than one second.

Compared level 2 with level 1 and the jack-knife cross-validation results (Table 4), the results suggest that the proposed method has a good generalization ability. Thus, in practice, a reduced amount of training data can be used when a large amount of data is not available. It can be seen from Figure 7 that EAV has the best prediction results in all joints compared with other inputs, which verifies the accuracy of the method proposed in this paper. Comparing our method with EA, the latter's $\overline{VAF}$ only reduced by 3.85%. Thus, it can be concluded that the effect of angular velocities on joint moment prediction is relatively small. Comparing the method with E, the latter's $\overline{VAF}$ reduced by 82.83%. This indicates that: (1) the EMG value alone cannot represent the value of the joint moment [60], and (2) the joint angle has a great influence on the joint moment prediction. From Figure 7, It can also be found that the EJAV has good prediction results, so it can be concluded that the effect of the joint moment's angle and angular velocity on joint moment prediction is very important. This is the reason why the musculoskeletal model use joint's angles and angular velocities as inputs to calculate joint moments. As the ANN model can adapt to the individual differences in the process of training and the muscle model is applicable to all muscles of any human body whether male or female, old or young and health or not, so the proposed method can also be applied to other joints of any human body theoretically.

It should be mentioned that the current study has some limitations. Firstly, there are only 10 muscles' EMG data of the right leg used in the method, which can't represent all muscles associated with the joint. our approach will be developed in a larger set in the future. Secondly, the gait patterns in the experimental only include run gait patterns, which is very limited. In the future study, more gait data will be collected, such as squatting, cutting and so on. Finally, the sample is only composed of young male subjects with similar anthropometry and age, which cannot ensure the diversity of the training samples. Data samples from different groups of people will be collected in the future, such as children, old people, women, patients and so on.

Author Contributions: B.X. and M.D. conceived the layout, the rationale, and the plan of this manuscript. B.X. wrote the first draft of the manuscript. N.Z., M.D., Y.L., M.H., G.M., W.S. and Y.Y. edited the manuscript. All authors have read and agreed to the published version of the manuscript.

Founding: This research received no external funding.

Acknowledgments: This work was supported in part by in part by National Nature Science Foundation of China (61773124, 61773415), in part by National Key Research and Development Program of China (2016YFE0122700), in part by UK-China Industry Academia Partnership Programmer\276, the Science and Tecohnology Project in Fujian Province Education Department (JT180344/ JT180320/JAT170398), and in part by the Scientific Fund Projects in Fujian University of Technology (GY-Z17151/GY-Z17144). Y.Y. is supported by the Dixon Translational Research Grants Initiative from the Northwestern Memorial Foundation.

Conflicts of Interest: The authors declare no conflict of interest.

References

1. Park, H.S.; Peng, Q.; Zhang, L.Q. A Portable telerehabilitation system for remote evaluations of impaired elbows in neurological disorders. *IEEE Trans. Neural Syst. Rehabil. Eng.* **2008**, *16*, 245–254. [CrossRef] [PubMed]

2. Zhang, S.; Guo, S.; Gao, B.; Hirata, H.; Ishihara, H. Design of a novel telerehabilitation system with a force-sensing mechanism. *Sensors* **2015**, *15*, 11511–11527. [CrossRef]

3. Song, Z.; Guo, S.; Pang, M.; Zhang, S.; Xiao, N.; Gao, B.; Shi, L. Implementation of resistance training using an upper-limb exoskeleton rehabilitation device for elbow joint. *J. Med. Biol. Eng.* **2014**, *34*, 188–196. [CrossRef]

4. Pfeiffer, M.; Hohmann, A. Applications of neural networks in training science. *Hum. Mov. Sci.* **2012**, *31*, 344–359. [CrossRef] [PubMed]

5. Iyer, S.R.; Sharda, R. Prediction of athletes performance using neural networks: An application in cricket team selection. *Expert Syst. Appl.* **2009**, *36*, 5510–5522. [CrossRef]

6. Schmidt, A. Movement pattern recognition in basketball free-throw shooting. *Hum. Mov. Sci.* **2012**, *31*, 360–382. [CrossRef]

7. Au, S.; Berniker, M.; Herr, H. Powered ankle-foot prosthesis to assist level-ground and stair-descent gaits. *Neural Netw.* **2008**, *21*, 654–666. [CrossRef]

8. Rupérez, M.J.; Martín-Guerrero, J.D.; Monserrat, C.; Alcañiz, M. Artificial neural networks for predicting dorsal pressures on the foot surface while walking. *Expert Syst. Appl.* **2012**, *39*, 5349–5357. [CrossRef]

9. Joshi, D.; Mishra, A.; Anand, S. ANFIS based knee angle prediction: An approach to design speed adaptive contra lateral controlled AK prosthesis. *Appl. Soft Comput.* **2011**, *11*, 4757–4765. [CrossRef]

10. Schoen, M.; Rotter, R.; Schattner, S.; Mittlmeier, T.; Claes, L.; Vollmar, B.; Gradl, G. Introduction of a new interlocked intramedullary nailing device for stabilization of critically sized femoral defects in the rat: A combined biomechanical and animal experimental study. *J. Orthop. Res.* **2008**, *26*, 184–189. [CrossRef]

11. Sanders, D.; Bryant, D.; Tieszer, C.; Lawendy, A.; Macleod, M.; Papp, S.; Allan, L.; Darius, V.; Chad, C.; Kevin, G. A multi-centre randomized control trial comparing a novel intramedullary device (InterTAN) versus conventional treatment (sliding hip screw) of geriatric hip fractures. *J. Orthop. Trauma* **2016**, *31*, 1. [CrossRef]

12. Pascoletti, G.; Cianetti, F.; Putame, G.; Terzini, M.; Zanetti, E.M. Numerical simulation of an intramedullary elastic nail: Expansion phase and load-bearing behavior. *Front. Bioeng. Biotechnol.* **2018**, *6*, 174. [CrossRef] [PubMed]

13. Yeung, S.S.; Yeung, E.W. Acute effects of kinesio taping on knee extensor peak torque and stretch reflex in healthy adults. *Medicine* **2016**, *95*, e2615. [CrossRef]

14. Ajoudani, A.; Tsagarakis, N.G.; Bicchi, A. Choosing poses for force and stiffness control. *IEEE Trans. Robot.* **2017**, *33*, 1483–1490. [CrossRef]

15. Al-Quraishi, M.S.; Ishak, A.J.; Ahmad, S.A.; Hasan, M.K.; Al-Qurishi, M.; Ghapanchizadeh, H.; Alamri, A. Classification of ankle joint movements based on surface electromyography signals for rehabilitation robot applications. *Med. Biol. Eng. Comput.* **2016**, *55*, 1–12. [CrossRef] [PubMed]

16. Koopman, B.; Asseldonk, E.H.F.V.; Kooij, H.V.D. Estimation of human hip and knee multi-joint dynamics using the LOPES gait trainer. *IEEE Trans. Robot.* **2016**, *32*, 920–932. [CrossRef]

17. Huo, W.; Mohammed, S.; Amirat, Y.; Kong, K. Fast gait mode detection and assistive torque control of an exoskeletal robotic orthosis for walking assistance. *IEEE Trans. Robot.* **2018**, *34*, 1035–1052. [CrossRef]

18. Focchi, M.; Prete, A.; Havoutis, I.; Featherstone, R.; Caldwell, D.G.; Semini, C. High-Slope terrain locomotion for torque-controlled quadruped robots. *Auton. Robot.* **2017**, *41*, 259–272. [CrossRef]

19. Prete, A.D.; Mansard, N. Robustness to joint-torque-tracking errors in task-space inverse dynamics. *IEEE Trans. Robot.* **2016**, *32*, 1091–1105. [CrossRef]

20. Souron, R.; Bordat, F.; Farabet, A.; Belli, A.; Feasson, L.; Nordez, A.; Lapole, T. Sex differences in active tibialis anterior stiffness evaluated using supersonic shear imaging. *J. Biomech.* **2016**, *49*, 3534–3537. [CrossRef]

21. Zhang, H.; Ahmad, S.; Liu, G. Torque estimation for robotic joint with harmonic drive transmission based on position measurements. *IEEE Trans. Robot.* **2017**, *31*, 322–330. [CrossRef]

22. Fregly, B.J.; Besier, T.F.; Lloyd, D.G.; Delp, S.L.; Banks, S.A.; Pandy, M.G.; D'Lima, D.D. Grand challenge competition to predict in vivo knee loads. *J. Orthop. Res.* **2012**, *30*, 503–513. [CrossRef] [PubMed]

23. Jiang, X.; Zhang, Y.-D. Chinese sign language fingerspelling recognition via six-layer convolutional neural network with leaky rectified linear units for therapy and rehabilitation. *J. Med. Imaging Health Inform.* **2019**, *9*, 2031–2038. [CrossRef]

24. Wang, S.; Sun, J.; Mehmood, I.; Pan, C.; Chen, Y.; Zhang, Y.-D. Cerebral micro-bleeding identification based on a nine-layer convolutional neural network with stochastic pooling. *Concurr. Comput. Pract. Exp.* **2019**, *32*, e5130. [CrossRef]

25. Zeng, N.; Wang, Z.; Zhang, H.; Kim, K.-E.; Li, Y.; Liu, X. An improved particle filter with a novel hybrid proposal distribution for quantitative analysis of gold immunochromatographic strips. *IEEE Trans. Nanotechnol.* **2019**, *18*, 819–829. [CrossRef]

26. Zhang, Y.-D.; Govindaraj, V.V.; Tang, C.; Zhu, W.; Sun, J. High performance multiple sclerosis classification by data augmentation and AlexNet transfer learning model. *J. Med. Imaging Health Inform.* **2019**, *9*, 2012–2021. [CrossRef]

27. Wasserman, P.D. *Neural Computing: Theory and Practice*; Van Nostrand Reinhold: New York, NY, USA, 1989.

28. Wang, S.-H.; Xie, S.; Chen, X.; Guttery, D.S.; Tang, C.; Sun, J.; Zhang, Y.-D. Alcoholism identification based on an AlexNet transfer learning model. *Front. Psychiatry* **2019**, *10*, 205. [CrossRef]

29. Uchiyama, T.; Bessho, T.; Akazawa, K. Static torque-angle relation of human elbow joint estimated with artificial neural network technique. *J. Biomech.* **1998**, *31*, 545–554. [CrossRef]

30. Luh, J.-J.; Chang, G.-C.; Cheng, C.-K.; Lai, J.-S.; Kuo, T.-S. Isokinetic elbow joint torques estimation from surface EMG and joint kinematic data: Using an artificial neural network model. *J. Electromyogr. Kinesiol.* **1999**, *9*, 173–183. [CrossRef]

31. Song, R.; Tong, K.Y. Using recurrent artificial neural network model to estimate voluntary elbow torque in dynamic situations. *Med. Biol. Eng. Comput.* **2005**, *43*, 473–480. [CrossRef]

32. Hahn, M.E. Feasibility of estimating isokinetic knee torque using a neural network model. *J. Biomech.* **2007**, *40*, 1107–1114. [CrossRef]

33. Ardestani, M.M.; Zhang, X.; Wang, L.; Lian, Q.; Liu, Y.; He, J.; Li, D.; Jin, Z. Human lower extremity joint moment prediction: A wavelet neural network approach. *Expert Syst. Appl.* **2014**, *41*, 4422–4433. [CrossRef]

34. Xiong, B.; Zeng, N.; Li, H.; Yang, Y.; Li, Y.; Huang, M.; Shi, W.; Du, M.; Zhang, Y. Intelligent prediction of human lower extremity joint moment: An artificial network approach. *IEEE Access.* **2019**, *7*, 29973–29980. [CrossRef]

35. Delp, S.L.; Loan, J.P.; Hoy, M.G.; Zajac, F.E.; Topp, E.L.; Rosen, J.M. An interactive graphics-based model of the lower extremity to study orthopaedic surgical procedures. *IEEE Trans. Biomed. Eng.* **1990**, *37*, 757–767. [CrossRef] [PubMed]

36. Putame, G.; Terzini, M.; Bignardi, C.; Beale, B.; Hulse, D.; Zanetti, E.; Audenino, A. Surgical treatments for canine anterior cruciate ligament rupture: assessing functional recovery through multibody comparative analysis. *Front. Bioeng. Biotechnol.* **2019**, *7*, 180. [CrossRef] [PubMed]

37. Hill, A.V. The heat of shortening and the dynamic constants of muscle. *Proc. R. Soc. Lond. Ser. B Biol. Sci.* **1938**, *126*, 136–195.

38. Zajac, F.E. Muscle and tendon: Properties, models, scaling, and application to biomechanics and motor control. *Crit. Rev. Biomed. Eng.* **1989**, *17*, 359–411.

39. Meyer, A.J.; Patten, C.; Fregly, B.J. Lower extremity EMG-driven modeling of walking with automated adjustment of musculoskeletal geometry. *PLoS ONE* **2017**, *12*, e0179698. [CrossRef]

40. Huang, G.-B.; Zhu, Q.-Y.; Siew, C.-K. Extreme learning machine: Theory and applications. *Neurocomputing* **2006**, *70*, 489–501. [CrossRef]

41. Yang, Y.; Wu, Q.M.J.; Wang, Y. Autoencoder with invertible functions for dimension reduction and image reconstruction. *IEEE Trans. Syst. Man Cybern. Syst.* **2016**, *48*, 1065–1079. [CrossRef]

42. Kasun, L.L.C.; Yang, Y.; Huang, G.B.; Zhang, Z. Dimension reduction with extreme learning machine. *IEEE Trans. Image Process.* **2016**, *25*, 3906–3918. [CrossRef]

43. Huang, G.-B.; Chen, L. Enhanced random search based incremental extreme learning machine. *Neurocomputing* **2008**, *71*, 3460–3468. [CrossRef]

44. Hamner, S.R.; Delp, S.L. Muscle contributions to fore-aft and vertical body mass center accelerations over a range of running speeds. *J. Biomech.* **2013**, *46*, 780–787. [CrossRef] [PubMed]

45. Sherman, M.A.; Seth, A.; Delp, S.L. Simbody: Multibody dynamics for biomedical research. *Procedia Iutam* **2011**, *2*, 241–261. [CrossRef] [PubMed]

46. Lloyd, D.G.; Besier, T.F. An EMG-driven musculoskeletal model to estimate muscle forces and knee joint moments In Vivo. *J. Biomech.* **2003**, *36*, 765–776. [CrossRef]

47. Li, L.; Tong, K.Y.; Hu, X.L.; Hung, L.K.; Koo, T.K.K. Incorporating ultrasound-measured musculotendon parameters to subject-specific EMG-driven model to simulate voluntary elbow flexion for persons after stroke. *Clin. Biomech.* **2009**, *24*, 101–109. [CrossRef]

48. Nikooyan, A.A.; Veeger, H.E.J.; Westerhoff, P.; Bolsterlee, B.; Graichen, F.; Bergmann, G.; van der Helm, F.C.T. An EMG-driven musculoskeletal model of the shoulder. *Hum. Mov. Sci.* **2012**, *31*, 429–447. [CrossRef]

49. Corcos, D.M.; Gottlieb, G.L.; Latash, M.L.; Almeida, G.L.; Agarwal, G.C. Electromechanical delay: An experimental artifact. *J. Electromyogr. Kinesiol.* **1992**, *2*, 59–68. [CrossRef]

50. Tanzil, S.M.S.; Hoiles, W.; Krishnamurthy, V. Adaptive scheme for caching YouTube content in a cellular network: Machine learning approach. *IEEE Access* **2017**, *5*, 5870–5881. [CrossRef]

51. Filatova, O.G.; Yang, Y.; Dewald, J.P.A.; Tian, R.; Maceira-Elvira, P.; Takeda, Y.; Kwakkel, G.; Yamashita, O.; van der Helm, F.C.T. Dynamic information flow based on EEG and diffusion MRI in stroke: A Proof-of-Principle Study. *Front. Neural Circuits* **2018**, *12*, 79. [CrossRef]

52. McRoberts, R.E.; Magnussen, S.; Tomppo, E.O.; Chirici, G. Parametric, bootstrap, and jackknife variance estimators for the k-Nearest Neighbors technique with illustrations using forest inventory and satellite image data. *Remote Sens. Environ.* **2011**, *115*, 3165–3174. [CrossRef]

53. Schöllhorn, W.I. Applications of artificial neural nets in clinical biomechanics. *Clin. Biomech.* **2004**, *19*, 876–898. [CrossRef] [PubMed]

54. Kipp, K.; Giordanelli, M.; Geiser, C. Predicting net joint moments during a weightlifting exercise with a neural network model. *J. Biomech.* **2018**, *74*, 225–229. [CrossRef] [PubMed]

55. Wang, L.; Buchanan, T.S. Prediction of joint moments using a neural network model of muscle activations from EMG signals. *IEEE Trans. Neural Syst. Rehabil. Eng.* **2002**, *10*, 30–37. [CrossRef]

56. de Vries, W.H.K.; Veeger, H.E.J.; Baten, C.T.M.; van der Helm, F.C.T. Can shoulder joint reaction forces be estimated by neural networks? *J. Biomech.* **2016**, *49*, 73–79. [CrossRef]

57. Kim, Y.-H.; Phuong, B.T.T. Estimation of joint moment and muscle force in lower extremity during sit-to-stand movement by inverse dynamics analysis and by electromyography. *Trans. Korean Soc. Mech. Eng. A* **2010**, *34*, 1345–1350. [CrossRef]

58. Happee, R.; Van der Helm, F.C.T. The control of shoulder muscles during goal directed movements, an inverse dynamic analysis. *J. Biomech.* **1995**, *28*, 1179–1191. [CrossRef]

59. Gardinier, E.S.; Manal, K.; Buchanan, T.S.; Snydermackler, L. Minimum detectable change for knee joint contact force estimates using an EMG-driven model. *Gait Posture* **2013**, *38*, 1051–1053. [CrossRef]

60. Henriksen, M.; Aaboe, J.; Bliddal, H.; Langberg, H. Biomechanical characteristics of the eccentric Achilles tendon exercise. *J. Biomech.* **2009**, *42*, 2702–2707. [CrossRef]

Review

Pain and Stress Detection Using Wearable Sensors and Devices—A Review

Jerry Chen [1], Maysam Abbod [2,*] and Jiann-Shing Shieh [1,*]

[1] Department of Mechanical Engineering, Yuan Ze University, Taoyuan 32003, Taiwan; s1088701@mail.yzu.edu.tw
[2] Department of Electronic and Computer Engineering, Brunel University London, Uxbridge UB8 3PH, UK
* Correspondence: Maysam.Abbod@brunel.ac.uk (M.A.); jsshieh@saturn.yzu.edu.tw (J.-S.S.)

Abstract: Pain is a subjective feeling; it is a sensation that every human being must have experienced all their life. Yet, its mechanism and the way to immune to it is still a question to be answered. This review presents the mechanism and correlation of pain and stress, their assessment and detection approach with medical devices and wearable sensors. Various physiological signals (i.e., heart activity, brain activity, muscle activity, electrodermal activity, respiratory, blood volume pulse, skin temperature) and behavioral signals are organized for wearables sensors detection. By reviewing the wearable sensors used in the healthcare domain, we hope to find a way for wearable healthcare-monitoring system to be applied on pain and stress detection. Since pain leads to multiple consequences or symptoms such as muscle tension and depression that are stress related, there is a chance to find a new approach for chronic pain detection using daily life sensors or devices. Then by integrating modern computing techniques, there is a chance to handle pain and stress management issue.

Keywords: pain detection; stress detection; wearable sensor; physiological signals; behavioral signals

Citation: Chen, J.; Abbod, M.; Shieh, J.-S. Pain and Stress Detection Using Wearable Sensors and Devices—A Review. *Sensors* **2021**, *21*, 1030. https://doi.org/10.3390/s21041030

Academic Editor: Ki H. Chon
Received: 25 December 2020
Accepted: 2 February 2021
Published: 3 February 2021

Publisher's Note: MDPI stays neutral with regard to jurisdictional claims in published maps and institutional affiliations.

1. Introduction

Pain is a highly inter-variated and subjective feeling. What makes one person feel excessive pain may not be exactly same for another. In order to reach a general perception of pain, people have been constantly looking for a relevant scale or index try to quantify this sensation objectively for hundreds of years [1]. To extract more information that helps better understanding of pain, required numerous studies based on experiments and clinical observations. Since pain generated in both types of scenarios is linked to the same original sensation that is embedded in the human body, the mechanism of the pain is being clarified by conducting more and more experiments or observing the symptoms in the clinic. Back in 1846, when the first anesthetic (ether) was publicly demonstrated for general anesthesia by Morton at Massachusetts General Hospital in Boston (MA, USA), the majority thought the agony of pain has become history. However, looking back from now, that event might just be the beginning of our understanding of the mechanisms of pain. In general anesthesia during surgery, anesthesiologists use their knowledge of anesthetics to make subjects go into unconsciousness and block the sensation of pain for the purpose of performing the surgery more smoothly, but pain is a strong sensation that not only exists during surgery but also can exist potentially in any moment of our lives. It acts as more than just an unpleasant experience to everyone but also plays the role of a useful reminder to avoid potential injuries or tissue damage. Thus, pain research is not only about how to stop it, but more importantly what is the problem that this sensation is implicitly pointing to. Their causation of some pains is easy to identify, and thus can easily taken care of by treating the causative wounds or injuries. However, not all types of pain have a clear or obvious reason responsible for it. Other kinds of pain have no clear correlated injuries or wounds that need to be cured. Sometimes this type of pain is observed even after the original injuries have healed. Unfortunately, pain is also tricky to study for two reasons:

pain-inducing tests are rarely viable in research; in fact, it is hard to design a pain-related experiment without ethical conflicts with human rights. For this reason existing pain-inducing tests (e.g., hot plate test, tail flick test, etc.) solely use animals as experimental subjects or focus on the relation between a pre-existing pain condition with some specific movement (e.g., back pain-inducing test [2] examines the pain condition during movements of lying supine, rolling over, and sitting up). The challenge of designing a proper pain-inducing test leads to the second difficulty, which is that pain detection is thus limited in the medical field or clinical aspects. Most pain-related research is done by observing patients in the clinics or during surgery. The devices used in measuring pain are usually expensive and only viable in hospitals and mostly used in surgery rooms. Thus, this review article aims first to emphasize the association between pain and stress and then, by adapting resourceful tests and wearable sensors-based detection techniques for stress detection, it aims to overcome the bottleneck of the universal pain detection problem.

2. Review Scope

Different from previous reviews that were centered around the compliance and usability different among self-report pain scaling [3] such as Visual Analogue Scale (VAS), Verbal Rating Scale (VRS) and the Numerical Rating Scale (NRS) in clinical use, this review also brought up other pain scaling approaches that are based not solely on subjective self-reporting but rather objective physiological signals for pain detection. Such equipment is only applicable in hospitals for inter- or post-surgery studies. Its potential to be universalized by wearable sensors is discussed in this review. On the other hand, although reviews on the topic of stress detection [4] are often based on physiological signals and algorithmic approaches for feature extraction, wearable sensor systems for stress detection are rarely discussed with pain detection and its integration for health monitoring systems in mind. Furthermore, the difficulties of applying pain-inducing experiments urges this paper to dive into the relations between pain and stress, their mechanisms, correlations, assessment, and applied medical devices and wearable sensors. The purpose of this review is summing up the modern physiological and behavioral-based techniques for both pain and stress detection. Then we also discuss the demands for a wearable-based monitoring system, the evaluation of the system and its possibilities to overcome the issues of pain management and stress management.

3. Mechanism of Pain

The mechanisms of pain are being clarified by more and more studies and research. The pain process is coming to be understood as a dynamic phenomenon [5]. The nociceptive signal travels from receptors (nociceptors) to peripheral nerves then to the spinal cord and then to cerebral structures where the thalamus transmits the signals to the somatosensory cortex, frontal cortex and limbic system. Although the sensation of pain is being carried by nerve fibers [6], different types of nerve are used for different sensations as shown in Table 1. The first kind of nerve fiber is A-alpha nerve fibers; its diameter is about 13−20 μm long. Its signal conduction speed is about 80−120 m/s and it is in charge of carrying information related to position and spatial awareness. The second type of nerve fiber are called A-beta; its diameter is 6−13 μm and it conducts touching signals at a speed of 35−75 m/s. The third type of nerve fibers are A-delta ones; despite the fact they have a smaller diameter (around 1−5 μm) and slower speed for conducting signals (5−35 m/s), information such as sharp pain and temperature are delivered through them. The last type of nerve fiber is C fibers which have the smallest diameter at around 0.2−1.5 μm and the slowest signal conducting speed at 0.5−2.0 m/s but they can carry information such as dull pain, temperature, and itching. The different speeds of signal conduction may cause the sensation of a sequence of signals in subjects. For example, since t A-delta fibers are larger and surrounded by myelin (a lipid-rich substance that acts as an insulator for nerve cell axons), when someone pricks their finger, they are expected to sense the sharp sensation followed by a slower ache due to the difference of speed between A-delta fibers and C fibers.

Table 1. Different kinds of nerve fiber.

	A-alpha	A-beta	A-delta	C
Myelinated/unmyelinated	Myelinated	Myelinated	Myelinated	Unmyelinated
Size (diameter)	13–20 µm	6–13 µm	1–5 µm	0.2–1.5 µm
Speed of signal transmission in meter per second	80–120 m/s	35–75 m/s	5–35 m/s	0.5–2.0 m/s
Related perception	Position and spatial awareness	touching	Sharp pain and temperatures sensation	Dull pain temperatures and itches

On the path of transmitting pain to the brain, nerve fibers go through the dorsal horn that acts as a relay station or gateway for the signals [7]. Inside the spinal cord, the dorsal horn intervenes in the transmission of nerve signals; it either amplifies the nociceptive signal and pass it through or decreases the amplitude of the signal and ends it. This gate-like behavior, first proposed by Melzack and Wall, is known as the "gate control theory of pain" [8]. According to this theory, when a pain signal reaches the spinal cord and the central nervous system (CNS) it could be either amplified, reduced, or blocked by the system. This kind of condition is commonly observed in cases after subjects have experienced a severe injury and suffering paralysis of the lower limbs. The intervention of pain signals is also related to the types of nerve fiber. That is, signals carried by different nerve fibers have different priority in the sensation mechanisms at the spinal cord [9]. One instance is when people rub a wounded body part, which seems to attenuate the sensation of pain. This is due to the modulating effect of the counter-mechanism on large-diameter afferent fibers inhibiting the transmission and small-diameter afferent fibers facilitating the transmission [10]. Thus, when the A-beta fibers synapse is activated, it has the tendency to close the gate then mediate the sensation of C nerve fibers. However, there do exist some cases (e.g., phantom limb pain) that gate control theory alone cannot explain. It also involves the mechanism of the brain [8].

Finally, besides the interactions between nerve fibers and the spinal cord, there are other factors that may deviate the perception of pain. Factors that could affect the perception of pain are emotions [11] and psychological state [12]. Different mindsets and expectations toward the pain could either enhance the pain experience or reduce it. Personal beliefs and values under social or cultural influences may alter the perception of pain or vice versa [13]. Physical state changes (e.g., age, health status, etc.) could also worsen the perception of pain [14].

4. Classification of Pain

Generally, pain is evaluated in multiple aspects such as the location of the pain, the possible causes, the frequency of the pain occurrence, its intensity and the period of the pain. Classifying pain benefits the communication between patients and clinicians which hence facilitates the assessment task, helps formulate treatment planning and increases the precision of diagnoses in the clinic. In reality, however, not all pains have clear causes linked to them or have an adequate treatment for the pain. Some pains might have apparent causes but no adequate treatment (e.g., deep tissue disorders, peripheral nerve disorders, etc.) while pains like trigeminal neuralgia have adequate treatment without the causation being known. Then there are other pains that neither have clear causes nor treatment such as back pain and fibromyalgia [15]. Due to the complexity of pain causes and adequate treatments, there exists numerous methods for the classification of pain. Such classifications are expanding and new types of pain are being overserved in the clinic. The classification itself can sometimes ironically be confusing to clinicians [16]. However, there is a general consensus on pain classification that is agreed upon by a majority of researchers and clinicians.

4.1. Classification of Pain by Its Mechanisms

One of the most common classification techniques is based on the different mechanisms that originate the pain [17], which classifies it into the following types:

4.1.1. Nociceptive Pain

Pain caused by injuries to body tissues is classified as nociceptive pain. This is the pain that comes after a cut, burn or fracture type of body tissue injury. Other pain of this type can commonly be observed in subjects who have undergone surgery during the postoperative period. This type of pain is described as aching, sharp or throbbing. Since the pain is caused by a body tissue injury, any movement (e.g., coughing, touching etc.) related to the injured part can amplify the pain sensation.

4.1.2. Neuropathic Pain

Neuropathic pain originally meant the pain caused by a primary lesion, the dysfunction or transitory perturbation of nerves or the peripheral or central nervous system until it was redefined by International Association for the Study of Pain (IASP) taxonomy as "pain that caused by a lesion or disease of the somatosensory system". Neuropathic pain is not a single disease; it is a syndrome caused by different diseases and lesions for which some of the underlying mechanisms might be unknown [18]. Sometimes neuropathic pain is depicted as a burning, tingling, and numbness sensation. People suffering neuropathic pain can also feel excessive pain from minor stimuli such as a light touch.

4.1.3. Nociplastic Pain

Nociplastic pain are defined in the IASP 2017 taxonomy as "pain that arises from altered nociception despite no clear evidence of actual or threatened tissue damage causing the activation of peripheral nociceptors or evidence for a disease or lesion of the somatosensory system causing the pain". Nociplastic pain is a relatively new term compared with nociceptive pain and neuropathic pain. In fact, there was only nociceptive pain and neuropathic pain before the IASP added the third mechanistic descriptor to its taxonomy. The call for the third mechanistic descriptor was to fill the lack of a proper valid pathophysiological descriptor among patient groups having fibromyalgia, complex regional pain syndrome (CRPS) type 1, or other instances of "musculoskeletal" pain and functional visceral pain disorders. As stated in [19]: "This group comprises people who have neither obvious activation of nociceptors nor neuropathy but in whom clinical and psychophysical findings suggest altered nociceptive function". However, signs of this altered nociception have not yet been characterized by IASP [20] and it requires more studies on patients suffering from chronic pain. Furthermore, there is also a proposal for definition modification of nociplastic pain as "pain that arises from altered nociceptive function" in [21].

4.2. Classification of Pain by Its Time Period

The most common classification is concerned with the time duration of the pain. That is, by observing how long the symptom lasts, the pain could also be divided in two types: acute pain and chronic pain [22,23].

4.2.1. Acute Pain

The term acute pain often refers to the occurrence of damage to tissues. It is a short-lived pain that works as a warning sign from the body. In most cases (i.e., broken bones, surgery, dental work, labor and childbirth, cuts, burns), the pain last fewer than six months and disappears once the injury or disease is cured or healed.

4.2.2. Chronic Pain

The features of chronic pain (e.g., osteoarthritis, frequent headaches, low back pain, etc.) are its long duration and its complicated mechanism. It usually lasts for more than six months, even after the original injury was healed. Due to its stubbornness, people living with chronic pain may develop symptoms of anxiety, depression, or other conditions (e.g., tense muscles, lack of energy, limited mobility, etc.).

5. What Is Stress and what is Its Correlation with Pain?

Pain and stress are interleaved and connected in many ways and the consensus of the highly intertwined relations between these two mechanisms has been established in [24]. Stress is a feeling of emotional strain and psychological stress and a state of threatened homeostasis and a reaction that breaks the balance of physiological processes. The sources of stress could be the experience of the pain or the consequences of ongoing pain or other psychological reasons. Experiencing pain is stressful enough, especially when it is not negligible, but what is worse is when the pain lasts for a longer time, then it could lead to a vicious cycle. For example, people suffering back pain could easily develop stress by further induced muscle tension or spasms [25]. The muscle tension produces more pressure on the nerves, not only causing more pain and stress but also squashing the nerve harder and exacerbating the pain. On the other hand, the consequences of ongoing pain usually last longer than half a year and this kind of chronic pain would have a greater impact to the patient's quality of life [26]. The stress and frustrated mood brought by the chronic pain are already tough, but the restricted movement or low physical activity in fear of amplifying the pain furthermore are even worse in these situations. People in the fear of being in pain tend to avoid any potential movement that does or may induce the pain. The avoidance and anticipation of pain that causes a lot of stress is the beginning of the vicious cycle. This kind of symptom are called "pain catastrophizing" [27]. It is a negative cognitive-affective response toward actual pain or the anticipation of pain. Furthermore, experiencing stress could also affect the endocrine system balance and then induce endocrine disorders which are linked back to chronic pain [28].

Upon encountering stress, the human body would respond with three components: adrenal medulla, hypothalamus and pituitary gland. These three components constitute the so called hypothalamic-pituitary-adrenal axis (HPA) which react to the stress by releasing hormones (the adrenal medulla could release norepinephrine) helping or exciting other parts of the organism through the sympathetic nervous system (SNS) [29]. When the SNS is activated, the subject's heart rate and blood pressure would increase in a short period of time; their breathing may get faster, adrenalin levels raise as do the blood sugar and cholesterol levels. The blood flow would also be redirected from lower priority organs such as the organs in the digestive system to higher priority vital organs such as the heart and the brain. The function of the immune system would be driven up since there is an immediate danger and the body needs to handle the "fight or flight" situation [30]. However, if this situation cannot be resolved immediately (e.g., chronic pain), this self-protecting mechanism might harm the body instead and becomes maladaptive in the long term. Excessive or prolonged activation of the SNS causes muscle tension, headaches, high blood pressure or even promotes the development of cancer [31]. People who are physically inactive due to a stress state could rather end up with depression.

6. Assessment for Pain and Stress

6.1. Pain Assessment

Pain is gradually being accepted as the fifth vital sign [32] since this was firstly proposed by American Pain Society (APS) in 1996. Different kinds of pain (i.e., acute pain or chronic pain) are assessed separately and serve different purposes. The assessment for acute pain is to avoid provoking the pain onset and to monitor the effect of the suppressant that is used. Contrarily, the goal of assessing chronic pain is collecting related signs in the early stages or to gather enough symptoms to track down the origin of the pain. In practice, there are multiple scales and measures that are helpful for tracking pain-related treatment outcomes. These kinds of measurement are resources for clinicians to select a treatment plan and validate the treatment effects. Commonly used measures for pain are: (1) Self-report measures: self-report measurement is a subjective score related to pain given by the subject ranging from 0 (no feeling of pain) to 10 (extreme pain). It usually refers to a numerical pain rating scale, and similar measurements are VAS [33]; (2) Physical performance tests: the 5-minutes walking, stair-climbing task, 15 meters walking, sit-to-stand and loaded forward-reach test [34] and the Abbey Pain Scale for the non-verbal individuals (e.g., patients with dementia) [35]; (3) Physiological response measures: the physiological and autonomic response measures are the most objective and physiological approach to pain. By observing the changes of multiple physiologic signals such as skin conductance and heart rate and other signals, researchers can formulate a valid index for pain evaluation (e.g., analgesia nociception index [36]). However, the correlation of such measurement with pain are still under debate. Since the idea of using such measurement is to apply the physiological signals that are the subject of the activation of the automatic nerve system. However, the activity of the automatic nerve system (mainly about the balance of sympathetic nervous system and parasympathetic nervous system) may be reacting not only to pain but also other factors. Thus, the physiological measurement in most cases is used in the surgical room where the subject is unconscious so the physiological approach is the only way to obtain any relevant information for pain monitoring.

To date, numeric pain scales based on the patient's self-reporting is still the easiest and most popular assessment for pain. However, the lack of an objective assessment for pain may cause the overuse of opioids and to their addiction in clinic. This problem could further lead to opioid-related unintended deaths [37].

6.2. Stress Assessment

Similar to pain assessment, one way to assess stress is by a self-report scale in a clinical environment. Rather than a subject filling out a questionnaire, the VAS provides a rapid quantitative assessment in a 10-points range [38]. Since stress is defined as a state in which homeostasis is threatened, the adaptive processes that are activated would cause both physiological and behavioral changes. In order to comprehend this stress response mechanism, numerous studies have been conducted observing the physiological and behavioral changes in the body under stress induction tests. The observation of physiological or pathophysiological changes in response to stress is fundamental to the development of novel pharmacological agents for stress management [39]. In the rapid development of modern society, people are dealing with stress and work fatigue on a daily basis; thus, like pain assessment, stress assessment could also benefit from observation during daily life. Without further inducing stress to the subjects, such cases are associated to fatigue and work-related stress [40] in the concern about the mental and physical health of employees.

Stress Induction Tests

Setting up stress-inducing scenarios helps researchers collect and validate the stress-correlated physiological signals or behavioral signals. These stress induction tests usually involve asking the subject to finish a certain task or perform a certain action in a specific condition designed by researchers; then the researchers could conclude which signals are related to stress by monitoring the changes of signals during the tests.

- Trier Social Stress Test

The original Trier Social Stress (TSST) consists of an anticipation period and a test period for 10 minutes each [41]. During the test, subject is told to take role of a job applicant and prepare for a 5-minutes speech. An audience of three persons plays as interviewers and managers. The subject must convince the interviewers of his/her suitability for the imaginary job without touching any topics that is previously noted before the test. If the subject finishes his/her presentation early, he/she will be asked to continue by the interviewers. Then after the speech period is over, the subject is asked to do a mental arithmetic which is counting down numbers from 1,022 in steps of 13. Once the subject makes a mistake, he/she will need to start all over from number 1,022 again. Following the mental arithmetic test, the subject is given details of the experiment and will be allowed to take a rest while his/her physiological signals are still under monitoring. There are other kinds of TSST variations such as using the virtual reality (VR) to reduced cost [42].

- Stroop Color-Word Inference Test

The idea of the Stroop color-word inference is asking the subject to read out the color of the words while that word is printed in different color of the word is represented literally. The stress is induced by the contradiction between the linguistic and visual perceptions. The Stroop color-word test has been widely used in psychology for a long time [43]. It has the advantages of high reliability and stability in measuring for individual differences with only relatively simple rules. The performance of Stroop color-word test also has a positive outcome in VR environment [44].

- Cold Pressor Test/Hot Water Immersion Test

In the cold pressor test, the subjects are asked to put their hand into a bucket of cold water and keep there it as long as they can. The subjects should notify the researcher when they first feel that the cold water starts causing pain to their hands. Then at any time after the first notification is given, the subjects are free to remove their hands when they feel the pain is unendurable. Then according to the timing of two notifications (i.e., when does the subject start feeling pain and when they remove their hands due to the intolerable pain) and the continually collected blood pressure and heart rate, researchers can further analyze the physiological features of stress. The cold pressor/hot water immersion test are basically the same and the only different is the temperature of the water that is being used in the test. The cold pressor test is efficient experimental stress induction [45] which has been observed to reliably increase HPA activity [46].

- International Affective Picture System Test

In psychological studies, one of the most common tests for emotion and attention research is the International Affective Picture System (IAPS). By providing pictures ranging from simple daily objects to extreme pictures that involve violent or erotic contents, the test induces stress or emotion in the subject. A relevant application is used to detect IAPS stress levels in human pilots [47].

6.3. Physiological Signals for Assessment

6.3.1. Heart Activity

Since stress causes fundamental disturbances in the autonomic nervous system (ANS) which has major effects on heart activity [48], some useful detection methods for stress are based on heart-related signals [49]. Heart activity could be represented by an electrocardiogram (ECG), which is recorded by measuring the electrical activity of heartbeats. Usually, a normal heartbeat includes three distinguishable waves: the P wave, QRS complex wave and T wave. Most of the studies on heart activity are related to three aspects of the heart: time domain, frequency domain and non-linear features of heart. The research in the time domain focuses on parameters such as heart rate (HR), inter-beat (RR) intervals and heart rate variability (HRV) [50]. For RR intervals, it could be further studied as its mean value, the standard deviation or root mean square. Frequency domain studies analyze the components in the low-frequency (LF), high-frequency (HF) or the LF/HF ratio. As for the non-linear features there are algorithms such as entropy, complexity, Poincare Plots [51], recurrence and fluctuation slopes.

6.3.2. Brain Activity

The brain activity is recorded as the electroencephalogram (EEG) for brain-related research (e.g., emotion changes, stress-related studies [52] or consciousness studies for anesthesia, etc.). The four bands of the EEG signal are alpha (8−13 Hz) which indicates the sign of calmness and balanced state of mind; beta (13−30 Hz), which is related to emotional and cognitive processes which correlates to stress; delta (0.1−4 Hz) which is associated with deep sleep stages (e.g., high brain activity in this range are being viewed as a sign of unconsciousness) and theta (4−8 Hz) that generates the theta rhythm which is a neural oscillation in the brain that is linked to interpretation of cognition [53] and behavior such as learning, memory and spatial navigation.

6.3.3. Muscle Activity

Muscle tension usually comes along with stress. Researchers are studying changes of muscles activities in human body under certain stress-presented activities [54]. The state of muscles like stretching and releasing could be monitored by electromyogram measurements. Electrodes placed on certain areas of muscle could detect the potential changes due to the locomotion of the body. After obtaining the measurement of muscle activity statistical techniques can be used to enhance the understanding of the signals. Such applications are often referred to "myomonitoring" and can be adopted by studies in the interest of muscle tension (e.g., monitoring mandibular closure maximum intercuspation of the teeth [55]), and muscle fatigue (with a sonic approach [56]).

6.3.4. Electrodermal Activity

EDA is a useful indicator for neurocognitive stress by giving the change of electrical properties of skins. When a stress-inducing scenario is applied on a subject, the body is expected to start sweating; this further increases the skin conductance [57]. The long-term shifts in tonic level are called skin conductance level (scl); and the transient responses within seconds are the galvanic skin response (GSR). The tonic and phasic measurement are the two main aspects of EDA. One example of using EDA to validate pain stimulation is given in [58]. Posada-Quintero et al. proved that thermal grill stimulation is highly correlated with VAS. In this research, the observed EDA also shows significant increases as the stimulation level goes up. A systematic review of EDA data collection and signal processing presented in [59] by Posada-Quintero and Chon provides a summary of EDA recording devices, signal analysis methods, and the synthesis framework for EDA-related research.

6.3.5. Blood Volume Pulse

Blood volume pulse (BVP) provides the changes of volume in blood between each heartbeat and it fluctuates along with the changes in heartbeats. BVP is measured by optical, non-invasive sensors by comparing the light absorbed by the blood. Xie et al. used BVP for identifying strong stress and weak stress [60].

6.3.6. Skin Temperature

Stress influences both the core and peripheral body temperature [61]. While the core temperature tends to rise in response to stress, the distal skin locations tends to decrease. When acute stress is present, it triggers peripheral vasoconstriction which causes a rapid drop in skin temperature [62].

6.4. Behavioral Signals for Assessment

6.4.1. Speech

Voice patterns can be quite different for a person under stress or not. Multiple features related to the voice patterns could be altered when stress is present such as the changes in pitch [63], tone and speaking rate, or even the words they choose in the speech.

6.4.2. Facial Expressions

Natural habits, such as facial expressions are the reflection of the psychological state and the indication of emotion that a person is experiencing. Lots of researchers are trying to capture these subtle-signs and correlate them with stress situations through facial electromyography (EMG) [64] or image recognition based on facial expression [65].

6.4.3. Keystroke and Mouse Dynamics

In modern times, most people use computers on a daily basis. The way they use them could provide clues about their mental state or emotions in the present. The speed of typing and mouse dynamics could provide useful clues to indicate whether a person is stressed or not [66,67]. The use of excessive strength when hitting a keyboard and clicking a mouse is no doubt an obvious sign of an upset mind.

6.4.4. Body Gestures and Movements

A stressful person also displays signs of their state of mind with their body actions such as jaw clenching, constant finger rubbing, or even posture changes while they are standing or sitting. Multiple behavioral features are provided in [68] for stress detection.

6.4.5. Mobile Phone Usage

A person experiencing stress could either choose to fight it or flee from it. Using a cell phone might provide an easy way to forget about the stress. By distracting themselves with various features built into a smartphone, the stress may seem to fade away temporarily. Mobile phone addiction could also be a sign of anxiety [69] which is common symptom for people under stress. One study [70] finds a significant correlation between mobile phone use and stress.

6.4.6. Questionnaires and Surveys

Questionnaires and surveys are already being widely used in psychological research for assessment of psychological state. By asking subjects questions that serve to identify a specific mental approach, subjects might expose their deepest concerns or stress on their minds. Sometimes, even the subject could not know the source of their own stress which requires questionnaires or consultation by a psychologist to unravel.

7. Medical Devices or Wearable Sensors used in Pain and Stress Detection

In this section, the common devices and wearable sensors that are suitable for detecting pain and stress are organized. Pain within a short time-period usually can be located by the person themselves or be detected by a clinician in a clinical environment. Even in surgery when the subject is unconscious, there are devices to provide a valid index for the degree of pain being experienced [71]. However, sometimes pain like chronic pain or stress are intermittent rather than constant. This features highly irregular seizure timings which are hard to detect in a specific time window using traditional medical devices or clinical assessments; in fact, relevant diagnoses are mostly dependent on self-reporting which is based on the subject's memory to get any relevant information. In this case, wearable sensors could be used to collect data when subjects are not in the clinic and monitor both physiological and behavioral signals for longer periods which provides useful information for clinicians [72].

7.1. Medical Devices Used in Pain Detection

Nociception is the most relevant and effective approach for pain detection. Even though pain is a subjective perception, nociception is a physiological reaction to the nociceptive stimuli; this nociceptive stimulus is based on the reaction of the autonomic nervous system (i.e., the balance of the sympathetic nervous system and parasympathetic nervous system). Then by analyzing the corresponding physiological signal variation caused by the activity of autonomic nervous system, researchers manage to formulate an index useful as a pain reference. A popular way to monitor the balance of the sympathetic nervous system and parasympathetic nervous system is by analyzing the heart rate variability. Other methods use the number of skin conductance fluctuations per second (NFSC), the size of pupil and its variability when illuminated, blood vessel contraction, EMG, EEG, and changes of body temperatures, etc. Heart rate variability and plethysmography are widely used in both types of research since these two signals are easier to obtain during surgery and they are highly sensitive to the activity of autonomic nervous system. The two most common medical devices used in assessing pain are the analgesia nociception index that is based on the heart rate variability and the surgical pleth index that is based on plethysmography.

7.1.1. Analgesia Nociception Index

The analgesia nociception index (ANI) is a technology that provides a measurement of the parasympathetic tone on a scale from 0 to 100. It has been used in the surgical room or post-operative room for pain assessment. By analyzing the electrocardiographic data which reflecting parasympathetic activity, the ANI provides a reference to the sympathetic/parasympathetic balance that allow doctors to control surgical stress. The ANI has shown a correlation to the self-rating system in the postoperative period after volatile agent and opioid-based anesthesia in [73]. Since this index is solely based on the physiological signals rather than being a self-rating system, it could be applied to patients under general anesthesia or in critically ill condition who have communication problems. Besides, in a study of ANI in pain-related surgical conditions (e.g., during labor [74], laparoscopic abdominal surgery [71]), there is also other research such as [75] that find the relations between ANI and emotional status. Basically, mechanisms that are correlated to parasympathetic changes could adopt ANI as a measurement tool.

7.1.2. Surgical Pleth Index

The surgical pleth index (SPI) is a digital monitor based on the patient's hemodynamic responses to surgical stimuli and analgesic medications during general anesthesia [76]. SPI measures the sympathetic activity as a reaction to painful stimuli; it creates a single index (using a scale from 0 to 100) by integrating the photoplethysmographic amplitude and photoplethysmographic pulse interval with algorithms. The SPI has been proved to correlate with self-reporting systems [77]; however, the appropriate selection of SPI target values has not been established yet. The representation of different score ranges of SPI (e.g., prediction of moderate-to-severe postoperative pain [78]) is still a major topic in the research field.

7.2. Wearable Sensors Used in Stress Detection

Assessments for stress detection could benefit from the proper use of wearable sensors for data collection based on the physiological/behavioral signal of interest. The useful physiological signals that are of interest in stress detection research are heart activity (ECG), brain activity (EEG), muscle activity (EMG), skin conductance (EDA), BVP, and skin/body temperatures and relevant wearable sensors and devices for stress detection are organized in Table 2. Most of the behavioral signals could be obtained by smartphone sensors [79] or recorded by video cameras for image analysis and voice analysis. Each sensor used in stress detection is listed in the following paragraphs, with additional details of the sensor placement illustrated in Figure 1.

- Empatica E4 wrist band: this device is a wrist band is a real-time physiological data streaming and visualization sensor. As a medical-grade wearable device, it enables researchers to collect multiple physiological data such as BVP for HRV analysis, and EDA that reflects the constantly fluctuating electrical properties of a certain area of skin and peripheral skin temperature. Besides, it also captures motion activity with a 3-axis accelerometer [80–83].
- AutoSense: this is a wireless sensor suite that packs six sensors in a small form factor which are capable of collecting cardiovascular, respiratory and thermoregularity measurements through radio transmission and processes collected signals for detecting the general stress state of subjects. The wearable sensor has advantages of excessive lifetime while fully charged which allows prolonging its use for constant data collection [84–86].

- SleepSense: this is a belt-like sensor which adopts a piezoelectric film sensor for converting chest or abdominal respiration motions to analog voltages and thus, provides an indication of respiration waveforms [87,88].
- BN-PPGED: this is a physiological sensor for measuring BVP via optical plethysmographic methods and EDA activity. The sensor could be worn as a wristband with an additional two electrodes situated on two fingers [89].
- Cardiosport TP3: this is also a belt-like wearable sensor. By attaching the sensor pod to the chest strap, the TP3 will be activated to collect HR and millisecond RR intervals as long as the HR is detected [90].
- Q-sensor: this is a wireless sensor designed by the Massachusetts Institute of Technology that aimed to "detect and record physiological signs of stress and excitement by measuring slight electrical changes in the skin." The emotion detection sensor could benefit individuals with autism who usually do not show his/her stress outward and helping to manifest the emotions before breakdown. The sensor could obtain the accelerometer data and skin conductance by measuring inner wrists of subject's hand [70].
- Wahoo chest belt: Wahoo chest belt is equipped with a sensor which collects HRV data on a chest belt. Besides provides the heart rate and calorie burn data for workout evaluation, the HRV data could also be an indicator of the autonomic nervous system activity [91].
- BioHarness 3: this is physiological monitoring telemetry device that are usable for subjects in the workplace. The device can store and transmit data such as HR, HRV, respiration rate, and 3-axis accelerometer data through Bluetooth [92].
- Shimmer sensor: the shimmer sensor is a monitoring wearable sensor for EDA. Composed of two finger electrodes and a main unit, the shimmer sensor can transmit data to personal computer or other devices through Bluetooth connections [92].
- MindWave mobile EEG headset: it is an EEG headset capable of logging single channel EEG raw data at a 512 Hz sampling rate then provides index of attention and meditation of the user after power spectral density analysis [92].
- DataLOG: this is a portable EMG signal collection and monitoring devices designed by Biometrics. It could be placed on the arm, the leg or waist for various fields studies like human performance, sports science, medical research, industrial ergonomics, gait laboratories, and educational settings [93].

Table 2. Wearable sensors used in stress detection.

Type of Signal	Commercialized Wearable Sensors Used in Relevant Research	Wearable Sensors Not Yet Commercialized but Used in Relevant Research
Heart activity	Empatica E4 wrist band, AutoSense, Cardiosport TP3, Wahoo chest belt, BioHarness 3	Device 1, Device 2
Brain activity	MindWave mobile EEG headset	
Muscle activity	DataLOG	Device 3
Electrodermal activity	Empatica E4 wrist band, BN-PPGED, Q-sensor, Shimmer sensor	
Respiratory	AutoSense, SleepSense	
Blood volume pulse/pulse plethysmograph	Empatica E4 wrist band, BN-PPGED	
Body/skin temperature	Empatica E4 wrist band, AutoSense	
Three-axis accelerometer data	Empatica E4 wrist band, Q-sensor	

Notes: Empatica E4 wrist band is used in [80–83]; AutoSense is used in [84–86]; SleepSense is used in [87,88]; BN-PPGED is used in [89]; Cardiosport TP3 is used in [90]; Q-sensor is used in [70]; Wahoo chest belt is used in [91]; BioHarness 3, Shimmer sensor, and MindWave mobile EEG headset are being used as an integrated system for stress monitoring in [92]; DataLOG is used in [93]; Device 1 is a EEG wearable sensor developed in Online Predictive Tools for Intervention in Mental Illness (PTIMI) project funded by European Union [94]; Device 2 is a noninvasive physiological sensor for stress assessment presented in [95]; Device 3 is used in [96] which they collect the EMG signals of the left trapezius muscle and then remove the contained ECG signal components.

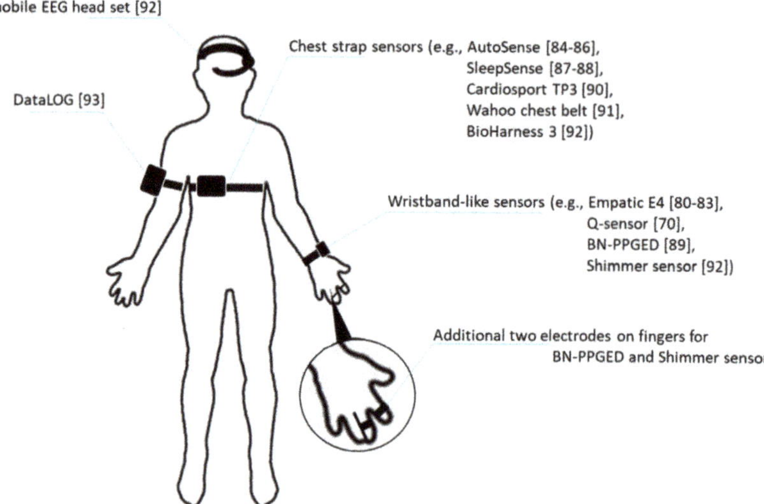

Figure 1. Illustration of sensors placement on human body.

8. Wearable Sensors in Healthcare

Wearable sensors are under rapid development and have been applied in many fields through the years, especially in healthcare. Recently, due to the severe pandemic, wearable sensors are also being used for COVID-19 detection [97]. The expanding needs of wearable sensors in healthcare domain is motivated by the increasing healthcare costs, the success application of wearable sensors in such domain is the result of advanced technologies of microelectronics and wireless communication [98]. The increasing healthcare costs are highly related to the ageing world population [99]. Constant monitoring for physiological and psychological advantages chronic diseases prognosis and detection in the early stage. A real-time feedback information about subject's health status could greatly improve the accuracy and capability to identify abnormal condition and prevent it in advance. Wearable sensors are the applicable solution for such tasks; in fact, a type of system based on wearable sensors are called the wearable health–monitoring systems (WHMS) which can be adopted for supervising subjects that are elderly people, have chronic disease or special abilities [100].

According to Alexandros et al. [98], WHMS are founded with a base of various types of miniature sensors, wearable or implantable for the purpose of measuring physiological signals. The collected parameters are transmitted through wireless or wired link to a central node (e.g., microcontroller board) to processing and display the information to the users. Then, these aggregated vital signs can be sent to the medical center for further analyses and diagnoses by medical professionals. A WHMS has multiple components (i.e., sensors, wearable materials, smart textiles, power supplies, communication modules, central processing unit (CPU), software and advanced algorithms, etc.). Yet, it also has to meet several criteria for practical use such as light weight and small size in order to enhance its comfortableness, and low radiation and mild heat dissipating to ensure safety. Finally, appropriate security during the data transmission for privacy concerns. After all of the above requirements are satisfied, the overall WHMS also has to be low-cost and affordable for majorities or even the underprivileged minority. Several available WHMS were reviewed by Alexandros and Nikolaos [98] according to the maturity level of the system. The maturity level of system is evaluated to their ability to measure multiple parameters, the detail level of the data documentation, the popularity of the system based on its citation number, the applied hardware technologies in the system and the algorithms used for feature extraction and decision support. A maximum maturity level WHMS also must face problems like battery lifetime shortage and private information security. However, integrating with modern techniques such as cloud computing, fog computing resources, the WHMS can further improve their performance with minimum cost on electronic components, processing unit and even save more space for individual sensors. Such integrations [101] release the frontier data collecting unit from massive computing tasks which may free more space and drop power consumption to the sensors. Then using smart phone as data transmission and processing relay station to construct a more comprehensive and compatible platform, and with federated learning an explicit data could be reproduced to implicit data which reduce the risk of data leakage and accelerate the process for optimizing diagnosis model.

9. Discussion

Pain and stress are useful mechanisms that help humans survive and is also a part of evolution. With the proper induction of the unpleasant feeling, individuals could sense the danger that is happening or might be harmful to its body; it is especially useful when the danger that caused the threat is beyond the individual's knowledge. Pain and stress serve as warning signals to acknowledge any incidents that are potentially harmful or fatal. This mechanism also prevents the same or similar incidents that might happen again by introducing stress before the harmful incident can cause further damages. This kind of experience based on threat learning could help to address threats and are useful but not harmful to the body in the short term. However, under the rapid development of human society, these mechanisms for survival instinct are no longer needed as much as they used to be; contrarily, the downside of these two mechanisms in the long term has gradually surfaced. Pain-inducing tests are much less effective compared to stress-inducing tests, and any experiment that may induce pain affects human rights. Thus, a possible solution for pain monitoring might rely on more research to find out more details about the relationship between pain and stress, then using WHMS is expected to resolve the pain management issue.

Pain management is a raising issue worldwide [102]. The access to pain management has been defined as a human right, despite the differences in social status and economic condition. Everyone should be able to be free from suffering pain, but in reality, it is truly sad that not everyone could afford a physical examination in the hospital or treatment resources. Pain management is also a public health issue [103]. In addition, in developed countries, the aging population also must face chronic age-related diseases [104]. On the other hand, stress management also demands people's attention as it is a common issue in modern society. Stress management techniques and relevant education are all necessary for students [105] and for workers [106], for everyone to meet their needs.

Fortunately, with the help of well-developed techniques, devices for monitoring pain or stress are becoming more and more accessible. The use of wearable sensors may allow the diagnose of pain and stress no longer restricted to hospital but everywhere by online doctors or artificial intelligence (AI) models.

10. Conclusions

Pain is an annoying feeling that everyone must have experienced; yet it is a subjective sensation to everyone. What is considered painful by one person might not be interpreted the same by others. The mechanisms of pain are so far being understood as a signal that travels from receptors to peripheral nerves, to the spinal cord, then to cerebral structures. Different nerve fibers have their own priority and duty for carrying sensations. There are a few classifications for pain according to their characteristics and mechanisms. One way to classify pain is by using the time-period of the pain duration. If the pain lasts less than 6 months, it is called acute pain; otherwise, if it lasts more than 6 months, it is known as chronic pain. Acute pain usually comes with a specific causation and the perception of it is constant until the causation to the pain has been removed or the causing injury healed.

Chronic pain, on the contrary, does not affects the subject for a much longer time but the symptoms are often intermittent, which raises challenges to detect or to find the origins of that pain. Moreover, having a persistent pain issue leads to the development of stress which is also hard to detect or treat. A person suffer either from pain or stress could end up with both, due to the vicious cycle. Luckily, with multiple physiological signals and behavioral signals collecting by wearable sensors, there is hope for detection to seek moderate treatment in the early stage. Available physiological signals for stress detection on wearable sensors are heart activity, brain activity, muscle activity, electrodermal activity, respiratory, blood volume pulse, skin temperature. Furthermore, wearable sensors canwork with multiple components (e.g., communication modules, CPU, advanced algorithms, etc.) to construct a wearable health-monitoring system for chronic disease detection and health status monitoring.

This article has presented the mechanisms of pain and stress, the correlation between them, their assessment, and their detection devices as well. Finally, wearable sensor-based health-monitoring systems are presented and discussed in the hope of solving the imbalanced resources global-wide for diagnosing pain and pain treatment issues. The low cost and easy to use features of wearable sensors might provide a perfect solution for this. Awareness about the importance of pain management is rising along with the promotion among humanity. Integrated with AI algorithms and cloud computing resources, wearable sensors could act as more than a component that collects data but as a foundation of a health monitoring and treatment system. Furthermore, by analyzing and quantifying pain and stress, they provide an opportunity to deal with the worldwide issues of pain t and stress management.

Funding: This research was funded by the Ministry of Science and Technology (MOST) of Taiwan (grant number: MOST 107-2221-E-155-009-MY2)

Conflicts of Interest: The authors declare no conflict of interest.

References

1. Beecher, H.K. The measurement of pain. *Pharmacol. Rev.* **1957**, *9*, 59–209. [PubMed]
2. Jin, H.; Ma, X.; Liu, Y.; Yin, X.; Zhu, J.; Wang, Z.; Fan, W.; Jin, Y.; Pu, J.; Zhao, J.; et al. Back Pain-Inducing Test, a Novel and Sensitive Screening Test for Painful Osteoporotic Vertebral Fractures: A Prospective Clinical Study. *J. Bone Miner. Res.* **2019**, *35*, 488–497. [CrossRef] [PubMed]
3. Karcioglu, O.; Topacoglu, H.; Dikme, O.; Dikme, O. A systematic review of the pain scales in adults: Which to use? *Am. J. Emerg. Med.* **2018**, *36*, 707–714. [CrossRef] [PubMed]
4. Panure, T.; Sonawani, S. Stress Detection Using Smartphone and Wearable Devices: A Review. *Asian J. Converg. Technol.* **2019**, *5*, 1–4. [CrossRef]
5. Marchand, S. The Physiology of Pain Mechanisms: From the Periphery to the Brain. *Rheum. Dis. Clin. North Am.* **2008**, *34*, 285–309. [CrossRef]
6. Sasaki, H.; Kishimoto, S. Diagnostic strategy for diabetic polyneuropathy: Focus on nerve fiber type and magnetic resonance neurography. *J. Diabetes Ingestig.* **2020**. [CrossRef]
7. Sadashivaiah, V.; Sacré, P.; Guan, Y.; Anderson, W.S.; Sarma, S.V. Selective relay of afferent sensory-induced action poten-tials from peripheral nerve to brain and the effects of electrical stimulation. In Proceedings of the 40th Annual International Conference of the IEEE EMBC, Honolulu, Hl, USA, 18–21 July 2018.
8. Melzack, R.; Wall, P.D. Pain mechanism: A new theory. *Science* **1965**, *150*, 971–979. [CrossRef]
9. Raffa, R.B.; Ossipov, M.H.; Porreca, F. Opiod analesicas and antagonists. In *Pharmacology and Therapeutics for Dentistry*, 7th ed.; Mosby: St. Louis, MI, USA, 2017; pp. 241–256.
10. Krabbenbos, I.P.; van Dongen, E.P.A.; Nijhuis, H.J.A.; Liem, A.L. *Mechanisms of Spinal cord Stimulation in Neuropathic Pain*; IntechOpen: London, UK, 2012.
11. Keefe, F.J.; Lumley, M.; Anderson, T.; Lynch, T.; Carson, K.L. Pain and emotion: New research directions. *J. Clin. Psychol.* **2001**, *57*, 587–607. [CrossRef]
12. Loggia, M.L.; Schweinhardt, P.; Villemure, C.; Bushnell, M.C. Effects of psychological state on pain perception in the dental environment. *J. Canadian Dent. Assoc.* **2008**, *74*, 651–656.
13. Williams, D.A.; Robinson, M.E.; Geisser, M.E. Pain beliefs: Assessment and utility. *Pain* **1994**, *59*, 71–78. [CrossRef]
14. Nakagami, Y.; Sugihara, G.; Takei, N.; Fujii, T.; Hashimoto, M.; Murakami, K.; Furu, M.; Moritoshi, F.; Uda, M.; Torii, M.; et al. Effect of Physical State on Pain Mediated Through Emotional Health in Rheumatoid Arthritis. *Arthritis Rheum.* **2019**, *71*, 1216–1223. [CrossRef] [PubMed]

5.   Arnold, L.M.; Clauw, D. Challenges of implementing fibromyalgia treatment guidelines in current clinical practice. *Postgrad. Med.* **2017**, *129*, 709–714. [CrossRef] [PubMed]

16.   Thienhaus, O.; Cole, B.E. The classification of pain. In *Pain Management a Practical Guide for Clinicians*; Weiner, R.S., Ed.; CRC Press LLC: London, UK, 2002.

17.   Shraim, M.A.; Massé-Alarie, H.; Hall, L.M.; Hodges, P.W. Systematic review and synthesis of mechanism-based classifica-tion systems for pain experienced in the musculoskeletal system. *Clin. J. Pain* **2020**, *36*, 793–812. [CrossRef]

18.   Jensen, T.S.; Baron, R.; Haanpää, M.; Kalso, E.; Loeser, J.D.; Rice, A.S.C.; Treede, R.D. A new definition of neuropathic pain. *Pain* **2011**, *152*, 2204–2205. [CrossRef] [PubMed]

19.   Kosek, E.; Cohen, M.; Baron, R.; Gebhart, G.F.; Mico, J.-A.; Rice, A.S.; Rief, W.; Sluka, A.K. Do we need a third mechanistic descriptor for chronic pain states? *Pain* **2016**, *157*, 1382–1386. [CrossRef]

20.   Trouvin, A.-P.; Perrot, S. New concepts of pain. *Best Pr. Res. Clin. Rheumatol.* **2019**, *33*, 101415. [CrossRef]

21.   Aydede, M.; Shriver, A. Recently introduced definition of "nociplastic pain" by the International Association for the Study of Pain needs better formulation. *Pain* **2018**, *159*, 1176–1177. [CrossRef]

22.   Grichnik, K.P.; Ferrante, F.M. The difference between acute and chronic pain. *Mt. Sinai J. Med. A J. Transl. Pers. Med.* **1991**, *58*, 217–220.

23.   Auvenshine, R.C. Acute vs. chronic pain. *Tex. Dent. J.* **2000**, *117*, 14–20.

24.   Melzack, R. Pain and stress: A new perspective. In *Psychosocial Factors in Pain: Critical Perspectives*; Gatchel, R.J., Turk, D.C., Eds.; The Guilford Press: New York, NY, USA, 1999; pp. 89–106.

25.   Flor, H.; Turk, D.C.; Birbaumer, N. Assessment of stress-related psychophysiological reactions in chronic back pain patients. *J. Consult. Clin. Psychol.* **1985**, *53*, 354–364. [CrossRef]

26.   Kühl, L.K. Effects of Stress Mechanisms on Pain Processing. Ph.D. Thesis, University of Trier, Trier, Germany, September 2010.

27.   Quartana, P.J.; Campbell, C.M.; Edwards, R.R. Pain catastrophizing: A critical review. *Expert Rev. Neurother.* **2009**, *9*, 745–758. [CrossRef]

28.   Tsigos, C.; Kyrou, I.; Kassi, E.; Chrousos, G.P. Stress: Endocrine physiology and pathophysiology. In *Endotext*; MDText: South Dartmouth, MA, USA, 2020.

29.   Guilliams, T.G.; Edwards, L. *Chronic Stress and the HPA Axis: Clinical Assessment and Therapeutic Considerations*; Point Institute of Nutraceutical Research: Wilmington, NC, USA, 2010; Volume 9.

30.   Dhabhar, F.S. A hassle a day may keep the pathogens away: The fight-or-flight stress response and the augmentation of immune function. *Integr. Comp. Biol.* **2009**, *49*, 215–236. [CrossRef] [PubMed]

31.   Dai, S.; Mo, Y.; Wang, Y.; Xiang, B.; Liao, Q.; Zhou, M.; Li, X.; Li, Y.; Xiong, W.; Li, G.; et al. Chronic stress promotes cancer development. *Front. Oncol.* **2020**, *10*, 1492. [CrossRef] [PubMed]

32.   Levy, N.; Sturgess, J.; Mills, P. "Pain as the fifth vital sign" and dependence on the "numerical pain scale" is being aban-doned in the US: Why? *Br. J. Anaesth.* **2018**, *120*, 435–438. [CrossRef]

33.   Hawker, G.A.; Mian, S.; Kendzerska, T.; French, M.R. Measures of adult pain: Visual analog scale for pain (vas pain), numeric rat-ing scale for pain (nrs pain), mcgill pain questionnaire (mpq), short-form mcgill pain questionnaire (sf-mpq), chronic pain grade scale (cpgs), short form-36 bodily pain scale (sf-36 bps), and measure of intermittent and constant osteoarthritis pain (icoap). *Arthritis Rheum.* **2011**, *63*, S240–S252. [CrossRef]

34.   Smeets, R.J.E.M.; Hijdra, H.J.M.; Kester, A.D.M.; Hitters, M.W.G.C.; Knottnerus, J.A. The usability of six physical performance tasks in a rehabilitation population with chronic low back pain. *Clin. Rehabil.* **2006**, *20*, 989–997. [CrossRef] [PubMed]

35.   Abbey, J.; Piller, N.; De Bellis, A.; Esterman, A.; Parker, D.; Giles, L.; Lowcay, B. The Abbey pain scale: A 1-minute numerical indicator for people with end-stage dementia. *Int. J. Palliat. Nurs.* **2004**, *10*, 6–13. [CrossRef] [PubMed]

36.   Chanques, G.; Tarri, T.; Ride, A.; Prades, A.; De Jong, A.; Carr, J.; Molinari, N.; Jaber, S. Analgesia nociception index for the assessment of pain in critically ill patients: A diagnostic accuracy study. *Br. J. Anaesth.* **2017**, *119*, 812–820. [CrossRef]

37.   Morone, N.E.; Weiner, D.K. Pain as the Fifth Vital Sign: Exposing the Vital Need for Pain Education. *Clin. Ther.* **2013**, *35*, 1728–1732. [CrossRef] [PubMed]

38.   Lesage, F.-X.; Berjot, S.; Deschamps, F. Clinical stress assessment using a visual analogue scale. *Occup. Med.* **2012**, *62*, 600–605. [CrossRef]

39.   Bali, A.; Jaggi, A.S. Clinical experimental stress studies: Methods and assessment. *Rev. Neurosci.* **2015**, *26*, 555–579. [CrossRef] [PubMed]

40.   Rose, D.M.; Seidler, A.; Nübling, M.; Latza, U.; Brähler, E.; Klein, E.M.; Wiltink, J.; Michal, M.; Nickels, S.; Wild, P.S.; et al. Associations of fatigue to work-related stress, mental and physical health in an employed community sample. *BMC Psychiatry* **2017**, *17*, 1–8. [CrossRef] [PubMed]

41.   Kirschbaum, C.; Pirke, K.M.; Hellhammer, D.H. The "trier social stress test"—A tool for investigating psychobiological stress responses in a laboratory setting. *Neuropsychobiology* **1993**, *28*, 76–81. [CrossRef] [PubMed]

42.   Allen, A.J.; Kennedy, P.J.; Dockray, S.; Cryan, J.F.; Dinan, T.G.; Clarke, G. The Trier Social Stress Test: Principles and practice. *Neurobiol. Stress* **2017**, *6*, 113–126. [CrossRef] [PubMed]

43.   Jensen, A.R.; Rohwer, W.D., Jr. The stroop color-word test: A review. *Acta Psychol.* **1996**, *25*, 36–93. [CrossRef]

44. Poguntke, R.; Wirth, M.; Gradl, S. Same same but different: Exploring the effects of the stroop color word test in virtual reality Human-computer interaction. In Proceedings of the 17th IFIP TC 13 International Conference, Paphos, Cyprus, 2–6 September 2019; pp. 699–708.

45. Schwabe, L.; Schächinger, H. Ten years of research with the Socially Evaluated Cold Pressor Test: Data from the past and guidelines for the future. *Psychoneuroendocrinology* **2018**, *92*, 155–161. [CrossRef]

46. McRae, A.L.; Saladin, M.E.; Brady, K.T.; Upadhyaya, H.; Back, S.E.; Timmerman, M.A. Stress reactivity: Biological and subjec-tive responses to the cold pressor and trier social stressors. *Hum. Psychopharmacol Clin. Exp.* **2006**, *21*, 377–385. [CrossRef]

47. Rhudy, M.B.; Dolan, S.K.; Wagner, A.R. A Pilot Study on Monitoring Airline Pilot Stress Levels. *AIAA Scitech 2020 Forum* **2020**. [CrossRef]

48. Van Bilsen, M.; Patel, H.C.; Bauersachs, J.; Böhm, M.; Borggrefe, M.; Brutsaert, D.; Coats, A.J.; De Boer, R.A.; De Keulenaer, G.W. Filippatos, G.S.; et al. The autonomic nervous system as a therapeutic target in heart failure: A scientific position statement from the Translational Research Committee of the Heart Failure Association of the European Society of Cardiology. *Eur. J. Hear. Fail.* **2017**, *19*, 1361–1378. [CrossRef]

49. Kim, H.-G.; Cheon, E.-J.; Bai, D.-S.; Lee, Y.H.; Koo, B.-H. Stress and Heart Rate Variability: A Meta-Analysis and Review of the Literature. *Psychiatry Investig.* **2018**, *15*, 235–245. [CrossRef]

50. Hernando, D.; Roca, S.; Sancho, J.; Alesanco, Á.; Bailón, R. Validation of the apple watch for heart rate variability measure-ments during relax and mental stress in healthy subjects. *Sensors* **2018**, *18*, 2619. [CrossRef] [PubMed]

51. Bu, N. Stress evaluation index based on Poincaré plot for wearable health devices. In Proceedings of the 2017 IEEE 19th International Conference on e-Health Networking, Applications and Services (Healthcom), Dalian, China, 12–15 October 2017; pp. 1–6.

52. Weygandt, M.; Meyer-Arndt, L.; Behrens, J.R.; Wakoning, K.; Bellmann-Strobl, J.; Ritter, K.; Scheel, M.; Brandt, A.U.; Labadie, C.; Hetzer, S.; et al. Stress-induced brain activity, brain atrophy, and clinical disability in multiple sclerosis. *Proc. Natl. Acad. Sci. USA* **2016**, *113*, 13444–13449. [CrossRef]

53. Buzzell, G.A.; Barker, T.V.; Troller-Renfree, S.V.; Bernat, E.M.; Bowers, M.E.; Morales, S.; Bowman, L.C.; Henderson, H.A.; Pine, D.S.; Fox, N.A. Adolescent cognitive control, theta oscillations, and social observation. *NeuroImage* **2019**, *198*, 13–30. [CrossRef] [PubMed]

54. Tsai, C.-M.; Chou, S.-L.; Gale, E.N.; McCall, W. Human masticatory muscle activity and jaw position under experimental stress. *J. Oral Rehabil.* **2002**, *29*, 44–51. [CrossRef] [PubMed]

55. George, J.P.; Boone, M.E. A clinical study of rest position using the kinesiograph and myomonitor. *J. Prosthet. Dent.* **1979**, *41*, 456–462. [CrossRef]

56. Song, X.; Li, H.; Gao, W. MyoMonitor: Evaluating Muscle Fatigue with Commodity Smartphones. *Smart Health* **2020**, 100175. [CrossRef]

57. Wickramasuriya, D.S.; Qi, C.; Faghih, R.T. A state-space approach for detecting stress from Electrodermal activity. In Proceedings of the 40th Annual Internationa Conference of the IEEE Engineering in Medicine and Biology Society, Honolulu, HI, USA, 18–21 July 2018.

58. Posada-Quintero, H.F.; Kong, Y.; Nguyen, K.; Tran, C.; Beardslee, L.; Chen, L.; Guo, T.; Cong, X.; Feng, B.; Chon, K.H. Using electrodermal activity to validate multilevel pain stimulation in healthy volunteers evoked by thermal grills. *Am. J. Physiol. Integr. Comp. Physiol.* **2020**, *319*, R366–R375. [CrossRef]

59. Posada-Quintero, H.F.; Chon, K.H. Innovations in Electrodermal Activity Data Collection and Signal Processing: A Systematic Review. *Sensors* **2020**, *20*, 479. [CrossRef]

60. Xie, J.; Wen, W.; Liu, G.; Chen, C.; Zhang, J.; Liu, H. Identifying strong stress and weak stress through blood volume pulse. In Proceedings of the 2016 International Conference on Progress in Informatics and Computing (PIC), Shanghai, China, 23–25 December 2016; pp. 179–182.

61. Vinkers, C.H.; Penning, R.; Hellhammer, J.; Verster, J.C.; Klaessens, J.H.G.M.; Olivier, B.; Kalkman, C.J. The effect of stress on core and peripheral body temperature in humans. *Stress* **2013**, *16*, 520–530. [CrossRef]

62. Herborn, K.A.; Graves, J.L.; Jerem, P.; Evans, N.P.; Nager, R.G.; McCafferty, D.J.; McKeegan, D.E. Skin temperature reveals the intensity of acute stress. *Physiol. Behav.* **2015**, *152*, 225–230. [CrossRef]

63. Simantiraki, O.; Giannakakis, G.; Pampouchidou, A.; Tsiknakis, M. Stress detection from speech using spectral slope measure-ments. In Proceedings of the Pervasive Computing Paradigms for Mental Health, FABULOUS 2016, Belgrade, Serbia, 24–26 October 2016; Volume 207.

64. Zhang, J.; Mei, X.; Liu, H.; Yuan, S.; Qian, T. Detecting Negative Emotional Stress Based on Facial Expression in Real Time. In Proceedings of the 2019 IEEE 4th International Conference on Signal and Image Processing (ICSIP), Wuxi, China, 19–21 July 2019; pp. 430–434.

65. Orguc, S.; Khurana, H.S.; Stankovic, K.M.; Lee, H.-S.; Chandrakasan, A.P. EMG-based Real Time Facial Gesture Recognition for Stress Monitoring. In Proceedings of the 2018 40th Annual International Conference of the IEEE Engineering in Medicine and Biology Society (EMBC), Honolulu, HI, USA, 18–21 July 2018; Volume 2018, pp. 2651–2654.

66. Lau, S.H. Stress Detection for Keystroke Dynamics. Ph.D. Thesis, Carnegie Mellon University, Pittsburgh, PA, USA, May 2018.

67. Carneiro, D.; Novais, P.; Sousa, N.; Pêgo, J.M.; Neves, J. Mouse dynamics correlates to student behavior in computer-based exams. *Logic J. IGPL* **2017**, *25*, 967–978. [CrossRef]

68. Aigrain, J.; Dubuisson, S.; Detyniecki, M.; Chetouani, M. Person-specific behavioral features for automatic stress detection. In Proceedings of the 11th IEEE International Conference and Workshops on Automatic Face and Gesture Recognition, Ljubljana, Slovenia, 4–8 May 2015.

69. Gao, T.; Li, J.; Zhang, H.; Gao, J.; Kong, Y.; Hu, Y.; Mei, S. The influence of alexithymia on mobile phone addiction: The role of depression, anxiety and stress. *J. Affect. Disord.* **2018**, *225*, 761–766. [CrossRef] [PubMed]

70. Sano, A.; Picard, R.W. Stress Recognition Using Wearable Sensors and Mobile Phones. In Proceedings of the 2013 Humaine Association Conference on Affective Computing and Intelligent Interaction, Geneva, Switzerland, 2–5 September 2013; pp. 671–676.

71. Jeanne, M.; Clément, C.; De Jonckheere, J.; Logier, R.; Tavernier, B. Variations of the analgesia nociception index during general anaesthesia for laparoscopic abdominal surgery. *J. Clin. Monit.* **2012**, *26*, 289–294. [CrossRef] [PubMed]

72. Can, Y.S.; Arnrich, B.; Ersoy, C. Stress detection in daily life scenarios using smart phones and wearable sensors: A survey. *J. Biomed. Informatics* **2019**, *92*, 103139. [CrossRef] [PubMed]

73. Abdullayev, R.; Uludağ, Ö.; Celik, B. Analgesia Nociception Index: Assessment of acute postoperative pain. *Braz. J. Anesthesiol. (English Ed.)* **2019**, *69*, 396–402. [CrossRef]

74. Le Guen, M.; Jeanne, M.; Siever, K.; Al Moubarik, M.; Chazot, T.; Laloë, P.A.; Dreyfus, J.F.; Fischler, M. The analgesia nociception index: A pilot study to evaluation of a new pain parameter during labor. *Int. J. Obstet. Anesth.* **2012**, *21*, 146–151. [CrossRef]

75. Abdullayev, R.; Yildirim, E.; Celik, B.; Sarica, L.T. Analgesia nociception index: Hear rate variability analysis of emotional status. *Cureus* **2019**, *11*, e4365. [CrossRef]

76. Huiku, M.; Uutela, K.; Van Gils, M.; Korhonen, I.; Kymäläinen, M.; Meriläinen, P.; Paloheimo, M.; Rantanen, M.; Takala, P.; Viertiö-Oja, H.; et al. Assessment of surgical stress during general anaesthesia. *Br. J. Anaesth.* **2007**, *98*, 447–455. [CrossRef]

77. Thee, C.; Ilies, C.; Gruenewald, M.; Kleinschmidt, A.; Steinfath, M.; Bein, B. Reliability of the surgical Pleth index for assessment of postoperative pain. *Eur. J. Anaesthesiol.* **2015**, *32*, 44–48. [CrossRef]

78. Ledowski, T.; Schneider, M.; Gruenewald, M.; Goyal, R.; Teo, S.; Hruby, J. Surgical pleth index: Prospective validation of the score to predict moderate-to-severe postoperative pain. *Br. J. Anaesth.* **2019**, *123*, e328–e332. [CrossRef]

79. Sağbaş, E.A.; Korukoglu, S.; Ballı, S. Stress Detection via Keyboard Typing Behaviors by Using Smartphone Sensors and Machine Learning Techniques. *J. Med. Syst.* **2020**, *44*, 1–12. [CrossRef] [PubMed]

80. Ollander, S.; Godin, C.; Campagne, A.; Charbonnier, S. A Comparison of Wearable and Stationary Sensors for Stress Detection. In Proceedings of the 2016 IEEE International Conference on Systems, Man, and Cybernetics (SMC), Budapest, Hungary, 9–12 October 2016; pp. 4362–4366.

81. Indikawati, F.I.; Winiarti, S. Stress Detection from Multimodal Wearable Sensor Data. In Proceedings of the 2nd International Conference on Engineering and Applied Sciences (2nd InCEAS), Yogyakarta, Indonesia, 16 November 2019; Volume 771.

82. Carreiro, S.; Chintha, K.K.; Shrestha, S.; Chapman, B.; Smelson, D.; Indic, P. Wearable sensor-based detection of stress and craving in patients during treatment for substance use disorder: A mixed methods pilot study. *Drug Alcohol Depend.* **2020**, *209*, 107929. [CrossRef] [PubMed]

83. Kaczor, E.; Carreiro, S.; Stapp, J.; Chapman, B.; Indic, P. Objective Measurement of Physician Stress in the Emergency Department Using a Wearable Sensor. In Proceedings of the 53rd Hawaii International Conference on System Sciences, Honolulu, HI, USA, 7–10 January 2020; pp. 3729–3738. [CrossRef]

84. Ertin, E.; Raij, A.; Stohs, N.; Al'Absi, M.; Kumar, S.; Mitra, S. An unobtrusively wearable sensor suite for inferring the onset, causality, and consequences of stress in the field. In Proceedings of the 9th ACM Conference on Recommender Systems, Seattle, WA, USA, 1–4 November 2011; pp. 437–438.

85. Kennedy, A.P.; Epstein, D.H.; Jobes, M.L.; Agage, D.; Tyburski, M.; Phillips, K.A.; Ali, A.A.; Bari, R.; Hossain, S.M.; Hovsepian, K.; et al. Continuous in-the-field measurement of heart rate: Correlates of drug use, craving, stress, and mood in polydrug users. *Drug Alcohol Depend.* **2015**, *151*, 159–166. [CrossRef] [PubMed]

86. Nakajima, M.; Lemieux, A.M.; Fiecas, M.; Chatterjee, S.; Sarker, H.; Saleheen, N.; Ertin, E.; Kumar, S.; Al'Absi, M. Using novel mobile sensors to assess stress and smoking lapse. *Int. J. Psychophysiol.* **2020**, *158*, 411–418. [CrossRef] [PubMed]

87. Wijsman, J.; Grundlehner, B.; Liu, H.; Hermens, H.J.; Penders, J. Towards mental stress detection using wearable physiological sensors. In Proceedings of the 2011 Annual International Conference of the IEEE Engineering in Medicine and Biology Society, Boston, MA, USA, 30 August–3 September 2011; pp. 1798–1801. [CrossRef]

88. Mühl1, C.; Jeunet, C.; Lotte, F. EEG-based workload estimation across affective contexts. *Front. Neurosci.* **2014**, *8*, 114.

89. Sandulescu, V.; Andrews, S.; Ellis, D.A.; Bellotto, N.; Mozos, O.M. Stress Detection Using Wearable Physiological Sensors. Proceedings of International Work-Conference on the Interplay Between Natural and Artificial Computation, IWINAC 2015, Elche, Spain, 1–5 June 2015; pp. 526–532.

90. Salai, M.; Vassányi, I.; Kósa, I. Stress Detection Using Low Cost Heart Rate Sensors. *J. Health Eng.* **2016**, *2016*, 1–13. [CrossRef]

91. Muaremi, A.; Arnrich, B.; Tröster, G. Towards Measuring Stress with Smartphones and Wearable Devices During Workday and Sleep. *BioNanoScience* **2013**, *3*, 172–183. [CrossRef]

92. Betti, S.; Lova, R.M.; Rovini, E.; Acerbi, G.; Santarelli, L.; Cabiati, M.; Del Ry, S.; Cavallo, F. Evaluation of an Integrated System of Wearable Physiological Sensors for Stress Monitoring in Working Environments by Using Biological Markers. *IEEE Trans. Biomed. Eng.* **2018**, *65*, 1748–1758. [CrossRef]

93. Pourmohammadi, S.; Maleki, A. Stress detection using ECG and EMG signals: A comprehensive study. *Comput. Methods Programs Biomed.* **2020**, *193*, 105482. [CrossRef]

94. Hu, B.; Peng, H.; Zhao, Q.; Hu, B.; Majoe, D.; Zheng, F.; Moore, P. Signal Quality Assessment Model for Wearable EEG Sensor on Prediction of Mental Stress. *IEEE Trans. NanoBioscience* **2015**, *14*, 553–561. [CrossRef]
95. Ahn, J.W.; Ku, Y.; Kim, H.C. A Novel Wearable EEG and ECG Recording System for Stress Assessment. *Sensors* **2019**, *19*, 1991. [CrossRef] [PubMed]
96. Wijsman, J.; Grundlehner, B.; Penders, J.; Hermens, H.J. Trapezius muscle EMG as predictor of mental stress. *ACM Trans. Embed. Comput. Syst.* **2013**, *12*, 1–20. [CrossRef]
97. Quer, G.; Radin, J.M.; Gadaleta, M.; Baca-Motes, K.; Ariniello, L.; Ramos, E.; Kheterpal, V.; Topol, E.J.; Steinhubl, S.R. Wearable sensor data and self-reported symptoms for COVID-19 detection. *Nat. Med.* **2021**, *27*, 73–77. [CrossRef]
98. Pantelopoulos, A.; Bourbakis, N.G. A Survey on Wearable Sensor-Based Systems for Health Monitoring and Prognosis. *IEEE Trans. Syst. Man, Cybern. Part C (Applications Rev.)* **2010**, *40*, 1–12. [CrossRef]
99. Hao, Y.; Foster, R. Wireless body sensor networks for health-monitoring applications. *Physiol. Meas.* **2008**, *29*, R27–R56. [CrossRef]
100. Bonato, P. Advances in wearable technology and applications in physical medicine and rehabilitation. *J. Neuroeng. Rehabil.* **2005**, *2*, 1–4. [CrossRef]
101. Chen, J.; Abbod, M.F.; Shieh, J.-S. Integrations between Autonomous Systems and Modern Computing Techniques: A Mini Review. *Sensors* **2019**, *19*, 3897. [CrossRef]
102. Brennan, F.; Carr, D.B.; Cousins, M. Pain Management: A Fundamental Human Right. *Anesth. Analg.* **2007**, *105*, 205–221. [CrossRef]
103. Brennan, F.; Carr, D.; Cousins, M. Access to Pain Management—Still Very Much a Human Right. *Pain Med.* **2016**, *17*, 1785–1789. [CrossRef]
104. Noroozian, M.; Raeesi, S.; Hashemi, R.; Khedmat, L.; Vahabi, Z. Pain: The neglect issue in old people's life. *J. Med. Sci.* **2018**, *6*, 1773–1778. [CrossRef]
105. Gulzhaina, K.K.; Aigerim, K.N. Stress management techniques for students. Advances in social Science, Education and Humanities Research. In Proceedings of the International Conference on the Theory and Practice of Personality Formation in Modern Society (ICTPPFMS 2018), Yurga, Russia, 20–22 September 2018; Volume 198, pp. 47–56.
106. Petković, A.I.; Nikolić, V. Educational needs of employees in work-related stress management. *Work* **2020**, *65*, 661–669. [CrossRef] [PubMed]

Review

A Survey of Heart Anomaly Detection Using Ambulatory Electrocardiogram (ECG)

Hongzu Li * and Pierre Boulanger

Computing Science Department, University of Alberta, Edmonton, AB T6G 2R3, Canada; pierreb@ualberta.ca
* Correspondence: hongzu@ualberta.ca

Received: 31 January 2020; Accepted: 2 March 2020; Published: 6 March 2020

Abstract: Cardiovascular diseases (CVDs) are the number one cause of death globally. An estimated 17.9 million people die from CVDs each year, representing 31% of all global deaths. Most cardiac patients require early detection and treatment. Therefore, many products to monitor patient's heart conditions have been introduced on the market. Most of these devices can record a patient's bio-metric signals both in resting and in exercising situations. However, reading the massive amount of raw electrocardiogram (ECG) signals from the sensors is very time-consuming. Automatic anomaly detection for the ECG signals could act as an assistant for doctors to diagnose a cardiac condition. This paper reviews the current state-of-the-art of this technology discusses the pros and cons of the devices and algorithms found in the literature and the possible research directions to develop the next generation of ambulatory monitoring systems.

Keywords: Review; ECG; Signal Processing; Machine Learning; Cardiovascular Disease; Anomaly Detection

1. Introduction

According to the World Health Organization (WHO), cardiovascular diseases (CVDs) are the number one cause of death globally. An estimated 17.9 million people die from CVD, representing 31% of all global deaths. Four out of five CVD deaths are due to heart attacks and strokes, and one third of these deaths occurs prematurely in people under 70 years of age [1]. An electrocardiogram (ECG) can record a patient's heart electrical signal activities over a long period [2] by measuring voltages from electrodes attached to the patient's chest, arms, and legs. ECGs are a quick, safe, and painless way to check for heart rate, heart rhythm, and signs of potential heart disease.

A twelve lead ECG is today's standard tool and is used by cardiologists for detecting various cardiovascular abnormalities. However, heart problems may not always be observed on a standard 10-second recording from the 12-lead ECG measurements performed in hospitals or clinics. Therefore, long term ECG monitoring that tracks the patient's heart condition at all times and under any circumstance has become possible with the development of new sensing technologies. Portable ECG recording devices such as the Apple Watch [3], AliveCor [4], Omron HeartScan [5], QardioMD [6], and, more recently, the Astroskin Smart Shirt [7] are revolutionizing cardiac diagnostics by measuring a patient's 24/7 cardiac activities and transmitting this information to a cloud service to be stored and processed remotely.

By itself, this massive data set is not very useful to the medical community as they usually do not have enough time or resources to read through long ECG recordings (two to three weeks) to detect any possible heart problems. For this technology to work, new automatic and reliable heart anomaly detection algorithms must be developed to assist doctors in coping with this massive data set. Our aim in this paper is to review the current state-of-the-art that address the challenges of performing ambulatory ECG anomaly detection and to highlight possible solutions. In order to do so, we will first review the medical background associated to ECG analysis and then review the state-of-the-art of automated anomaly detection in ambulatory and non-ambulatory contexts. We will then conclude by discussing the possible research directions to develop the next generation of ambulatory monitoring systems.

2. ECG Monitoring and Its Signals

2.1. Standard 12-lead ECG

A standard 12-lead electrocardiogram provides views of the heart in both the frontal and horizontal planes and views the surfaces of the left ventricle from 12 different angles. A 12-lead ECG has six limb leads (I, II, III, aVF, aVL, and aVR), and six chest leads (V1–V6). The standard 12-lead ECG is used as a standard clinical dysrhythmia analysis tool for chest pain or discomfort, electrical injuries, electrolyte imbalances, medication overdoses, ventricular failure, stroke, syncope, and unstable patients. It is widely used in clinics and hospitals for heart disease diagnosis [8]. However, when the patient needs to be monitored continuously, a 12-lead ECG is impractical as the patient needs to be attached to 10 electrodes.

2.2. Three-lead vs. 12-lead ECGs

Due to the fact that the standard 12-lead ECG is impractical for continuous ECG recording, therefore, three (3)-lead ECGs are widely used in portable ECG devices for a 24-h recording. Frank's lead system [9] is a 3-lead system that is practical for clinical use. In addition, much research has been done to show that a 3-lead ECG is useful to make a valid diagnosis. Antonicelli [10] was able to validate the accuracy of a 3-lead telecardiology (tele)-ECG compared to a 12-lead tele-ECG in an older population. Their study demonstrated a high level of concordance between the ECG diagnosis using a simple home telecardiology device (3-lead tele-ECG) and more complex instruments, like the 12-lead tele-ECG, as well as the standard 12-lead ECG. The study also demonstrated that a simple 3-lead tele-ECG could be used to detect cardiac alterations, such as arrhythmias, atrioventricular blocks, and re-polarization abnormalities, with good agreement with the observations measured by a 12-lead tele-ECG and the standard 12-lead ECG.

Kristensen et al. also evaluated how well an inexpensive portable three-lead ECG monitor (PEM) can detect patients with atrial fibrillation (AF) compared to a standard 12-lead ECG [11]. In their study, the results demonstrated that the sensitivity of diagnosing AF using PEM recordings was 86.7% and the specificity was 98.7% when compared to a 12-lead ECG. According to cardiologists, the misclassification of three PEM recordings was due to interpretation errors and not related to the PEM recording. In their article, they concluded that PEM devices could be used to diagnose AF. Dehnavi et al. performed an analysis of 3-lead vectorcardiogram (VCG) signals for the detection of cardiovascular diseases [12]. In the study, the authors experimented with detecting ischemia using a VCG algorithm using 3-lead and 12-lead ECGs and demonstrated a similar performance in both cases.

Furthermore, many researchers have tried to reconstruct a 12-lead ECG signal from 3-lead signals. Piotr Augustyniak reviewed and compared two transformation functions between 3-lead VCG and 12-lead ECGs [13]: the Dower and Levkov transformations. The author then tested how a synthesized 12-lead ECG and a VCG compared to an actual 12-lead ECG. The results showed that the synthesized 12-lead ECG was 10.08% distorted and the synthesized vectocardiogram was 6.347% distorted. Atoui [14] introduced a neural network-based model that could derive a standard 12-lead ECG from a serial 3-lead ECG. As a result, the derived 12-lead ECG from the ANN model has an average correlation coefficient of 0.93 compared to the actual 12-lead ECG.

Figueriedo et al. proposed using a 3-signal-lead sensor to synthesize a 12-lead ECG [15]. The authors used a linear equation to combine the collected signals from a 3-signal-lead sensor to output the 12-lead ECG. H. Zhu et al. proposed a novel, lightweight synthetic method [16], which could reconstruct the standard 12-lead ECG from 3-leads: I, II, and V2. The proposed method is called the adaptive region segmentation-based piece wise linear (APSPL) method. It consists of adaptive region segmentation, linear regression operation, and ECG sequence restoration.

Moreover, Nelwan et al. [17] and Drew et al. [18,19] have done several studies demonstrating that it is possible to reconstruct a standard 12-lead ECG from a reduced lead set ECG. I. Tomasic et al. performed a study [20] to investigate how a regression trees algorithm can be used to transform a 3-lead ECG into a synthesized 12-lead ECG. Their study demonstrated that the regression trees algorithm can synthesized an accurate 12-lead reconstruction and that the reduced ECG lead set, contains enough information to detect most heart anomalies.

2.3. Normal ECG Signals

To detect anomalies on ECG signals, one must first know what a normal heartbeat looks like. In [8], a normal rhythm (see Figure 1) is defined as the result of an electrical impulse that starts from the sinoatrial (SA) node, propagates through the heart muscles, and then to the patient's chest. A normal rhythm is composed of the following segments in sequence: a P wave generated by the atrial depolarization, the QRS complex generated by the ventricular depolarization, and a T wave and U wave generated by ventricular re-polarization. In normal ECG signals, the P wave, QRS complex, and T wave should be similar over time at a frequency ranging from 60 to 100 bpm. A normal ECG signal should have paced rhythm (PR) intervals within 0.12–0.2 s, and QT intervals less than half of the corresponding RR interval. Also, in a normal ECG signal, the variation between the shortest PP interval/RR interval and the longest PP interval/RR interval should be less than 0.04 s (see Figure 2).

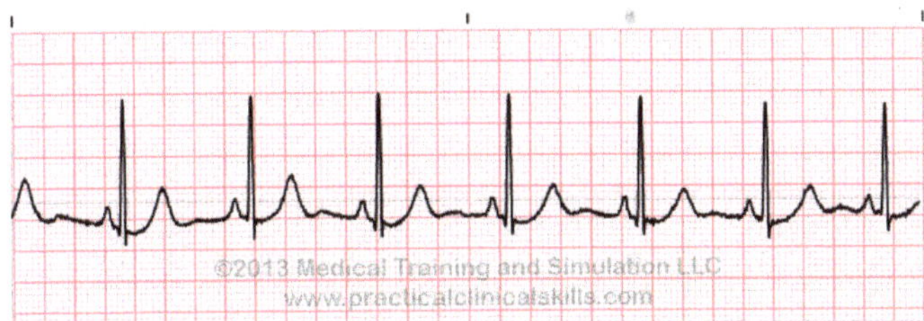

Figure 1. Normal sinus rhythm (NSR) [21].

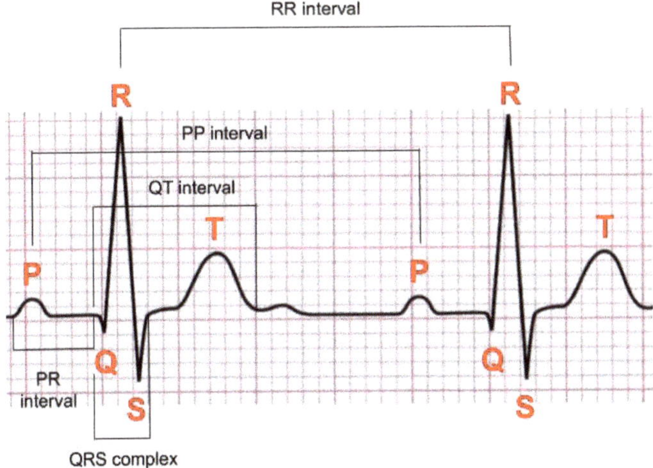

Figure 2. A normal electrocardiogram (ECG) signal and the corresponding notation [21].

2.4. Abnormal ECG Signals

The anomalies in ECG signals can be categorized into three subsets: irregular heart rate, irregular rhythm and ectopic rhythm. The heart rate could be counted by measuring the PP/RR intervals on the ECG. If the PP/RR interval is long, this indicates a low heart rate, otherwise, it indicates a high heart rate. If the heartbeats start from SA node, but the PP/RR intervals are longer than 1 s, this may indicate sinus bradycardia (Figure 3a), which indicates that the heart is pumping too slow. When the PP/RR intervals are shorter than 0.6 s, this may be the sign of Sinus Tachycardia (Figure 3b). Moreover, if the variations between the PP/RR intervals are too large, this may indicate Sinus Arrhythmia, Sinus Block, and Sinus Arrest (Figure 3c–e).

These ECG anomalies may indicate a patient's current conditions. For instance, Sinus Bradycardia may be associated with hypothyroidism, hyperkalemia, sick sinus syndrome, sleep apnea syndromes, carotid sinus hypersensitivity syndrome, and vasovagal reactions. Sinus Tachycardia is commonly associated with anxiety, excitement, pain, drug reactions, fever, congestive heart failure, pulmonary embolism, acute myocardial infarction, hyperthyroidism, pheochromocytoma, intravascular volume loss, and alcohol intoxication or withdrawal. Sinus Block, and Sinus Arrest can be caused by hypoxemia, myocardial ischemia or infarction, digitalis toxicity, and a toxic response to drugs [22].

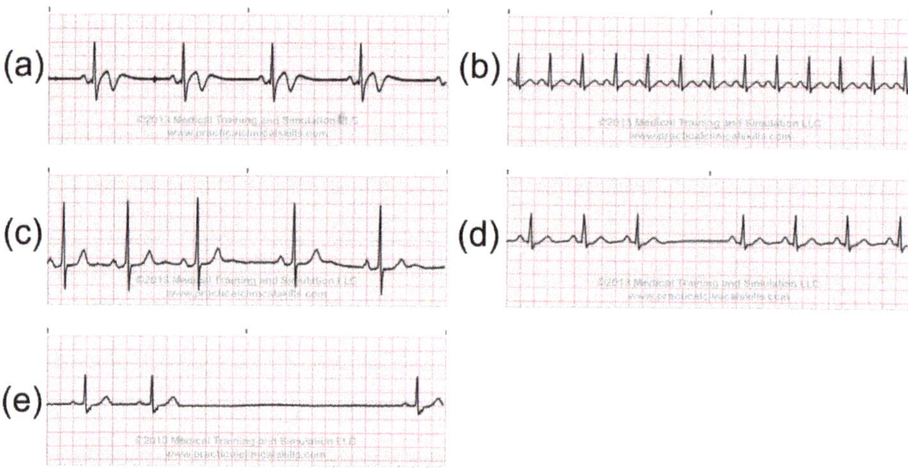

Figure 3. Abnormal sinus rhythms: (**a**) sinus bradycardia, (**b**) sinus tachycardia, (**c**) sinus arrhythmia, (**d**) sinus block, (**e**) sinus arrest [21].

Even if the heartbeat starts from the SA node, the heartbeat signal shape could be abnormal as well. For example, in the ECG signal, the ST segment and the T wave could have abnormal shapes; these are usually called ST-T changes. The ST-T changes could indicate hyperkalemia, ischemia, and so on [23]. Some examples of ST-T changes can be found in Figure 4.

Ectopic rhythms are started from a source other than the sinus node. For example, Atrial Rhythms begin in the atria. In this case, the P wave is shaped differently from the P wave beginning in the SA node. There are several abnormal rhythms that can occur when the Atria is firing the heartbeat: Premature Atrial Contraction, Wandering Atrial Pacemaker, Atrial Tachycardia, Atrial Flutter, Atrial Fibrillation. Examples are shown in Figure 5. The Premature Atrial Contraction is a very common heartbeat that could be caused by emotional stress, excessive intake of caffeine, and hyperthyroidism. If Premature Atrial Contraction consecutively occurs three or more times, the rhythm is considered as Atrial Tachycardia. It may cause light-headiness or even fainting. Atrial Flutter and Atrial Fibrillation are two distinct but closely related tachyarrhythmias. They could lead to many symptoms, such as palpitations, light-headiness, fainting, angina, and congestive heart failure.

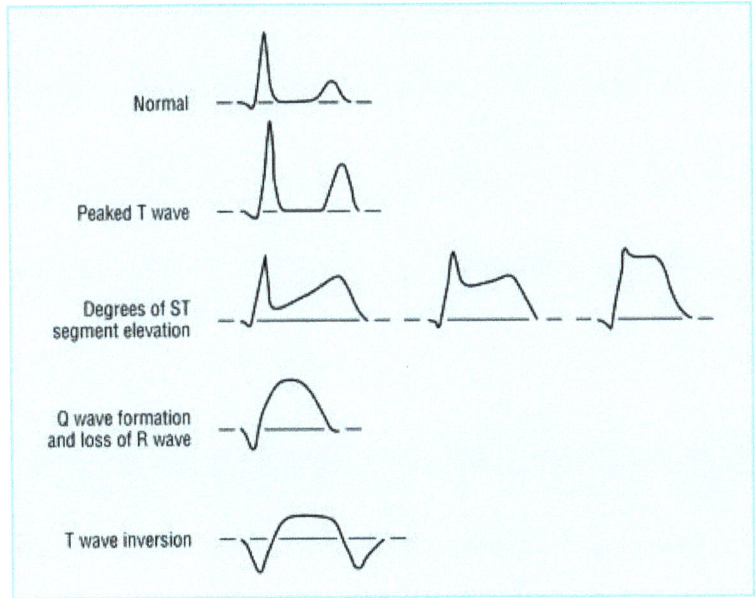

Figure 4. Examples of ST-T changes.

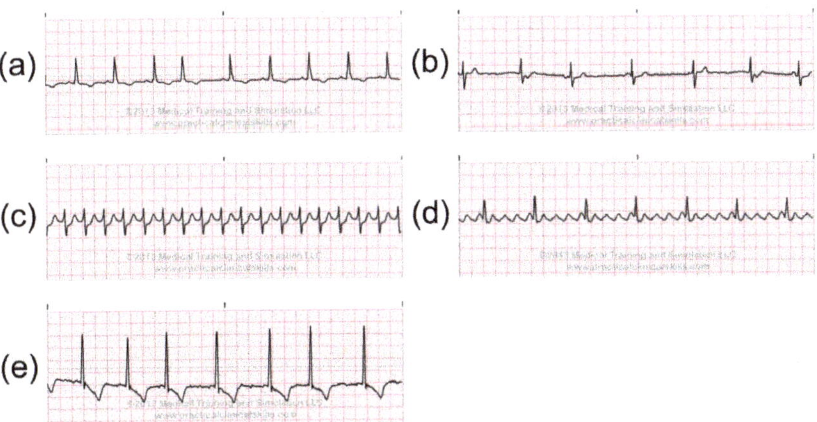

Figure 5. Abnormal Atrial Rhythms: (**a**) Premature Atrial Contraction, (**b**) Wandering Atrial Pacemaker, (**c**) Atrial Tachycardia, (**d**) Atrial Flutter, (**e**) Atrial Fibrillation [21].

Junctional Rhythms are another kind of ectopic rhythm. These occur when the atrioventricular (AV) junction paces the heart. In such a case, the P wave on the ECG signal may disappear or become negative. There are several anomaly examples shown in Figure 6: Premature Junctional Complex, Junctional Escape Rhythm, Junctional Tachycardia. The Premature Junctional Complex usually has the same cause as the Premature Atrial Contraction described previously. A Junctional Escape Rhythm could be caused by sick sinus syndrome, digitalis toxicity, excessive effects of beta-blockers or calcium channel blockers, acute myocardial infarction, hypoxemia, and hyperkalemia. One of the most common anomalies is the Junctional Tachycardia, the Atrioventricular nodal re-entrant tachycardia (AVNRT). This is an arrhythmia that results from a rapidly recirculating impulse in the nodal part of the AV junction, and could be caused by digitalis toxicity [22].

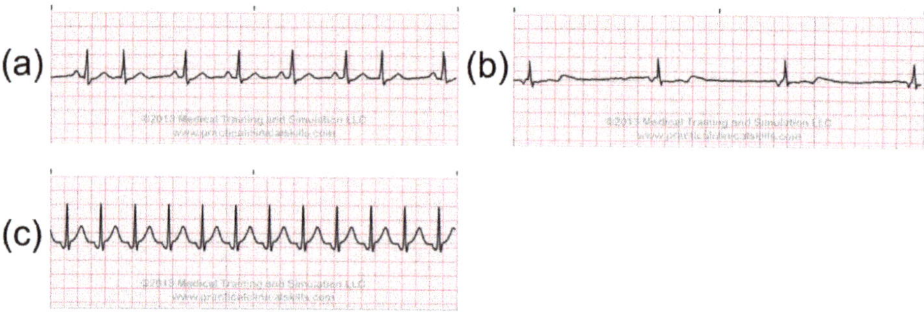

Figure 6. Abnormal Junctional Rhythms: (**a**) Premature Junctional Contraction, (**b**) Junctional Escaped Rhythm, (**c**) Junctional Tachycardia [21].

Ventricular Rhythms is another kind of ectopic rhythm. It occurs when an ectopic site within a ventricle assumes responsibility for pacing the heart. As a result, the ventricular heartbeats and rhythms usually have QRS complexes that have abnormal shapes and longer lengths. The following are the examples of abnormal Ventricular Rhythm: Premature Ventricular Contraction, Ventricular Escaped Rhythm, Accelerated Idioventricular Rhythm, Ventricular Tachycardia, and Ventricular Fibrillation, Ventricular Asystole. We can see the ECG signals in Figure 7. Individuals with Premature Ventricular Contraction may have the marker of severe organic heart disease associated with an increased risk of cardiac arrest and sudden death from Ventricular Fibrillation. Ventricular Tachycardia consists of three or more consecutive Premature Ventricular Contraction, and it could lead to more life-threatening Ventricular Fibrillation. With Ventricular Fibrillation, the ventricles do not heartbeat in any coordinated fashion but instead, fibrillate or quiver asynchronously and ineffectively. It will cause the patient to become unconscious immediately [22].

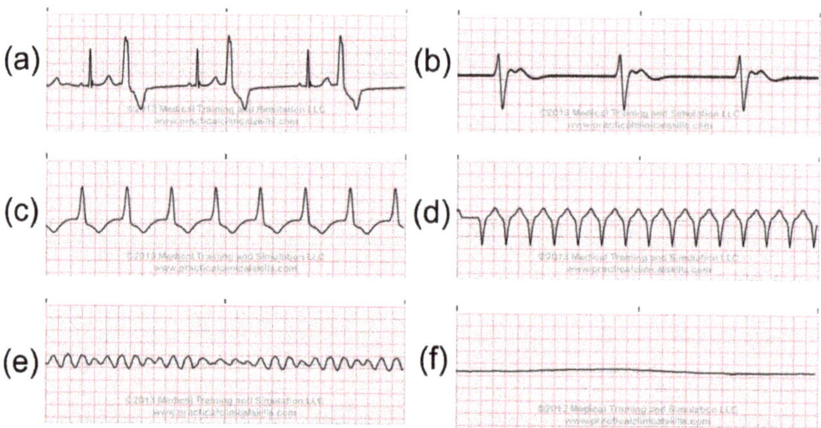

Figure 7. Abnormal Ventricular Rhythms: (**a**) Premature Ventricular Contraction, (**b**) Ventricular Escaped Rhythm, (**c**) Accelerated Idioventricular Rhythm, (**d**) Ventricular Tachycardia, (**e**) Ventricular Fibrillation, (**f**) Ventricular Asystole [21].

As depolarization and re-polarization are slow in the atrioventricular (AV) node, this area is vulnerable to blocks in conduction. Therefore, when a delay or interruption happens during impulse conduction from the atria to the ventricle, AV blocks may occur. AV blocks, also called Heart blocks, are classified into: First-degree AV blocks; Second-degree AV blocks (types I and II); Third-degree AV blocks (complete) see Figure 8. Among the heart blocks, the lower degree heart blocks could lead to Third-degree AV blocks, also called Complete Heart blocks, which are the most severe heart anomaly. With the Complete Heart blocks, the atria and ventricle are pacing independently, which could slow down the ventricular rate, and eventually lead to fainting [22].

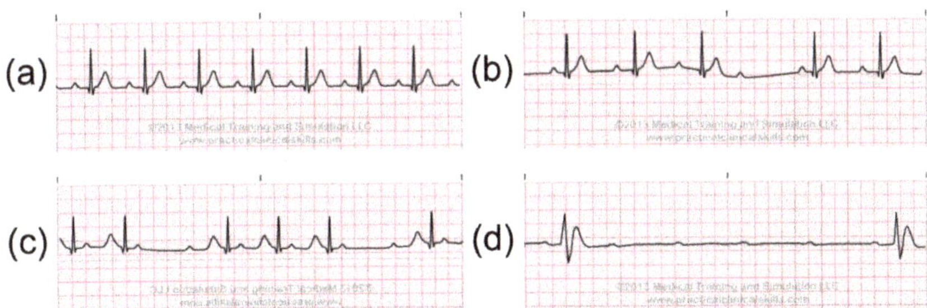

Figure 8. AV Blocks: (**a**) First-degree AV blocks, (**b**) Second-degree AV blocks type I, (**c**) Second-degree AV blocks type II, (**d**) Third-degree AV blocks [21].

3. Automatic Heart Anomaly Detection: A State-of-the-Art

3.1. Automatic Heart Anomaly Detection

The objective of detecting anomalies in ECG signals consists of finding the irregular heart rates, heartbeats, and rhythms. To achieve this goal, an anomaly detection system must be able to find them on all heartbeat sequences; therefore, to obtain the essential metrics as stated in Section 2. Also, the system looks at the entire recording to detect any irregular rhythm segments such as an inconsistent R-R interval and ectopic rhythms. Therefore, an anomaly detection system is composed of five different sub-systems: noise removal (Section 3.2), heartbeat detection (Section 3.3), heartbeat segmentation (Section 3.3), heartbeat classification (Section 3.4), and rhythm classification (Section 3.5).

A typical heartbeat anomaly detection system can be seen in Figure 9. The noise reduction process intends to minimize its effect on signal interpretation caused by the recording device or patient's movement. The heartbeat detection aims to find the location of the heartbeats to calculate the heart rate. The heartbeat segmentation extracts the entire heartbeat based on a known heartbeat location. The heartbeat classification checks for any abnormal heartbeat shape on the ECG signal. The irregular heart rhythm classification is similar to the heartbeat classification, but instead of checking only one heartbeat shape, it checks a period signal on the ECG record. Pertinent research found in the literature relating to the five sub-systems are introduced in the following sections.

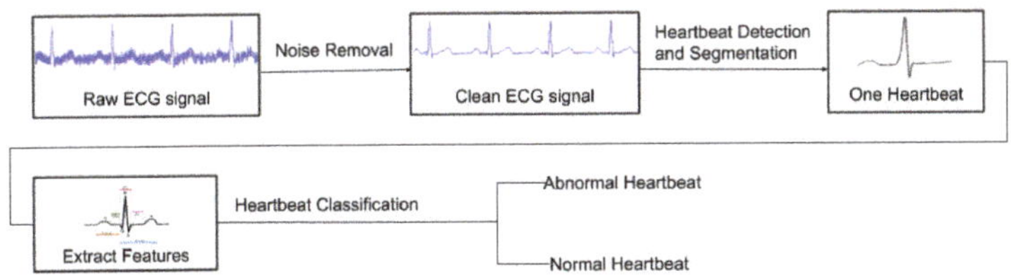

Figure 9. Typical Heartbeat Anomaly Detection.

MIT-BIH Database

Before explaining the sub-systems of the anomaly detection, the MIT-BIH Arrhythmia Database [24,25] need to be described first, as it is widely used in the ECG analysis related research. It was the first generally available set of standard test materials for the evaluation of the arrhythmia detector. It contains 48 half-hour excerpts of two-channel ambulatory ECG recordings from 47 subjects. The ECG data was collected with Del Mar Avionics model 445 two-channel reel-to-reel Holter recorders. The database has the annotation labels for 16 different heartbeat types and 15 different types of rhythms. All the selected research in this review used the MIT-BIH database, which allows us to test and compare the performance of the algorithms.

The annotation labels and the corresponding heartbeat types and rhythm types used in this database are listed below. The heartbeat types are:

1. Normal (N)
2. Left bundle branch block beat (LBBB)
3. Right bundle branch block beat (RBBB)
4. Atrial premature beat (PAC/APC)
5. Aberrated atrial rremature beat (a)
6. Nodal(junctional) premature beat (J)
7. Supraventricular premature beat (S)
8. Premature ventricular contraction (PVC)
9. Fusion of ventricular and normal beat (F)
10. Atrial escape beat (e)
11. Nodal (junctional) escape beat (j)
12. Ventricular escape beat (E)
13. Paced beat (P)
14. Fusion of paced and normal beat (f)
15. Classifiable beat (Q)
16. Atrial/Ventricular flutter beat (!).

The rhythm types are:

1. Atrial bigeminy (AB)
2. Atrial fibrillation (AF)
3. Atrial flutter (AFL)
4. Ventricular bigeminy (B)
5. $2°$ Heart block (BII)
6. Idioventricular rhythm (IVR)
7. Normal sinus rhythm (NSR)
8. Nodal (A-V junction) rhythm
9. Paced rhythm (PR)
10. Pre-excitation (PREX)
11. Sinus bradycardia (SBR)
12. Supraventricular tachyarrhythmia (SVTA)
13. Ventricular trigeminy (T)
14. Ventricular flutter (VFL)
15. Ventricular tachycardia (VT).

3.2. Noise Removal

ECG signals may be distorted by many other artifacts that have nothing to do with the heart functions. The ECG artifacts have various and uncertain forms. Some physiologic artifacts could mimic true dysrhythmia, leading to false diagnostics [26]. Therefore, noise removal is a necessary step for anomaly detection in ambulatory ECGs.

There are two main groups of artifacts: non-physiological and physiological artifacts. The first is caused by equipment problems, such as power-line interference, and the other one is caused by muscle activities, skin interference or body motion such as baseline wander, electromyogram, and motion artifacts. For example, the motion wander could significantly affect the measurement of the ST segment in an ECG signal [27]. Among all the artifacts, the motion artifact is the most challenging noise to remove as the noise spectrum overlaps the ECG signal [28]. Various ECG motion artifact examples are shown in Figures 10 and 11 [29].

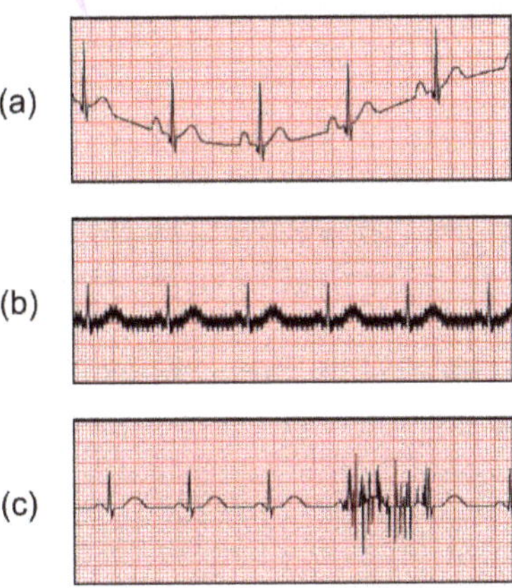

Figure 10. ECG Artifact examples: (**a**) Baseline Wander, (**b**) Power line Interference, (**c**) Muscle Interference.

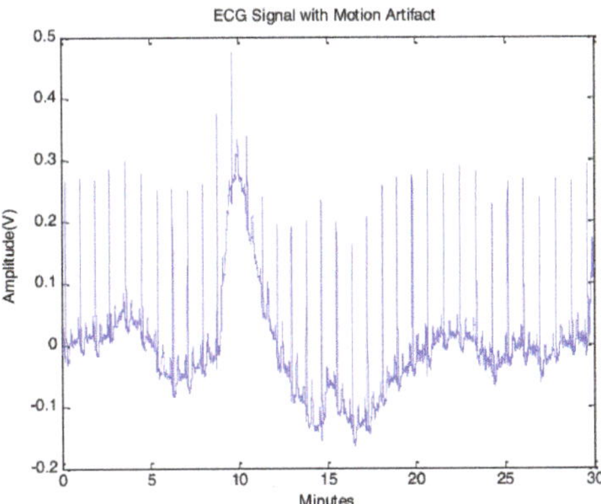

Figure 11. ECG Motion Artifact.

In this section, various noise removal algorithms in the research literature are categorized and compared. There are four conventional methods used for noise removal in ECG signals.

The first approach consists of using digital low-pass, high-pass, band-pass, and notch filters to remove the noise. Many studies, such as [30–35], use a combination of low-pass and high-pass filters to remove the corresponding noise on an ECG signal. The low-pass filter cut-off frequency is in the range of 11 Hz to 45 Hz, and it mainly suppresses the high-frequency noise. The high-pass filter cut-off frequency is in the range of 1 Hz to 2.2 Hz, and it focuses on removing the baseline wander in the signal. In [30,32], notch filters range from 50 Hz to 60 Hz and are used for removing the power-line interference. Band-pass filters with cut-off frequencies from 0.1 to 100 Hz are used by [36] to remove the noisy components of electronic noise. The advantage of using a fixed digital filter is that it is easy to implement and is highly efficient.

The second approach is to use a discrete wavelet transform (DWT) to remove the noise components from a signal. Wavelet transform is a powerful method for analyzing non-stationary signals, such as ECGs [37]. The DWT noise removal method is used in [38–40]. This method decomposes the signal into the approximation and detail coefficients by using a wavelet function. The selection of the wavelet function in the wavelet transform is the most important task, which depends upon the type of signal [41]. The commonly used Mother Wavelet basis functions are Daubechies filters (Db), Symmlet filters (Sym), Coiflet filters (C), Battle-Lemarie filters (Bt), Beylkin filters (Bl), and Vaidyanathan filters (Vd) [42].

According to studies in [41–43], the Daubechies filters of order 4 and 8 (Figure 12), and the Symmlet filters of order 5 and 6 (Figure 13) are the best wavelet functions for ECG signal analysis due to their similar signal structure to the QRS complex. After decomposing the ECG signal, a threshold method is applied to the DWT coefficients. A clean ECG signal could be reconstructed from the thresholded DWT coefficients.

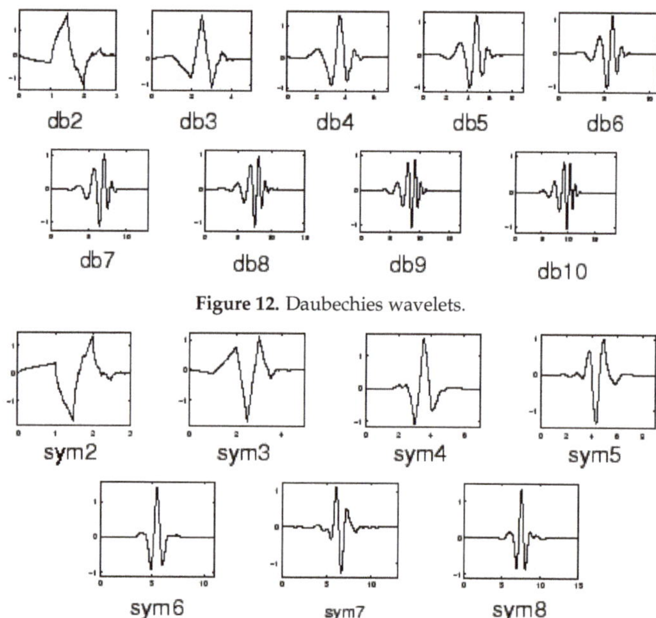

Figure 12. Daubechies wavelets.

Figure 13. Symlet wavelets.

DWT relies on the choice of the wavelet basis [44]. The level of DWT may be different between different data sets; therefore, re-implementation is needed. Another wavelet analysis method is the empirical mode decomposition (EMD). The EMD is an adaptive and fully data-driven technique that obtains the oscillatory modes present in the data [44]. The EMD, similar to the wavelet analysis, decomposes a time series signal into individual components without leaving the time domain. In EMD, the high-frequency components are called the intrinsic mode function (IMF), and the low-frequency part is called the residual. The procedure can be applied to residuals iteratively until no IMFs can be extracted. The IMFs must satisfy two conditions:

- The number of extremas and zero-crossings must be equal or differ at most by one;
- All local maximas and minimas must be symmetric to zero.

After decomposition using EMD, an IMFs and one residual signal will be obtained. Let $c(t)$ be the IMFs, we will have $c_1(t)$ to $c_n(t)$ from higher frequency components to lower frequency components. Then digital filters or thresholds can be applied to the IMFs that contain the noise. After processing, the signal can be reconstructed using the following equation:

$$x(t) = \sum_{i=1}^{n} c_i(t) + r(t) \tag{1}$$

where $x(t)$ is the reconstructed signal, $c(t)$ is the IMFs, and $r(t)$ is the residual signal.

In [45–47], the authors performed an EMD on the MIT-BIH database to suppress the high frequency noise and the baseline wander. Ensemble empirical mode decomposition (EEMD) [48] fixed the EMD shortcoming of mode mixing. The mode mixing can cause serious aliasing in the time-frequency distribution, and also makes the physical meaning of individual IMF unclear. The EEMD adds one extra step comparing to the EMD. By adding white noise to the original signal before decomposing the signal into IMFs using EMD. Many noise removal works were found using the EEMD, such as [49–51].

The previous approaches work well when the noise is in a fixed frequency range. However, there are some cases where these approaches could fail. The first one is in motion wander removal. Raimon Jane et al. stated in [27] that the motion wander frequency may not always be below 0.05 Hz. It could depend on the frequency of the heart rate, which could be less than 0.8 Hz. Also, a fixed digital filter could introduce nonlinear phase distortion and key point displacement [52]. These two approaches could not remove the motion artifact from the ECG signal, as its spectrum completely overlaps with the ECG signal. Therefore, many approaches use adaptive filtering to solve the proposed problem.

In 1991, Thakor et al. [28] introduced the least mean squares (LMS) adaptive filter (ARF) to reduce the baseline wander, 60 Hz power line noise, muscle noise, and motion wander. In their research, two adaptive filter structures were proposed. The first one has the primary input as $s_1 + n_1$, while the reference input is noise, n_2, which could be recorded from another generator that is correlated with n_1. The second one is an ECG that is recorded from several electrode leads, the primary input is $s_1 + n_1$ from one of the leads, the reference input is S_2 from another lead that is noise-free. In both cases, the signal s_1 can be extracted by recursively minimizing the mean squared error (MSE) between the primary and the reference inputs. The MSE can be calculated as:

$$E[\epsilon^2] = E[(s_1 - y)^2] + E[N_1^2]. \tag{2}$$

The least mean squares (LMS) algorithm was used to minimize the MSE. The LMS algorithm could be written as:

$$W_{k+1} = W_k + 2\mu\epsilon_k X_k \tag{3}$$

where W_k is a set of filter weights at time k, X_k is the input vector at time k of the samples from the reference signal, $\epsilon =$ primary input d_k- filter output y, and parameter μ is empirically selected to produce convergence at a desired rate. The error ϵ_k can be calculated as:

$$\epsilon_k = d_k - y_k \tag{4}$$

where d_k is the desired primary input from the ECG to be filtered, and y_k is the filter output that is the best least squares estimate of d_k.

As LMS adaptive filters are sensitive to scaling of the input, a power normalized least mean squares has been introduced to solve this problem [53]. Another convention adaptive filter type is the recursive least square (RLS) adaptive filter. The RLS algorithm recursively finds the filter coefficients that minimize a weighted linear least-squares cost function relating to the input signal. It is known for its excellent performance when working in time-varying environments but at the cost of increased computational complexity and some stability problems [54]. The algorithm updates the filter weight vector using the following equations:

$$w(n) = \overline{w}^T(n-1) + k(n)\overline{e}_{n-1}(n), \tag{5}$$

$$k(n) = u(n)/(\lambda + x^T(n)u(n)), \tag{6}$$

$$u(n) = \overline{w_\lambda^{-1}}(n-1)x(n), \tag{7}$$

where $w(n)$ is the weights vector of iteration n, $x(n)$ is the input signal, and λ is a small positive constant very close to but smaller than 1.

The filter output $\overline{y}_{n-1}(n)$ and the error signal $\overline{e}_{n-1}$ are calculated using the filter tap weights of the previous iteration and the current input vector as in the following equations:

$$\overline{y}_{n-1}(n) = \overline{w}^T(n-1)x(n), \tag{8}$$

$$\overline{e}_{n-1} = d(n) - \overline{y}_{n-1}(n). \tag{9}$$

An adaptive filtering approach could remove baseline wander, motion artifacts, power-line interference, and the muscle noise; however, it requires a reference input that is correlated to the original noisy input. Obtaining a clean ECG signal is very difficult to acquire. Due to the added complexity for the data collection, many studies have considered using an accelerometer as the reference noise signal for the adaptive filter. For example, in [55], Raya et al. explored the possibility of using both a signal axis and dual-axis accelerometer signal as the noise reference input to a least mean square (LMS) adaptive filter and a recursive least square (RLS) adaptive filter. As a result, the RLS adaptive filter outperformed the LMS adaptive filter. Using an accelerometer signal showed better results than using a dual-axis accelerometer signal. The authors believed that the use of one axis reference input, particularly the y-axis, was sufficient to minimize the noise.

3.3. Heartbeat Detection and Segmentation

Heartbeat detection is often related to the detection of an irregular heart rate and inconsistent RR-intervals, which are explained in Section 2. Heartbeat detection is also the key step to extract the heartbeats from the ECG signal to be used for classification. Heartbeat detection consists of three main parts: P wave detection, QRS complex detection, and T wave detection. Therefore, it is usually related to heartbeat segmentation. Heartbeat segmentation usually means segmenting a heartbeat from its start point (onsite) of P wave to its endpoint (offsite) of the T wave.

However, the P wave and T wave may not be detectable in certain types of abnormal heartbeat, and the QRS complex is the most obvious waveform. Thus the location of the QRS complex is often used to locate the origin of the heartbeat; see Figure 2. There are many studies that detect the R peak location in the QRS complex.

The Pan–Tompkins algorithm [56] is one of the most popular and earliest algorithms that has been implemented (Figure 14). It is widely used in many applications due to its robustness and computational efficiency. The algorithm uses a filter bank that consists of band-pass filters, a differentiator, a squaring filter, and a moving window integrator to reduce the signal noise so that only R wave information is present. Inspired by the Pan–Tompkins algorithm, many researchers, such as [57–60] developed their own filter banks to improve the accuracy of the detection. In order to reduce the detection of false positives, [58,60] used a predefined amplitude threshold, [59,60] used a predefined RR interval length threshold.

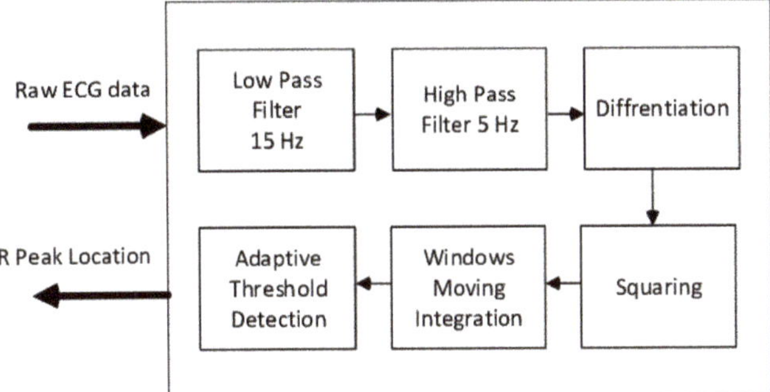

Figure 14. The Pan–Tompkins Algorithm.

Zidelmal et al. introduced a QRS detection method based on wavelet decomposition [61]. In the algorithm, the authors decomposed the raw ECG signal using a discrete wavelet transform, then reconstructed the signal by selecting only the sub-signals that contained ECG information. To detect the QRS complex, a threshold was set to select the peaks that have a large amplitude. Similar works have been done in [62].

Manikandan et al. introduced a new algorithm that uses a Shannon energy envelop and Hilbert-transform (SEEHT, Figure 15) to detect the QRS complex location [63]. In the preprocessing stage of their algorithm, a band-pass filter is applied to the raw ECG signal to remove the baseline wander and high-frequency noise. After that, a differentiator and normalizer is applied to the clean signal to highlight the QRS complex components. The Shannon energy of the processed signal is calculated using the following equation:

$$s[n] = -d^2[n]log(d^2[n]),$$ (10)

where d[n] is the processed signal. The calculated Shannon energy sequence is then processed by a zero-phase filter to preserve the sharp peaks around the QRS complex and smooth out the noisy peaks. In the peak finding algorithm, a Hilbert transform is applied on all the candidate R peaks to obtain the R wave envelope. In each R wave envelope, the zero-crossing locations indicate an R peak.

Inspired by SEEHT, [64] introduced An R-peak detection method based on peaks of Shannon energy envelope(PSEE) that improves the computational inefficiency of the Hilbert transform by using both predefined amplitude thresholds and predefined RR interval length thresholds. An improved R-peak detection method based on Shannon energy envelope (ISEE) [65] improved further the SEEHT and PSEE algorithms by using a filter bank consisting of a moving average filter, a differentiator, a normalizer, and a squaring filter to eliminate the noisy peaks. The filter bank computational costs is less than the Hilbert transform and does not use a predefined threshold. Most recently, Park combined discrete wavelet transform and ISEE to detect R peaks on the ECG signals [66].

As explained previously, the P and T waves represent important information and the heartbeat segmentation depends on the P and T wave detection. Therefore, a good detection of the P and T waves is critical for diagnosis. Pal and Mitra proposed an algorithm that could detect the PQRST peak points [67]. The algorithm is based on discrete wavelet decomposition. It reconstructs the signal from selected wavelet coefficients, which are related to peaks such as: R, QS, and PT. For example, when the algorithm is detecting the R peak, a signal is reconstructed with d3, d4, and d5 coefficients, and this preserves the information for the R peaks but diminishes the other peaks.

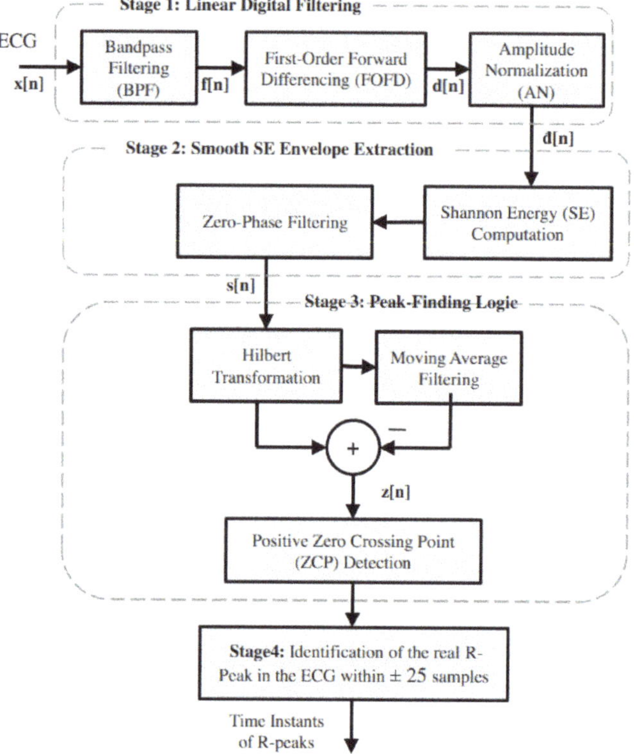

Figure 15. SEEHT R-peak detection algorithm.

A few years later, Banerjee also developed a T wave and QRS complex detection algorithm based on discrete wavelet decomposition and adaptive thresholding [68]. Karimipour uses discrete wavelet transform and adaptive thresholding to detect the QRS complex location, and give an estimate of the P-wave and T-wave locations [69]. In practice, many studies, such as [31,70] used the 'ecgpuwave' detector from PhysioNet for heartbeat segmentation [25]. However, because the P and T wave detection works well with normal heartbeats, but not for many abnormal heartbeat types. Many researchers choose manual annotation, such as [71], or a fixed window, such as [32,33,38,71,72], for their heartbeat segmentation.

In Table 1, we compare the performance of some of the heartbeat detection algorithms that have been tested on the MIT-BIH Arrhythmia database. [24].

The metrics used to compare each algorithm are calculated as follow:

- TP: Number of correctly detected heartbeats
- FP: Number of incorrectly detected heartbeats
- FN: Number of missed heartbeats
- Sensitivity (SEN) = TP / (TP+FN)
- Positive Detection (+P) = TP / (TP+FP)
- Detection Error Rate(DER) = (FP+FN) / TP
- Accuracy (ACC) = TP / (TP+FP+FN).

Table 1. The heartbeat detection performance comparison using the MIT-BIH data set.

Method	Year	Total Heartbeats	TP	FP	FN	SEN	+P	DER	ACC
Pan–Tompkins [56]	1985	116,137	115,860	507	277	99.76%	99.56%	0.68%	99.33%
FBBBD [57]	1999	91,283	90,909	406	374	99.59%	99.56%	0.86%	99.15%
S.W.Chen [58]	2006	102,654	102,195	529	459	99.55%	99.49%	0.97%	99.04%
DOM [60]	2008	116,137	115,971	58	166	99.86%	**99.95%**	0.19%	99.81%
S.Choi [59]	2010	109,494	109,118	218	376	99.66%	99.80%	0.54%	99.46%
Z.Zidelmal [61]	2012	109,494	109,101	193	393	99.64%	99.82%	0.54%	99.47%
SEEHT [63]	2012	109,496	109,417	140	79	99.93%	99.87%	0.2%	99.80%
S.Banerjee [68]	2012	19140	19126	20	20	99.90%	99.90%	0.21%	99.79%
PSEE [64]	2013	109,494	109,401	91	93	99.92%	99.92%	0.17%	99.83%
F.Bouaziz [62]	2014	109,494	109,354	232	140	99.87%	99.79%	0.34%	99.66%
A.Karimipour [69]	2014	116,137	115,945	308	192	99.83%	99.74%	0.43%	99.57%
ISEE [65]	2016	109,532	109,474	116	58	**99.95%**	99.89%	**0.16%**	**99.84%**
WTSEE [66]	2017	109,494	109,415	99	79	99.93%	99.91%	**0.16%**	**99.84%**

3.4. Irregular Heartbeat Classification

Irregular heartbeat classification focuses on the shape of the heartbeats, and aims at classifying the type of a single heartbeat. As discussed previously, the heartbeat shape may vary when the heartbeat starts from an ectopic location. For example, a premature heartbeat may have a missing P wave. The abnormal shape of a heartbeat may indicate potential heart disease. By classifying and annotating the types for all the heartbeats on the ECG, one could easily notice the frequency of anomalies that happens in the heart to make an appropriate diagnosis and treatment. Heartbeat classification consists of two main parts: feature extraction and model training.

3.4.1. Feature Extraction

The feature extraction step converts the raw ECG signal to machine-readable information. Based on the existing research, there are two common features: morphological features and derived features. The morphological features describe the heartbeats based on the observations of the signal itself. There are many morphological features (see in Table 2) that have been used in various studies.

Other derived features are calculated from the ECG signal. There are many different methods are proposed in the literature:

- Vectorcardiography (VCG) vector;
- DWT coefficients produced by Discrete Wavelet Transform (DWT);
- Independent components from Independent Component Analysis (ICA);
- PCA components generated from Principal Component Analysis (PCA);
- IMFs from Empirical Mode Decomposition (EMD)/Ensemble EMD (EEMD);
- DTCWT coefficients from Dual Tree Complex Wavelet Transform;
- Eigenvector methods;
- Dynamic Time Warping (DTW) distance.

Table 2. Conventional Morphological Features of Heartbeats.

Features	Description	Reference
QRS complex duration	The time interval between the onsite of the Q wave and offsite of the S wave	[30,31,38] [73–75]
QRS velociy left	The QRS slope velocity calculated for the time-interval between the QRS complex onset and the first peak	[30,73]
QRS velociy right	The QRS slope velocity calculated for the time-interval between the first peak and the second peaks	[30,73]
QRS complex area	The sum of the positive area and absolute negative area in the QRS complex	[30,73]
QRS complex morphology	Sample points from the QRS onsite to the QRS offsite	[31]
QRS complex AC power	The total power content of the QRS complex signal	[32]
QRS complex Kurtosis	The kurtosis indicates the peakedness of the QRS complex	[32]
QRS complex Skewness	The skewness measures the symmetry of the distribution of the QRS complex	[32]
Q wave valley	The valley value of Q wave	[75]
S wave valley	The valley value of S wave	[75]
T wave peak	The peak value of T wave	[75]
T wave duration	The duration from the QRS offsite to the T wave offsite	[31]
T wave morphology	Sample points from the QRS offsite to the T wave offsite	[31]
P wave flag	A Boolean value indicates the presence or absence of the P wave	[31]
P wave duration	The duration from the P wave onsite to the P wave offsite	[74]
P wave morphology	Sample points from the P wave onsite to the P wave offsite	[34,74]
PR interval duration	The duration from the P wave onsite to the QRS complex onsite	[74]
PR interval morphology	Sample points from the P wave onsite to the QRS complex onsite	[34]
QT interval duration	The duration from the QRS complex onsite to the T wave offsite	[74]
QT interval morphology	Sample points from the QRS complex onsite to the T wave offsite	[34,75]
ST interval morphology	Sample points from the S wave valley to the T wave offsite	[75]

Table 2. *Cont.*

Features	Description	Reference
Max peak(R peak) value	The maximum amplitude of the heartbeat	[30,73,75]
Min peak value	The minimum amplitude of the heartbeat	[30,73]
Positive QRS complex area	The area of the positive sample points in the QRS complex	[30,73,74]
Negative QRS complex area	The area of the negative sample points in the QRS complex	[30,73,74]
Positive P wave area	The area of the positive sample points in the P wave	[74]
Negative P wave area	The area of the negative sample points in the P wave	[74]
Positive T wave area	The area of the positive sample points in the T wave	[74]
Negative T wave area	The area of the negative sample points in the T wave	[74]
Absolute velocity sum	Sum of the absolute velocities in the pattern interval	[30,73]
Ima	Time-interval from the QRS complex onset to the maximal peak	[30,73]
Imi	Time-interval from the QRS complex onset to the minimal peak	[30,73]
Pre-RR interval	The RR interval between the heartbeat and its previous heartbeat	[31,71,74]
Post-RR interval	The RR interval between the heartbeat and its following heartbeat	[31,71,74]
Post-PP interval	The PP interval between the heartbeat and its following heartbeat	[74]
Average-RR interval	The average value of all valid RR intervals in the ECG record	[31,71,74] [32,75]
Local Average-RR interval	The average value of ten valid RR intervals surrounding the heartbeat	[31,71,74]
Normalized signal	The heartbeat sample points are normalized and down-sampled to have a mean of zero and standard deviation of one	[76–78]
Raw/downsampled ECG signal	The unprocessed ECG signal or the only processing on the signal is downsampled	[36,79]

Vectorcardiography (VCG) is one of the ECG analysis tools. It displays the various complexes of the ECG. It provides the possibility to use vector analysis on the cardiac electric potentials [80].

Discrete Wavelet Transform (DWT) decomposes the signal into many sub-signals (detail coefficients) with different frequency ranges, as described in Section 3.2. Not only could the DWT method be used to remove unwanted noises, it could also find features for the heartbeats as the heartbeat waves are much clearer in the specific detail coefficients, such as D4 and D5. Therefore, much research, such as [38,71], uses features from the detail coefficients to classify the heartbeat.

The conventional DWT technique lacks the property of shift-invariance due to the downsampling operations at each stage of DWT implementation. Hence, the energy of the wavelet coefficient changes significantly for a small-time shift in the input pattern. The Dual-Tree Complex Wavelet Transform [81] is a simple technique that overcomes the DWT shortcomings. The DTCWT uses two sets of filters: one is used for level 1 decomposition, and the other one is used for the higher levels. In the first level decomposition, the original signal is decomposed into two Trees, and each Tree contains two sub-band signals. One tree could be interpreted as the real part of a complex wavelet, and the other tree could be the imaginary part. For each tree, the conventional DWT is applied for further decomposition [32]. The DTCWT method was used by Thomas to extract heartbeat features to classify the heartbeat type [32].

Similar to DWT and DWCWT, the ICA, PCA, and EMD/EEMD also decompose the signal into many sub-signals. The difference is that the ICA and PCA aims to reduce the input size to minimize the computation speed. The EMD/EEMD, as explained in Section 3.2, does not require the knowledge of the level of scale and the basis function that is needed in DWT. The ICA method has been used in [71] to produce the independent components to be part of the heartbeat feature set. The PCA method used in [82] reduces the input size for higher efficiency. Rajesh et al. computed the heartbeat features from IMFs by applying the EMD/EEMD method to the ECG signal.

Eigenvector methods are used for estimating the frequencies and powers of signals from noise-corrupted measurements. These methods are based on an eigendecomposition of the correlation matrix of the noise-corrupted signal [83]. In [83], Ubeyli et al. used three kinds of eigenvector methods to generate the feature set: Pisarenko, Multiple Signal Classification (MUSIC), and Minimum-Norm. The Pisarenko method is particularly useful for estimating a PSD that contains sharp peaks at the expected frequencies. The MUSIC method is a noise subspace frequency estimator and could eliminate the effects of spurious zero on the noise subspace. The Minimum-Norm method aims to differentiate spurious zeros from real zeros, and it uses a linear combination of all noise subspace eigenvectors.

Dynamic Time Warping measures the similarity between two heartbeat segments. It computes the distance between these two heartbeat segments. Therefore, if we let one of the heartbeat segments to be the sample heartbeat of a specific type, and the other one to be the test heartbeat, then the distance indicates the similarity score between the test heartbeat and the sample heartbeat. The similarity score could be used as a feature that represents the heartbeat, such as in work by [74,76]. Details of the features of each method reviewed can be seen in Table 3.

Table 3. Conventional Derived Features of the Heartbeats.

Features	Method	Description	Reference
VCG amplitude	VCG	Maximal amplitude of the VCG vector	[30,38]
VCG sine angle	VCG	Sine component of the angle of the maximal amplitude vector	[30,38]
VCG cosine angle	VCG	Cosine component of the angle of the maximal amplitude vector	[30,38]
DTW distance	DTW	The Dynamic Time Warping distance between a heartbeat segment and the median heartbeat segment of the recording	[74,76]
Positive peak of the QRS complex	DWT	The positive peak amplitude of QRS complex on the fourth scale of the DWT	[38]
Negative peak of the QRS complex	DWT	The absolute negative peak amplitude of QRS complex on the fourth scale of the DWT	[38]
Positive peak of T wave	DWT	The positive peak amplitude of the T wave on the fourth scale of the DWT	[38]
Absolute T wave offsite	DWT	The absolute amplitude of the T wave offsite on the fourth scale of the DWT	[38]
R-S interval distance	DWT	The relative distance between the R peak and S valley on the fourth scale of the DWT	[38]
S-T interval distance 1	DWT	The relative distance between the S valley to the T wave peak on the fourth scale of the DWT	[38]
S-T interval distance 2	DWT	The relative distance between the S valley to the T wave offsite on the fourth scale of the DWT	[38]
Absolute maximum	DWT	The absolute maximum value and location on the fourth scale of the DWT signal	[38]
Zero crossing	DWT	The zero crossing location on the fourth scale of DWT signal	[38]
Wavelet scale	DWT	Calculate which scale the QRS complex is centered on	[38]
DWT coefficients	DWT	The down-sampled third and fourth detail coefficients and the fourth approximation coefficients	[71]
Independent Components	ICA	Independent components calculated with a fast fixed point algorithm	[71]

Table 3. *Cont.*

Features	Method	Description	Reference
Fourier spectrum	DTCWT	Compute the absolute value of fourth and 5th scale DTCWT detail coefficients(dc). Then 1D FFT is applied to the selected DC to obtain the Fourier spectrum. Then take logarithm value of the Fourier spectrum	[32]
IMF sample entropy	EMD/EEMD	The sample entropy is measured of regularity of a time series used to quantify the complexity of heartbeat dynamics	[33]
IMF variation coefficient	EMD/EEMD	The coefficient of variation is a statistical parameter defined as σ^2 / μ^2. [1]	[33]
IMF singular values	EMD/EEMD	The singular value decomposition	[33]
IMF band power values	EMD/EEMD	The band power is the average power of each IMF	[33]
PCA components	PCA	PCA components for size reduction	[82]
Pisarenko PSD	Eigenvector	Power spectral density estimates generated with Pisarenko method	[83]
MUSCI PSD	Eigenvector	Power spectral density estimates generated with Multiple signal classification method	[83]
Minimum-Norm PSD	Eigenvector	Power spectral density estimates generated with Minimum-Norm methods	[83]

[1] σ is the standard variation of the selected IMF, μ is the mean of the selected IMF.

3.4.2. Model Training

Once the feature vectors are extracted from the raw ECG signal, then they can be used by a model for training and classification. There are several methods that have been proven to be valid for identifying heartbeat types. They are clustering, traditional machine learning classification, and deep learning classification.

The clustering aims to find the similarity between the two groups (heartbeat segments) by computing the distance between the two groups. The conventional distances for ECG signals are the Euclidean Distance and Dynamic Time Warping Distance. [84].

The Euclidean distance is the most common distance when comparing two groups with the same dimensions. An example of using Euclidean distance for abnormal heartbeat detection can be found in Chuah and Fu's [76]. They introduce an adaptive window discord discovery (AWDD) to detect the anomaly in ECG recordings. It was developed from a brute force discord discovery (BFDD) algorithm [85]. The algorithm finds candidates with an abnormal heartbeat by selecting the largest Euclidean distance when comparing the heartbeats to each other. Also, they have set a threshold for the Euclidean distance to reduce the false alarm rate. The Euclidean distance only works when both heartbeat segments are the same length.

K-mean clustering is a popular clustering method that builds on the Euclidean distance. The K-mean clustering clusters the heartbeat segments into many different clusters. Veeravalli et al. developed an algorithm for real-time and personalized anomaly detection from wearable health care ECG devices [86]. The K-means cluster algorithm is used to cluster all the heartbeat classes. To avoid calibration of the technique for individual users, they assigned the most frequent heartbeat segments as the normal heartbeat segments. The authors tested their algorithm on the MIT-BIH database and the European ST-T Database. They were able to achieve 97.1% sensitivity and 99.5% specificity.

Sivarake and Ratanamahatana proposed a robust and accurate anomaly detection algorithm (RAAD) that reduced the false alarm detection rate on ECG anomaly detection [34]. They extracted heartbeat morphological features to be their input feature vectors. Then, they calculated the dynamic time warping distance to measure the similarity between two variable-length heartbeats. In their experiment, they tested their algorithm on INCARTDB01-05 [25], the MIT-BIH arrhythmia database [24,25], and the MIT-BIH long term database [25]. Overall, their algorithm achieved 94.35% accuracy and a 0% false alarm rate.

Another major method is the traditional machine learning classification algorithms: Kth nearest neighbor(KNN), Linear Discriminant Analysis(LDA), Quadratic Discriminant Functions(QDF), Support Vector Machine(SVM), and Multilayer perceptron neural network(MLPNN). These algorithms build a mathematical model based on the provided training data. The trained model could correlate the input data with its corresponding label. Many research could be found in this field.

Ivaylo Christov et al. used both the ECG morphology features and VCG features to represent the heartbeat, and then train the feature vectors and its labels with Kth nearest neighbor. As a result, the classification performance on both feature sets is over 96% for five heartbeat types (N, PVC, LBBB, RBBB, and P) [30].

Philip de Chazal et al. used linear discriminant analysis as a classification algorithm. The input feature vectors are ECG morphology features. As a result, this algorithm could perform around 97% accuracy on MIT-BIH database with five heartbeat types (N, S, V, F, and Q) classification [31].

Mariano Llamedo et al. validated a heartbeat classification method for Normal, Supra-ventricular, and Ventricular heartbeats based on ECG interval features, morphological features, and DWT features [38]. The feature vectors are trained with quadratic discriminant functions. The model had a 94% overall classification accuracy on the test dataset.

Li et al. uses the concept of transductive transfer learning to detect the abnormal instance on an ECG signal. They trained a model to learn from a labeled data set to detect irregular heartbeats, and then they use a kernel mean matching (KMM) algorithm [87] to enable knowledge transferring between a labeled data set and unlabeled data set. The model they used was a weighted transductive one-class support vector machine, which could solve the problem of imbalanced data set [78]. The authors performed experiments on records 100, 101, 103, 105, 109, 115, 121, 210, 215, and 232 from the MIT-BIH database. They achieved a 87.89% average accuracy.

Ye et al. classified 16 heartbeat types by using both morphological and dynamic features of ECG signals. Then, both morphological and dynamic features were trained by the support vector machine for the classification. Also, two channels of the ECG signal in the database were trained separately and generated two models. Both models were used for the final classification part. The authors introduced two ways of making a final decision: one is rejection, which requires both models to make the same decision, and the other one is Bayesian, which is based on the fusion of both model's results [71]. The experiment result of this research is compared in Table 4.

Zhang et al. built 46 feature vectors to represent the heartbeat to classify the abnormal heartbeat shape on MIT-BIH database [74]. In the study, the authors apply the ecgpuwave tool from PhysioNet [25] to detect the boundaries of the P wave, QRS complex, and ST waves. Then they have collected five types of features, which are five inter-heartbeat intervals, five intra-heartbeat intervals, 29 morphological amplitudes, six morphological areas, and morphological distance. The five types of features could generate a feature vector with 46 morphological features. In the classification step, the author used the support vector machine to learn the patterns of the feature vectors. Additionally, both channels of the ECG signal have a trained support vector machine model. The results of both models are considered in the final classification result. The result table of the paper shows that the algorithm has nearly 90% accuracy for four heartbeat types (N, F, V, and S) classification.

Thomas et al. introduced an automatic ECG arrhythmia classification idea using dual-tree complex wavelet-based features to detect normal, paced, RBBB, LBBB, and PVC heartbeats. The authors proposed a feature extraction technique based on a dual-tree complex wavelet transform (DTCWT) technique. Then the feature vectors were input to a multilayer perceptron neural network for abnormal heartbeat detection [32]. The experimental results of this research are compared in Table 4.

Kandala Rajesh et al. used ensemble empirical mode decomposition (EEMD) features to classify normal PVC, PAC, LBBB, and RBBB heartbeats. For the classification tool, a sequential minimal optimization SVM was used to train and classify the different heartbeat types [33]. The experimental results of this research are compared in Table 4.

Wess et al. implemented a multi-layer perceptron (MLP) classifier to detect anomalies in ECG signal. To reduce the size of feature vectors, the author applied PCA on the extracted heartbeats. Finally, the processed feature vectors were used as inputs to train an MLP neural network. The trained model could be used for classifying the anomalies in the ECG signal [82]. The authors were able to test their model on the MIT-BIH database with an overall accuracy of 99.82%.

Most researchers have used traditional methods to solve the problem. Traditional machine learning classification methods do not require a considerable amount of training data, and they do not need a lot of computational power. Recently, due to the development of GPUs, deep learning has been proven to be reliable and fast for classification problems. Compared to traditional algorithms, deep learning does not require cardiology experts to extract features since the network can extract the features automatically. Instead, a deep learning model needs many labeled data for training. Luckily, public data sets could be easily found on the Internet. Therefore, many studies using various deep learning architecture have published new algorithms to classify heartbeats.

The Ubeyli algorithm uses Eigenvectors as the feature vectors and a recurrent neural network as the classification tool. In the experiment, normal, congestive heart failure, VT, and AFIB rhythms were trained and tested [83]. The experiment result of this research is compared in Table 4.

Table 4. Heartbeat classification performance on the MIT-BIH dataset.

Method	Year	Abnormal/Normal	Heartbeat Types	TP	FP	TN	FN	Sensitivity	False Alarm	Accuracy
Christov et al. [30]-morphology	2006	18,378/47,239	5	180,42	1604	45,635	336	98.17%	3.40%	97.04%
Christov et al. [30]-frequency	2006	18,378/47,239	5	17,590	1459	45,780	788	95.71%	3.09%	96.58%
Chazal et al. [31]-frequency	2006	4317/34,394	5	4108	1962	32,432	209	95.16%	5.70%	94.39%
Ubeyli et al. [83]	2009	269/90	4	268	2	88	2	99.26%	2.22%	99.89%
Llamedo et al. [38]	2010	5441/44,188	3	4752	2238	41,950	689	87.34%	5.06%	94.10%
Ye et al. [71]-rejection	2012	19,913/64,042	16	19,815	93	63,949	98	99.51%	0.15%	99.77%
Ye et al. [71]-bayesian	2012	20,745/65,264	16	20,557	286	64,978	188	99.09%	0.44%	99.45%
Zhang et al. [74]	2014	5653/44,011	4	5248	4869	39,142	405	92.84%	11.06%	89.38%
Thomas et al. [32]	2015	26,626/672,68	5	22,900	1300	65,968	3726	86.01%	1.93%	94.65%
Kiranyaz et al. [36]	2015	7366/42,191	5	6539	1228	40,963	827	88.77%	2.97%	95.85%
Rajesh et al. [33]	2017	8000/2000	5	7677	33	1967	323	95.96%	1.65%	96.44%
Sahoo et al. [75]	2017	807/244	4	798	5	239	9	98.88%	2.04%	98.67%

Chauhan and Vig developed a predictive algorithm that could detect normal, PVC, PAC, paced heartbeats via deep LSTM (long short-term memory) neural network (Figure 16). In their algorithm, the features extraction/selection step is neglected, raw ECG data, and corresponding labels are used as inputs to the stacked (two-layer) LSTM neural network. In the experiment, they split the MIT-BIH database into four sets: a non-anomalous training set (S_N), non-anomalous validation set (V_N), mixture of both abnormal and normal validation sets (S_{N+A}) and the test sets (t_{N+A}). The LSTM network was trained on S_N, and used V_N for early stopping. The trained LSTM network was then applied to S_{N+A} to find the threshold for detecting abnormal heartbeats. Finally, the chosen threshold was used on t_{N+A} to discriminate regular and anomalous heartbeats while predicting [79]. The presented model was able to achieve a 97.5% precision with a 46.47% recall on the test set (t_{N+A}).

Kiranyaz et al. presented a fast and accurate patient-specific ECG classification and monitoring system. In their experiment setup, they picked five heartbeat types, N, V, S, F, and Q, from 20 ECG records (100–124) from the MIT-BIH database as the training samples. The raw heartbeat segments were submitted to a 1-D adaptive convolutional neural network (CNN) for pattern recognition. The 1-D convolutional neural network acted as a feature extraction tool as well as a classification tool. The classification times for this model is 0.58 and 0.74 ms for 64 and 128 sample heartbeat resolutions, respectively. The speed is more than 1000x faster than the real-time requirement [36]. The experiment result of this research is compared in Table 4.

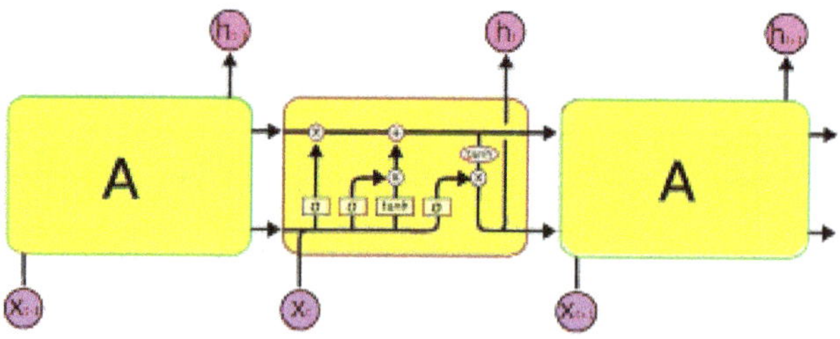

Figure 16. Long short-term memory layers.

Sahoo et al. made an improvement to Rai's algorithm [39] by using multi-resolution wavelet transform and machine learning to detect Normal, LBBB, RBBB, and Paced heartbeats [75]. The authors used Q-peak, R-peak, S-peak, T-peak, QR-interval, ST-interval, RR-interval, and QRS duration as the input feature vector and used a MLP and a SVM classifier as the classification tool. In their experimental results, the overall classification accuracy of normal, LBBB, RBBB, and Paced heartbeats were 96.67% for the SVM classifier and 98.39% for the MLP classifier. The algorithm was tested on the MIT-BIH database [24].

In addition to training with a public data set, some researchers used a patient-specific approach to train the model. The first step of a patient-specific approach is to train an initial classifier with the public data set. Then the second step requires a local cardiologist to review and correct the produced labels by the initial classifier. The final step consists of training the initial classifier with corrected labels to produce the final classifier to this specific patient. The patient-specific approach could eliminate the inter-patient variations of the ECG signals. Biel et al.'s research shows that the variance in different human heartbeats can be very high [88]. Many research works, [31,89–93], have proven that by using a patient-specific model, the detection algorithms have a higher accuracy than the traditional systems in practical cases.

3.5. Irregular Rhythm Classification

Different irregular heartbeat classifications can be found in the literature. Rhythm classification focuses on finding abnormal rhythm among normal rhythms. To find a rhythm anomaly, the algorithm needs to process more than one heartbeat.

Ge et al. [94] used an auto-regressive (AR) modeling technique to classify the Normal, PAC, PVC, SVT, VT, and VF rhythms. The algorithm uses Burg's algorithm to compute the AR coefficients X. In their paper, the authors have attempted two ways to classify the AR coefficients of X: a generalized linear model (GLM) and multi-layer feed-forward neural network. The GLM equation is:

$$Y = X\beta + \varepsilon, \tag{11}$$

where $Y = [y_1, y_2, ..., y_N]$ is an N-dimensional vector of the observed responses, X is the N * P matrix of the AR coefficients, β is a P-dimensional vector, ε is an N-dimensional error vector. The GLM outputs, y_1 to y_N, compared to predefined conditions to classify various heartbeat types. An artificial neural network with the AR coefficients as inputs was used for training and classification. Their experimental results show that artificial neural networks perform better than GLM.

Ozbay et al. integrated a type-2 fuzzy clustering and discrete wavelet transform in order to build a neural network-based ECG classifier to detect Normal, Br, VT, SA, PAC, P, RBBB, LBBB, AF, and AFI results [95]. The proposed diagnostic algorithm can distinguish 10 different rhythm types. The system was formed by combining fuzzy clustering layers, feature extraction layers, and a final classifier layer. The fuzzy clustering layer select segments represents the arrhythmia class in the ECG. A wavelet transformation was applied to the ECG segments to generate features. The authors have trained three Type-2 Fuzzy Clustering Neural Network models (T2FCWNN-1, T2FCWNN-2, and T2FCWNN-3) with three different training data sets. The three training data sets have the same amount of ECG segments. However, the length of each ECG segment is 101 sample points, 52 sample points, and 27 sample points. As a result, the T2FCWNN-3 had the lowest training time, which is 4.86 s and test error rate, which is 0.23% among all three models.

Patel et al. used a thresholding technique to detect arrhythmias on ECGs collected from a mobile platform [35]. In the paper, they first used the Pan–Tompkins [56] algorithm to detect the R peaks on the ECG recordings. Then they characterized SB, ST, PVC, PAC, and Sleep Apnea using a predefined threshold to classify different rhythms. Their system a 97.3% detection accuracy.

Rajpurkar et al. developed an algorithm that could out-perform a board-certified cardiologist in the detection of 12 types of arrhythmia using a 34-layer CNN [96]. The network took a 30 s long raw ECG signal recording as input, and the output was a sequence of label prediction. The model output a new prediction every second. The training data set contained 64,121 ECG records from 29,163 patients, and the testing data set contained 336 records from 328 patients. The model performed with 80.9% precision, 82.7% sensitivity, and a 0.809 F1 score.

Acharya et al. used two 11-layer CNNs to detect AFIB, AFL, ad VF(VFL) from normal heartbeat rhythms [40]. The two networks, Net A and Net B, used a 2-second raw ECG recording and 5-second raw ECG recording as inputs, respectively and output the corresponding label. In the algorithm, no wave detection was performed on the input data. Before submission to the 1-D deep CNN, the ECG segments were Z-score normalized. The result of Net A and Net B are compared in Table 5.

Table 5. Rhythm classification performance on the MIT-BIH dataset.

Method	Year	Abnormal/Normal	Rhythm Types	Rhythm Length	TP	FP	TN	FN	Sensitivity	False Alarm	Accuracy
Ge et al. [94]	2002	713/143	6	1.2 s	706	10	133	7	88.77%	6.99%	98.01%
U. Acharya Net A [40]	2017	20807/902	4	2 s	19,160	62	840	1647	92.08%	6.87%	92.13%
U. Acharya Net B [40]	2017	8322/361	4	5 s	7946	376	294	67	95.48%	18.56%	94.9%

3.6. Heartbeat/Rhythm Classification Algorithm Comparison

In the previous sections, we reviewed many algorithms that classify the ECGs in various categories. We can see in Table 4 the classification results performed using the MIT-BIH database. In addition, some algorithms' performance metrics were converted to binary classification, which detects normal and abnormal heartbeats. The reason is that computer diagnoses are not 100% accurate. We still need doctors to make the final diagnosis as they are the only ones who know the context. The methods should be focusing on binary classification, which classifies all abnormal heartbeat as one class. The terms used in the table are explained:

- TP: Number of correctly detected abnormal heartbeats
- FP: Number of incorrectly detected abnormal heartbeats
- TN: Number of correctly detected normal heartbeats
- FN: Number of incorrectly detected normal heartbeats
- Sensitivity = TP/(TP+FN)
- False Alarm Rate= 1 - Specificity = FP/(FP+TN)
- Accuracy = (TP+TN)/(TP+FP+TN+FN)

Similarly, Table 5 compares all methods that classify the rhythms on MIT-BIH database. In addition, the table has only shown the algorithms that provided enough information to compute our metrics.

4. Discussion

4.1. Challenges for Heart Anomaly Detection with Ambulatory Electrocardiograms

There are still several challenges in heart anomaly detection:

- ECG signals may be contaminated with motion noise as the patient is constantly moving. The noisy signal may have a similar morphology to abnormal cardiac signals resulting in false positive. It is easy for the human eye to identify these conditions; however, for computers, it is much harder to separate the noise from the signal.
- The model training requires a labeled ECG signal. In order to label the ECG data set, trained personnel are needed. In addition, the labeling process is very time consuming. For example, a 10 s one ECG signal has 2500 data points, and the continuous monitoring usually takes 24–48 h.
- The ECG heartbeat data is highly imbalanced. Over 99% of the heartbeat data is the normal case and only 1% of the heartbeat data presents 16 abnormal cases. Therefore, the highly imbalanced dataset makes it more difficult to adjust the learning step. Several options could be explored to reduce the effect of imbalanced data, such as database re-sampling or using the cost-sensitive method, kernel based method, or active learning [97].

4.2. Future Works

The next generation of heart anomaly detection algorithms should be able to deal with ambulatory health measurements taking advantage of multiple synchronized measurements from: accelerometer, real-time blood pressure (based on pulse transit time), skin temperature, and upper and lower chest breathing sensors. An excellent review of the state-of-the-art of body sensor fusion work can be found in Gravina et al. [98]. One commercial example of such a data fusion system is Astroskin from Carre Technologies Inc. The Astroskin space-grade garments offer state-of-the-art continuous real-time monitoring for 48 h of blood pressure, pulse oximetry, 3-lead ECG, respiration, skin temperature, and activity. Using Astroskin, one can develop new fusion algorithms that can compensate for ECG motion artifacts by correlations with synchronized accelerometer and breathing data.

This can be accomplished by using advanced LSTM and recurrent neural networks (RNN). An example of this approach can be found in Shrimanti et al. [99] where ECG, peak blood oxygenation signal (PPG), and accelerometer measurements were combined using a LSTM and RNN to compute in real-time motion compensated blood pressure. Such technology could open the door to real-time patient-specific anomaly detection that goes far beyond simple ECG measurements, for example correlating cardiac and respiratory events with patient activities.

4.3. Conclusions

In this survey, we have first introduced the definition of anomaly detection on ambulatory electrocardiograms (ECG) and its importance. We then discussed the basic medical background (Section 2) of electrocardiogram interpretation and the type of anomalies that need to be detected. Most electrocardiogram anomalies can be categorized into two major categories: irregular heart rates and irregular heart rhythms. The irregular heart rates on ECG could indicate bradycardia, tachycardia, heart block, arrhythmia, and so on. The irregular heart rhythms could be ectopic heartbeat when checking a period of ECG signal.

Therefore, based on the different irregularities on the ECG, anomaly detection can be divided into several categories: heartbeat detection (Section 3.3) for detecting the location of each heartbeat; heartbeat segmentation (Section 3.3) for segmenting the heartbeats from the entire ECG signal; heartbeat classification (Section 3.4) for classifying the type of one heartbeat; and rhythm classification (Section 3.5) for classifying the type of a period of ECG signal. In addition, as the ECG signal is frequently contaminated with electrical noise and motion artifacts, noise removal (Section 3.2) is important for anomaly detection on the ambulatory ECG.

From the literature, we have reviewed the conventional methods for each part. For the noise removal on ECG, fixed digital filters, discrete wavelet transform, empirical mode decomposition, and adaptive filters have been used by many researchers. For heartbeat detection, many researchers used fixed digital filters, discrete wavelet transform, and Shannon energy envelopes to remove the noise and unwanted waves while preserving the R peak information. They then used the R peak location to compute the heartbeat. For heartbeat segmentation, the most common method was to use a predefined window to segment the heartbeat signal from the entire signal.

For the literature on heartbeat classification, authors used morphological features and derived features to represent the heartbeat signal. The morphological features were calculated from the ECG signal, and derived features were computed using other methods, such as discrete wavelet transform, independent component analysis, empirical model decomposition, and many more. Both morphological and derived features are then used for training in order to generate a mathematical model of the heartbeat signal.

The most popular models used k nearest neighbor, linear discriminant analysis, support vector machines, multilayer perceptron neural networks, and deep neural networks, such as CNN and RNN. Similarly for the rhythm classification, the algorithms take a period of ECG signal as the input to the model.

The current challenges of anomaly detection for ambulatory electrocardiograms are analyzed in this paper. We determined three major challenges. First, the reduction of motion artifacts on the ECG signal interferes with the anomaly detection. Second, model training requires a massive amount of labeled data that are had to come by. Third, ECG databases have very imbalanced data making it difficult for deep learning model training.

Author Contributions: H.L. did the literature review and analysis, and P.B. did editing. All authors have read and agreed to the published version of the manuscript.

Funding: Funded by the CISCO Chair in Healthcare.

References

1. World Health Organization. Cardiovascular Diseases. 2017. Available online: https://www.who.int/health-topics/cardiovasculardiseases/tab=tab3 (accessed on 4 March 2020).
2. Zipes, D.P.; Libby, P.; Bonow, R.O.; Mann, D.L.; Tomaselli, G.F. *Braunwald's Heart Disease E-Book: A Textbook of Cardiovascular Medicine*; Elsevier Health Sciences: Philadelphia, PA, USA, 2018.
3. Apple Inc. Apple Watch Series 4. 2018. Available online: https://www.apple.com/ca/apple-watch-series-4/health (accessed on 4 March 2020).
4. AliveCor Inc. Alive cor. 2020. Available online: https://www.alivecor.com/ (accessed on 4 March 2020).
5. Omron Healthcare Asia. Omron ecg Monitor hcg-801. 2020. Available online: https://zenicor.com/zenicor-ekg/ (accessed on 4 March 2020).
6. Qardio Inc. Qardiomd. 2018. Available online: https://www.getqardio.com/qardiomd-ecg/ (accessed on 4 March 2020).
7. Carre Technologies Inc. Hexoskin Smart Shirt. 2012. Available online: https://www.hexoskin.com/ (accessed on 4 March 2020).
8. Hampton, J. *The ECG Made Easy E-Book*; Elsevier Health Sciences: St. Louis, MO, USA, 2013.
9. Frank, E. An accurate, clinically practical system for spatial vectorcardiography. *Circulation* **1956**, *13*, 737–749. [CrossRef] [PubMed]
10. Antonicelli, R.; Ripa, C.; Abbatecola, A.M.; Capparuccia, C.A.; Ferrara, L.; Spazzafumo, L. Validation of the 3-lead tele-ECG versus the 12-lead tele-ECG and the conventional 12-lead ECG method in older people. *J. Telemed. Telecare* **2012**, *18*, 104–108. [CrossRef] [PubMed]
11. Kristensen, A.N.; Jeyam, B.; Riahi, S.; Jensen, M.B. The use of a portable three-lead ECG monitor to detect atrial fibrillation in general practice. *Scand. J. Primary Health Care* **2016**, *34*, 304–308. [CrossRef] [PubMed]
12. Mehri-Dehnavi, A.; Salehpour, N.; Rabbani, H.; Farahabadi, A.; Farahabadi, E. Automatic Analysis of Vectorcardiogram Signal for Detection of Cardiovascular Diseases. Ph.D. Thesis, Isfahan University of Medical Sciences, Isfahan, Iran, 2013.
13. Augustyniak, P. On the Equivalence of the 12-Lead ECG and the VCG Representations of the Cardiac Electrical Activity. In Proceedings of the 10th International Conference on System-Modelling-Control, Zakopane, Poland, 21–25 May 2001; pp. 51–56.
14. Atoui, H.; Fayn, J.; Rubel, P. A novel neural-network model for deriving standard 12-lead ECGs from serial three-lead ECGs: Application to self-care. *IEEE Trans. Inf. Technol. Biomed.* **2010**, *14*, 883–890. [CrossRef]
15. Figueiredo, C.; Mendes, P. Towards Wearable And Continuous 12-Lead Electrocardiogram Monitoring-Synthesis of the 12-lead Electrocardiogram using 3 Wireless Single-lead Sensors. In *International Conference on Biomedical Electronics and Devices*; SCITEPRESS: Setúbal, Portugal, 2012; Volume 2.
16. Zhu, H.; Pan, Y.; Cheng, K.T.; Huan, R. A lightweight piecewise linear synthesis method for standard 12-lead ECG signals based on adaptive region segmentation. *PLoS ONE* **2018**, *13*, e0206170. [CrossRef]
17. Nelwan, S.P.; Kors, J.A.; Meij, S.H.; van Bemmel, J.H.; Simoons, M.L. Reconstruction of the 12-lead electrocardiogram from reduced lead sets. *J. Electrocardiol.* **2004**, *37*, 11–18. [CrossRef]
18. Drew, B.J.; Adams, M.G.; Pelter, M.M.; Wung, S.F.; Caldwell, M.A. Comparison of standard and derived 12-lead electrocardiograms for diagnosis of coronary angioplasty-induced myocardial ischemia. *Am. J. Cardiol.* **1997**, *79*, 639–644. [CrossRef]
19. Drew, B.J.; Pelter, M.M.; Brodnick, D.E.; Yadav, A.V. Comparison of a new reduced lead set ECG with the standard ECG for diagnosing cardiac arrhythmias and myocardial ischemia. *J. Electrocardiol.* **2002**, *35*, 13. [CrossRef]
20. Tomasic, I.; Trobec, R.; Lindén, M. Can the regression trees be used to model relation between ECG leads? In *International Internet of Things Summit*; Springer: Rome, Italy, 2015; pp. 467–472.
21. M. Training and S. LLC. Ekg reference. 2019. Available online: https://www.practicalclinicalskills.com/ekg-reference (accessed on 4 March 2020).
22. Goldberger, A.L.; Goldberger, Z.D.; Shvilkin, A. *Clinical Electrocardiography: A Simplified Approach E-Book: A Simplified Approach*; Elsevier Health Sciences: Philadelphia, PA, USA, 2017.
23. Morris, F.; Brady, W.J. ABC of clinical electrocardiography: Acute myocardial infarction—Part I. *Bmj* **2002**, *324*, 831–834. [CrossRef]

24. Moody, G.B.; Mark, R.G. The impact of the MIT-BIH arrhythmia database. *IEEE Eng. Med. Biol. Mag.* **2001**, *20*, 45–50. [CrossRef]

25. Goldberger, A.L.; Amaral, L.A.N.; Glass, L.; Hausdorff, J.M.; Ivanov, P.C.; Mark, R.G.; Mietus, J.E.; Moody, G.B.; Peng, C.K.; Stanley, H.E. PhysioBank, PhysioToolkit, and PhysioNet: Components of a New Research Resource for Complex Physiologic Signals. *Circulation* **2000**, *101*, e215–e220. [CrossRef] [PubMed]

26. Harrigan, R.A.; Chan, T.C.; Brady, W.J. Electrocardiographic electrode misplacement, misconnection, and artifact. *J. Emerg. Med.* **2012**, *43*, 1038–1044. [CrossRef]

27. Jané, R.; Laguna, P.; Thakor, N.V.; Caminal, P. Adaptive baseline wander removal in the ECG: Comparative analysis with cubic spline technique. In *Proceedings Computers in Cardiology*; IEEE: Durham, NC, USA, 1992; pp. 143–146.

28. Thakor, N.V.; Zhu, Y.S. Applications of adaptive filtering to ECG analysis: Noise cancellation and arrhythmia detection. *IEEE Trans. Biomed. Eng.* **1991**, *38*, 785–794. [CrossRef] [PubMed]

29. MAUVILA.COM. Ecg Artifacts. 2018. Available online: http://www.mauvila.com/ECG/ecgartifact.htm (accessed on 4 March 2020).

30. Christov, I.; Gómez-Herrero, G.; Krasteva, V.; Jekova, I.; Gotchev, A.; Egiazarian, K. Comparative study of morphological and time-frequency ECG descriptors for heartbeat classification. *Med. Eng. Phys.* **2006**, *28*, 876–887. [CrossRef]

31. De Chazal, P.; Reilly, R.B. A patient-adapting heartbeat classifier using ECG morphology and heartbeat interval features. *IEEE Trans. Biomed. Eng.* **2006**, *53*, 2535–2543. [CrossRef]

32. Thomas, M.; Das, M.K.; Ari, S. Automatic ECG arrhythmia classification using dual tree complex wavelet based features. *AEU-Int. J. Electron. Commun.* **2015**, *69*, 715–721. [CrossRef]

33. Rajesh, K.N.; Dhuli, R. Classification of ECG heartbeats using nonlinear decomposition methods and support vector machine. *Comput. Biol. Med.* **2017**, *87*, 271–284. [CrossRef] [PubMed]

34. Sivaraks, H.; Ratanamahatana, C.A. Robust and accurate anomaly detection in ECG artifacts using time series motif discovery. *Comput. Math. Methods Med.* **2015**, 453214. [CrossRef]

35. Patel, A.M.; Gakare, P.K.; Cheeran, A. Real time ECG feature extraction and arrhythmia detection on a mobile platform. *Int. J. Comput. Appl.* **2012**, *44*, 40–45.

36. Kiranyaz, S.; Ince, T.; Gabbouj, M. Real-time patient-specific ECG classification by 1-D convolutional neural networks. *IEEE Trans. Biomed. Eng.* **2015**, *63*, 664–675. [CrossRef]

37. Velayudhan, A.; Peter, S. Noise Analysis and Different Denoising Techniques of ECG Signal-A Survey. *IOSR J. Electron. Commun. Eng. (IOSR-JECE)* **2016**, *2278–2834*, 40–44.

38. Llamedo, M.; Martínez, J.P. Heartbeat classification using feature selection driven by database generalization criteria. *IEEE Trans. Biomed. Eng.* **2010**, *58*, 616–625. [CrossRef] [PubMed]

39. Rai, H.M.; Trivedi, A.; Shukla, S. ECG signal processing for abnormalities detection using multi-resolution wavelet transform and Artificial Neural Network classifier. *Measurement* **2013**, *46*, 3238–3246. [CrossRef]

40. Acharya, U.R.; Fujita, H.; Lih, O.S.; Hagiwara, Y.; Tan, J.H.; Adam, M. Automated detection of arrhythmias using different intervals of tachycardia ECG segments with convolutional neural network. *Inf. Sci.* **2017**, *405*, 81–90. [CrossRef]

41. Rai, H.M.; Trivedi, A. De-noising of ECG Waveforms based on Multi-resolution Wavelet Transform. *Int. J. Comput. Appl.* **2012**, *45*, 25–30.

42. Singh, B.N.; Tiwari, A.K. Optimal selection of wavelet basis function applied to ECG signal denoising. *Digit. Signal Process.* **2006**, *16*, 275–287. [CrossRef]

43. Lin, H.Y.; Liang, S.Y.; Ho, Y.L.; Lin, Y.H.; Ma, H.P. Discrete-wavelet-transform-based noise removal and feature extraction for ECG signals. *Irbm* **2014**, *35*, 351–361. [CrossRef]

44. Labate, D.; La Foresta, F.; Occhiuto, G.; Morabito, F.C.; Lay-Ekuakille, A.; Vergallo, P. Empirical mode decomposition vs. wavelet decomposition for the extraction of respiratory signal from single-channel ECG: A comparison. *IEEE Sens. J.* **2013**, *13*, 2666–2674. [CrossRef]

45. Weng, B.; Blanco-Velasco, M.; Barner, K.E. ECG denoising based on the empirical mode decomposition. In Proceedings of the 2006 International Conference of the IEEE Engineering in Medicine and Biology Society, New York, NY, USA, 30 August–3 September 2006; pp. 1–4.

46. Blanco-Velasco, M.; Weng, B.; Barner, K.E. ECG signal denoising and baseline wander correction based on the empirical mode decomposition. *Comput. Biol. Med.* **2008**, *38*, 1–13. [CrossRef]

47. Chang, K.M. Arrhythmia ECG noise reduction by ensemble empirical mode decomposition. *Sensors* **2010**, *10*, 6063–6080. [CrossRef]

48. Wu, Z.; Huang, N.E. Ensemble empirical mode decomposition: A noise-assisted data analysis method. *Adv. Adapt. Data Anal.* **2009**, *1*, 1–41. [CrossRef]

49. Jenitta, J.; Rajeswari, A. Denoising of ECG signal based on improved adaptive filter with EMD and EEMD. In Proceedings of the 2013 IEEE Conference on Information & Communication Technologies, Thuckalay, Tamil Nadu, India, 11–12 April 2013; pp. 957–962.

50. Zhidong, Z.; Yi, L.; Qing, L. Adaptive noise removal of ECG signal based on ensemble empirical mode decomposition. In *Adaptive Filtering Applications*; IntechOpen: London, UK, 2011; ISBN 978-953-307-306-4. [CrossRef]

51. Singh, G.; Kaur, G.; Kumar, V. ECG denoising using adaptive selection of IMFs through EMD and EEMD. In Proceedings of the 2014 International Conference on Data Science & Engineering (ICDSE), Kochi, India, 26–28 August 2014; pp. 228–231.

52. Lisheng, X.; Kuanquan, W.; Zhang, D.; Cheng, S. Adaptive baseline wander removal in the pulse waveform. In Proceedings of the 15th IEEE Symposium on Computer-Based Medical Systems (CBMS 2002), Maribor, Slovenia, 4–7 June 2002; pp. 143–148.

53. Haykin, S.S.; Widrow, B.; Widrow, B. *Least-Mean-Square Adaptive Filters*; Wiley Online Library: Hoboken, NJ, USA, 2003; Volume 31.

54. Thenua, R.K.; Agarwal, S. Simulation and performance analysis of adaptive filter in noise cancellation. *Int. J. Eng. Sci. Technol.* **2010**, *2*, 4373–4378.

55. Raya, M.A.D.; Sison, L.G. Adaptive noise cancelling of motion artifact in stress ECG signals using accelerometer. In Proceedings of the Second Joint 24th Annual Conference and the Annual Fall Meeting of the Biomedical Engineering Society Engineering in Medicine and Biology, Houston, TX, USA, 23–26 October 2002; Volume 2, pp. 1756–1757.

56. Pan, J.; Tompkins, W.J. A real-time QRS detection algorithm. *IEEE Trans. Biomed. Eng.* **1985**, *32*, 230–236. [CrossRef]

57. Afonso, V.X.; Tompkins, W.J.; Nguyen, T.Q.; Luo, S. ECG beat detection using filter banks. *IEEE Trans. Biomed. Eng.* **1999**, *46*, 192–202. [CrossRef]

58. Chen, S.W.; Chen, H.C.; Chan, H.L. A real-time QRS detection method based on moving-averaging incorporating with wavelet denoising. *Comput. Methods Progr. Biomed.* **2006**, *82*, 187–195. [CrossRef]

59. Choi, S.; Adnane, M.; Lee, G.J.; Jang, H.; Jiang, Z.; Park, H.K. Development of ECG beat segmentation method by combining lowpass filter and irregular R–R interval checkup strategy. *Expert Syst. Appl.* **2010**, *37*, 5208–5218. [CrossRef]

60. Yeh, Y.C.; Wang, W.J. QRS complexes detection for ECG signal: The Difference Operation Method. *Comput. Methods Progr. Biomed.* **2008**, *91*, 245–254. [CrossRef]

61. Zidelmal, Z.; Amirou, A.; Adnane, M.; Belouchrani, A. QRS detection based on wavelet coefficients. *Comput. Methods Progr. Biomed.* **2012**, *107*, 490–496. [CrossRef]

62. Bouaziz, F.; Boutana, D.; Benidir, M. Multiresolution wavelet-based QRS complex detection algorithm suited to several abnormal morphologies. *IET Signal Process.* **2014**, *8*, 774–782. [CrossRef]

63. Manikandan, M.S.; Soman, K. A novel method for detecting R-peaks in electrocardiogram (ECG) signal. *Biomed. Signal Process. Control* **2012**, *7*, 118–128. [CrossRef]

64. Zhu, H.; Dong, J. An R-peak detection method based on peaks of Shannon energy envelope. *Biomed. Signal Process. Control* **2013**, *8*, 466–474. [CrossRef]

65. Rakshit, M.; Panigrahy, D.; Sahu, P. An improved method for R-peak detection by using Shannon energy envelope. *Sādhanā* **2016**, *41*, 469–477. [CrossRef]

66. Park, J.S.; Lee, S.W.; Park, U. R peak detection method using wavelet transform and modified Shannon energy envelope. *J. Healthc. Eng.* **2017**. [CrossRef] [PubMed]

67. Pal, S.; Mitra, M. Detection of ECG characteristic points using multiresolution wavelet analysis based selective coefficient method. *Measurement* **2010**, *43*, 255–261. [CrossRef]

68. Banerjee, S.; Gupta, R.; Mitra, M. Delineation of ECG characteristic features using multiresolution wavelet analysis method. *Measurement* **2012**, *45*, 474–487. [CrossRef]

69. Karimipour, A.; Homaeinezhad, M.R. Real-time electrocardiogram P-QRS-T detection–delineation algorithm based on quality-supported analysis of characteristic templates. *Comput. Biol. Med.* **2014**, *52*, 153–165. [CrossRef]

70. De Chazal, P.; O'Dwyer, M.; Reilly, R.B. Automatic classification of heartbeats using ECG morphology and heartbeat interval features. *IEEE Trans. Biomed. Eng.* **2004**, *51*, 1196–1206. [CrossRef]

71. Ye, C.; Kumar, B.V.; Coimbra, M.T. Heartbeat classification using morphological and dynamic features of ECG signals. *IEEE Trans. Biomed. Eng.* **2012**, *59*, 2930–2941.

72. Yang, S.; Shen, H. Heartbeat classification using discrete wavelet transform and kernel principal component analysis. In Proceedings of the IEEE 2013 Tencon-Spring, Sydney, Australia, 17–19 April 2013; pp. 34–38.

73. Jekova, I.; Bortolan, G.; Christov, I. Assessment and comparison of different methods for heartbeat classification. *Med. Eng. Phys.* **2008**, *30*, 248–257. [CrossRef]

74. Zhang, Z.; Dong, J.; Luo, X.; Choi, K.S.; Wu, X. Heartbeat classification using disease-specific feature selection. *Comput. Biol. Med.* **2014**, *46*, 79–89. [CrossRef]

75. Sahoo, S.; Kanungo, B.; Behera, S.; Sabut, S. Multiresolution wavelet transform based feature extraction and ECG classification to detect cardiac abnormalities. *Measurement* **2017**, *108*, 55–66. [CrossRef]

76. Chuah, M.C.; Fu, F. ECG anomaly detection via time series analysis. In Proceedings of the International Symposium on Parallel and Distributed Processing and Applications, Niagara Falls, ON, Canada, 29–31 August 2007; pp. 123–135.

77. Li, P.; Chan, K.L.; Fu, S.; Krishnan, S.M. An abnormal ecg beat detection approach for long-term monitoring of heart patients based on hybrid kernel machine ensemble. In Proceedings of the International Workshop on Multiple Classifier Systems, Seaside, CA, USA, 13–15 June 2005; pp. 346–355.

78. Li, K.; Du, N.; Zhang, A. Detecting ECG abnormalities via transductive transfer learning. In Proceedings of the 2012 ACM Conference on Bioinformatics, Computational Biology and Biomedicine, Orlando, FL, USA, 8–10 October 2012; pp. 210–217.

79. Chauhan, S.; Vig, L. Anomaly detection in ECG time signals via deep long short-term memory networks. In Proceedings of the IEEE International Conference on Data Science and Advanced Analytics (DSAA), Paris, France, 19–21 October 2015; pp. 1–7.

80. Burch, G.E. The history of vectorcardiography. *Med. History* **1985**, *29*, 103–131. [CrossRef] [PubMed]

81. Kingsbury, N. Complex wavelets for shift invariant analysis and filtering of signals. *Appl. Comput. Harmon. Anal.* **2001**, *10*, 234–253. [CrossRef]

82. Wess, M.; Manoj, P.S.; Jantsch, A. Neural network based ECG anomaly detection on FPGA and trade-off analysis. In Proceedings of the 2017 IEEE International Symposium on Circuits and Systems (ISCAS), Baltimore, MD, USA, 28–31 May 2017; pp. 1–4.

83. Übeyli, E.D. Combining recurrent neural networks with eigenvector methods for classification of ECG beats. *Digit. Signal Process.* **2009**, *19*, 320–329. [CrossRef]

84. Berndt, D.J.; Clifford, J. Using dynamic time warping to find patterns in time series. In Proceedings of the KDD Workshop, Seattle, WA, USA, 31 July –1 August 1994; Volume 10, pp. 359–370.

85. Pham, N.D.; Le, Q.L.; Dang, T.K. HOT aSAX: A novel adaptive symbolic representation for time series discords discovery. In Proceedings of the Asian Conference on Intelligent Information and Database Systems, Hue City, Vietnam, 24–26 March 2010; pp. 113–121.

86. Veeravalli, B.; Deepu, C.J.; Ngo, D. Real-time, personalized anomaly detection in streaming data for wearable healthcare devices. In *Handbook of Large-Scale Distributed Computing in Smart Healthcare*; Springer: New York, NY, USA, 2017; pp. 403–426.

87. Gretton, A.; Smola, A.; Huang, J.; Schmittfull, M.; Borgwardt, K.; Schölkopf, B. Covariate shift by kernel mean matching. *Dataset Shift Mach. Learn.* **2009**, *3*, 5.

88. Biel, L.; Pettersson, O.; Philipson, L.; Wide, P. ECG analysis: A new approach in human identification. *IEEE Trans. Instrum. Meas.* **2001**, *50*, 808–812. [CrossRef]

89. Hu, Y.H.; Palreddy, S.; Tompkins, W.J. A patient-adaptable ECG beat classifier using a mixture of experts approach. *IEEE Trans. Biomed. Eng.* **1997**, *44*, 891–900.

90. Jiang, W.; Kong, S.G. Block-based neural networks for personalized ECG signal classification. *IEEE Trans. Neural Netw.* **2007**, *18*, 1750–1761. [CrossRef]

91. Ince, T.; Kiranyaz, S.; Gabbouj, M. A generic and robust system for automated patient-specific classification of ECG signals. *IEEE Trans. Biomed. Eng.* **2009**, *56*, 1415–1426. [CrossRef]

92. Kiranyaz, S.; Ince, T.; Hamila, R.; Gabbouj, M. Convolutional neural networks for patient-specific ecg classification. In Proceedings of the 2015 37th Annual International Conference of the IEEE Engineering in Medicine and Biology Society (EMBC), Milano, Italy, 25–29 August 2015; pp. 2608–2611.

93. Luo, K.; Li, J.; Wang, Z.; Cuschieri, A. Patient-specific deep architectural model for ECG classification. *J. Healthc. Eng.* **2017**, *2017*. [CrossRef]

94. Ge, D.; Srinivasan, N.; Krishnan, S.M. Cardiac arrhythmia classification using autoregressive modeling. *Biomed. Eng. Online* **2002**, *1*, 5. [CrossRef]

95. Özbay, Y.; Ceylan, R.; Karlik, B. Integration of type-2 fuzzy clustering and wavelet transform in a neural network based ECG classifier. *Expert Syst. Appl.* **2011**, *38*, 1004–1010. [CrossRef]

96. Rajpurkar, P.; Hannun, A.Y.; Haghpanahi, M.; Bourn, C.; Ng, A.Y. Cardiologist-level arrhythmia detection with convolutional neural networks. *arXiv* **2017**, arXiv:1707.01836.

97. He, H.; Garcia, E.A. Learning from imbalanced data. *IEEE Trans. Knowl. Data Eng.* **2009**, *21*, 1263–1284.

98. Gravina, R.; Alinia, P.; Ghasemzadeh, H.; Fortino, G. Multi-sensor fusion in body sensor networks: State-of-the-art and research challenges. *Inf. Fusion* **2017**, *35*, 68–80. [CrossRef]

99. Ghosh, S.; Banerjee, A.; Ray, N.; Wood, P.W.; Boulanger, P.; Padwal, R. Using accelerometric and gyroscopic data to improve blood pressure prediction from pulse transit time using recurrent neural network. In Proceedings of the 2018 IEEE International Conference on Acoustics, Speech and Signal Processing (ICASSP), Calgary, AB, Canada, 15–20 April 2018; pp. 935–939.

MDPI

St. Alban-Anlage 66

4052 Basel

Switzerland

Tel. +41 61 683 77 34

Fax +41 61 302 89 18

www.mdpi.com

Sensors Editorial Office

E-mail: sensors@mdpi.com

www.mdpi.com/journal/sensors